Innovations in Fuel Cell Technologies

RSC Energy and Environment Series

Editor-in-Chief:
Professor Laurence Peter, University of Bath, UK

Series Editors:
Professor Heinz Frei, Lawrence Berkeley National Laboratory, USA
Professor Ferdi Schüth, Max Planck Institute for Coal Research, Germany
Professor Tim S. Zhao, The Hong Kong University of Science and Technology, Hong Kong

How to obtain future titles on publication:
A standing order plan is available for this series. A standing order will bring delivery of each new volume immediately on publication.

For further information please contact:
Book Sales Department, Royal Society of Chemistry, Thomas Graham House, Science Park, Milton Road, Cambridge, CB4 0WF, UK
Telephone: +44 (0)1223 420066, Fax: +44 (0)1223 420247, Email: books@rsc.org
Visit our website at http://www.rsc.org/Shop/Books/

Innovations in Fuel Cell Technologies

Edited by

Robert Steinberger-Wilckens and Werner Lehnert
Forschungszentrum Jülich, Jülich, Germany

RSC Publishing

RSC Energy and Environment Series No. 2

ISBN: 978-1-84973-033-4
ISSN: 2044-0774

A catalogue record for this book is available from the British Library

© Royal Society of Chemistry 2010

Published by The Royal Society of Chemistry,
Thomas Graham House, Science Park, Milton Road,
Cambridge CB4 0WF, UK

Registered Charity Number 207890

For further information see our web site at www.rsc.org

Preface

Fuel cells have evolved from an exotic technology only feasible under the constraints of space flight into a product addressing the 'everyman' consumer, although, at first, in niche markets only. The considerable level of technological readiness that has been reached today finally gives rise to hopes that fuel cells will eventually make it to larger markets within the decade leading up to the year 2020. Their high potential for emission-free energy supply, high electrical efficiency, modularity, low maintenance requirements and almost noiseless operation have encouraged researchers and developers worldwide to stubbornly and gradually improve performance, robustness and cost effectiveness, inching their way towards eventually technically realising the potential dormant in the concept. Many companies have been seen to invest in fuel cell technology, lose a lot of money and leave again, whereas the number of firms with commercially successful operations is small but – in more recent times – constantly growing.

Different fuel cell types are being discussed for different applications: stationary, mobile and portable. The requirements on the fuel cell systems differ in each application not only from the point of view of lifetime but also with respect to dynamic or stationary application and to power range. In the milliwatt range the efficiency is often not the main driver for fuel cell development whereas in the kilowatt range and in the future in the megawatt power range the use of fuel cells is mainly driven by the high efficiency of the evolving fuel cell technology.

The alkaline fuel cell (AFC) proved its reliability in demanding applications like space flight. Cost issues and the requirement for pure gases at both electrodes prevent a broad application outside some niche markets. Consequently, despite the proven technological readiness, most activities in the field of R&D were stopped, especially after Nafion was shown to be a good electrolyte for low temperature fuel cells. Now, worldwide R&D activities on Nafion-based fuel cells have shown a sharp rise in research activities and also in industry. Several automotive companies who have the vision of bringing fuel cell driven

RSC Energy and Environment Series No. 2
Innovations in Fuel Cell Technologies
Edited by Robert Steinberger-Wilckens and Werner Lehnert
© Royal Society of Chemistry 2010
Published by the Royal Society of Chemistry, www.rsc.org

cars into the market focus on these polymer electrolyte membrane fuel cells (PEFCs). The effort of the industry to develop a reliable and competitive product led to advanced fuel cell systems which were proven to fulfil the requirements of cars in daily use. The necessary hydrogen infrastructure and cost issues are the main reasons that the market entry is further delayed. In the slipstream of the automotive industry the development of PEFCs for other applications, especially for stationary power generation has increased. Based on natural gas, several companies started a development of PEFC combined heat and power systems. The reforming of the natural gas, the treatment of the reformate gas in order to remove the resulting CO from the fuel gas and the water management lead to complex system architectures and it seems to be difficult to reach the desired cost goals.

A few years ago high-temperature polymer electrolyte fuel cells (HT-PEFC) came into the focus of the fuel cell community. This fuel cell type is based on a phosphoric acid-doped polybenzimidazol-type electrolyte. The advantage of the operating temperature of about 160°C is the high CO tolerance of about 1 to 2%. Furthermore, the difficult water management is not a major issue as in classical PEFCs. This opened up the option to simplify the complex system when using natural gas as fuel. In addition, reformate gas from middle distillates like diesel or jet fuel can be used without the necessity of employing a selective oxidation or methanation step after the shift reactors. At the moment HT-PEFC auxiliary power unit (APU) systems are within the focus of developers. One drawback is the missing cold start capability of a HT-PEFC, the stack has to be heated to at least 120°C before current can be delivered by the system. Compared to classical PEFCs the development of HT-PEFCs is at an early stage but many improvements have been made in the last few years. The latest research activities focus on the oxygen reduction reaction (ORR) which is a major reason for the low power density. While lifetime is high at steady state operation, dynamic operation and thermal cycling leads to increased degradation.

These are also major issues for phosphoric acid fuel cells (PAFCs) which use the same electrolyte as HT-PEFCs. Industrial companies have invested a great deal of time and money in order to develop stationary PAFC systems. Despite the high costs of such systems it can be said that the PAFC is a product which has proven its functionality especially in the 200 kW range for stationary applications. Until the 1980s there were many publications concerning phosphoric acid fuel cells but in the last few years research and development has been performed mainly within industry and few results have been published. Due to the high interest in HT-PEFCs, research activities worldwide on fuel cells with phosphoric acid as electrolyte are growing again.

In the field of high-temperature fuel cells the technology of molten carbonate fuel cells (MCFCs) lies in the hands of few industrial companies. Systems of 250 kW capacity for stationary applications are state of the art. Much experience has been gained during the last few years in field tests. These fuel cell stacks have been operated for some tens of thousands of hours without failure in various system environments. The future transition to an automated assembly may lead to decreasing costs. As a result, the MCFC may also become attractive outside the niche markets where they are located today.

The solid oxide fuel cell (SOFC) also has a long tradition. The main advantage of the SOFC over other fuel cells is the ability of internally reforming methane, the main constituent of natural gas which results in a very high electric efficiency when operated with natural gas. Furthermore, the SOFC provides a high off-gas temperature which can be used for a large range of heat applications. An SOFC can also be coupled with steam and gas turbines which further increases the overall efficiency of the system. Intensive industrial and institutional research and development resulted in long-term stable materials and highly efficient systems. Operating times of 40 000 h were demonstrated under steady-state conditions. As with the MCFC, thermal cycles will induce stress and microcracking in the ceramic materials. Therefore applications are preferred where thermal cycling can be avoided or kept to a minimum. This is typical for stationary applications in the high-power range. In the low power range of about 1 kW SOFCs will be used as residential combined heat and power systems. In this application the main goal of the system is heating. The electricity is more or less a by-product. Nevertheless, the high electrical efficiency of up to 60% offers a high potential in de-central electricity generation. Major issues for further development are redox stability of the anode and the ability to withstand hundreds of thermal cycles. Furthermore, the sealing between the cells inside a stack has to be improved.

Last but not least the direct methanol fuel cell has to be mentioned. This low-temperature fuel cell can be operated with a mixture of methanol and water. The advantages are (1) the simple system layout, and (2) methanol can be converted in the fuel cell directly without a preceding reforming step. The disadvantages of the DMFC are the low efficiency and the limited lifetime of typically a few thousand hours. Typical applications are in the 100 W region, e.g. in recreational applications. Fuel cell systems in this power range can be purchased from the company Smart Fuel Cells. Instead of using a battery which has to be recharged after a while, the DMFC offers the ability to operate as long as methanol is present or can be refilled. Therefore continuous operation of electric devices is possible even when no electricity for recharging of a battery is available. Another typical application can be found in the kilowatt power range. In pallet trucks a DMFC can replace the battery pack. The advantage is the continuous operation with only short refuelling breaks. At the moment the research effort is focusing on new membrane materials and long-term stable catalysts.

Seeing this overview, it can well be stated that fuel cells are evolving to be a well developed technology with far-reaching R&D expertise. When and how the 'real' markets with 'real', 'typical' customers can be entered now remains to be seen. This will not only be a question of attractiveness of the technology and the products that are made thereof, but also of a variety of societal and economical conditions, regulations, political decisions on emission control *etc*. Generally, in technology development three phases can be discerned, and fuel cells are no exception to this rule: first, the technology is brought from scientific

principle to laboratory demonstration, then first demonstration and niche
market items are brought to the public, and, finally, the technology achieves a
break-through and wins a reasonable share in markets or at least firmly
establishes itself in niche areas. The third phase would especially be dominated
by considerations that have nothing much to do with the technology as such –
although customers would still tend to discuss such topics as the pros and cons
of diesel versus petrol vehicles – but rather with aspects like usefulness, durability, cost, added value *etc.* According to our assessment, fuel cells stand at
the beginning of phase 2. This is a good point in time to leave the major part of
further developments to the engineers and company laboratories and scientifically venture out to new shores!

In the volume you are just holding we therefore have attempted to look
beyond our own noses and try to identify new fields of high potential and
interest to the fuel cell community. This includes many fields where success is
still fragile or technical solutions are at a laboratory level, and in some cases
also theoretical concepts that sound worthwhile but still lack practical proof. In
all chapters of this book, though, we have tried to encourage authors to extend
today's knowledge to new concepts and ideas. *Innovations in Fuel Cell Technologies* does not address a single fuel cell technology – many topics covered
here are relevant to several types of fuel cells – but tries to direct the reader's
attention to the developments of tomorrow and the technology of the day after
tomorrow. The chapters may serve as an early warning to technology developers of the rewarding prospects on the horizon as well as orientation to stu-
dents and young researchers in guiding their studies.

The first group of two chapters describes the prospects of miniaturising fuel
cells. This topic has two intriguing aspects: first, that of extreme simplification
and (possibly) low cost; and second, that of integration into everyday life and
equipment. Miniature fuel cells can be part of food wrappings, can act as
environmental and medical sensors, be used as implants *etc.* In all these ways
they would become an integral part of our lives without our even noticing that
we are using 'fuel cell technology'.

The second set of two chapters looks at high-temperature polymer membrane fuel cells and their application as on-board electricity supply, even in very
large vehicles, such as aircraft. Although well in development, this membrane
type has its intrinsic problems and scaling up to large units is a huge challenge.

The following group of two chapters covers non-standard fuels like pure
carbon, and the handling of fuel impurities. For very different reasons, the use
of carbon offers a high potential for applications, although large technological
challenges still exist. This fuel is not conventionally considered for direct use in
fuel cells, but ridding oneself of excessive fuel processing would bring a high
potential for simplification of systems and subsequent cost reduction. The same
goes for a higher tolerance of fuel electrodes towards impurities. Admittedly,
these chapters predominately address SOFC technology, with some relevance
for high temperature fuel cells in general.

The twin set on degradation modelling and accelerated testing looks into a
very critical area of fuel cell research. Although vehicle applications call for no

more than 5000 to 10 000 h of operational lifetime, stationary applications for electricity production require up to 10 years of total component life. It is obviously impossible to test components for this length of time and there is a dire need for acceleration in testing lifetime limiting effects. The main prerequisite in accelerated lifetime testing, though, is the profound understanding of degradation issues.

Finally, we look into the prospects of reversing the fuel cell reactions towards producing instead of consuming hydrogen. This possibility has been under discussion for some years and the first technical units that can actually reverse their principle of operation seem to be under development. In line with the concept of this book, though, we look a little further at the concept of co-producing hydrogen in high-temperature fuel cells, at constructing closed loops of electrochemical conversion of chemicals to energy and vice-versa, and at the use of the SOFC principle to – in reverse – produce hydrogen at extremely high efficiencies.

The concluding chapter then inspects the pitfalls in bringing a technology from demonstration to technical maturity, including the issues of incumbent and concurrent technologies and introduction of products to consumer markets, an issue scientists and laboratory engineers may find strange and irrelevant to their work. Nevertheless, these aspects should be considered at an early stage in development and a careful assessment of efficiency, added value and usefulness aspects may have cut many technology developments short, long before they failed in the markets.

We do hope that readers will find this volume useful or at least interesting reading. We are fully aware that developments are not static and that many topics covered here will find their technical solution and market entry. We are convinced that the potential in fuel cell technologies is tremendous and that their commercial success is necessary in tailoring the worldwide energy supply systems towards efficiencies and emission levels that allow a long-term stable and sustainable development for the world economy and environment. Therefore we can at this point only hope that our book will be outdated soon! In which case we will gladly offer an update ...

Robert Steinberger-Wilckens and Werner Lehnert
Forschungszentrum Jülich, Germany

Contents

RSC Energy and Environment Series No. 2
Innovations in Fuel Cell Technologies
Edited by Robert Steinberger-Wilckens and Werner Lehnert
© Royal Society of Chemistry 2010
Published by the Royal Society of Chemistry, www.rsc.org

**Chapter 8   Accelerated Lifetime Testing for Phosphoric Acid
Fuel Cells   249**
John Donahue, Ned Cipollini and Robert Fredley

Part 5: Hydrogen Generation and Reversible Fuel Cells

Chapter 9   Electrolysis Using Fuel Cell Technology   267
A. Brisse, J. Schefold, C. Stoots and J. O'Brien

Part 6: Outlook

Part 1
Micro-applications and Micro-systems

Introduction

Generally, fuel cells are intended for applications in the range from a few kilowatts to several megawatts of electrical power. In recent years, though, applications have been suggested for 'micro-systems' of a few milliwatts to watts. These devices open up completely new markets and fields of application; for instance, with fuel cells that feed directly on body fluids and enzymatic processes that can be used for medical purposes. Another challenging idea is food wrappings with integrated fuel cells running on decay products (enzymes), thus automatically indicating degraded content.

As another example, micro fuel cells based on conventional electrochemical processes and fuels address the market for sensors and other small systems where it is desirable to integrate the power supply with the electronic device.

In Chapter 1, 'Printed Enzymatic Current Sources', recent research and development of biofuel cells is presented. These cells are capable of transforming chemical energy directly to electrical energy *via* electrochemical reactions involving enzymatic catalysis. The operating conditions and materials, as well as fuels utilised differ considerably from conventional fuel cells. The possibility to utilise biological catalysts, *i.e.* enzymes, as active components of a printed power source cell are reviewed. Research on biofuel cells started in the 1960s. Since the year 2000 there has been increasing interest for this technology.

Chapter 2, 'Potential of Multilayer Ceramics for Micro Fuel Cells', describes the challenges of this new technology and also its realisation. Like any fuel cell system, a micro fuel cell consists of sub-systems like the fuel cell, blowers, pumps, electrical converters, *etc.* Often all these components are shrunk versions of larger fuel cell systems. In order to reach the envisaged cost goals, the number of system components has to be reduced in a micro-system. Furthermore, a reduced number of parts will decrease the probability of a system failure. Multilayer ceramics open the possibility of a high integration of fuel cell components including, for example, flowfields and electronics. The principles of this new technology and also several applications are presented in this second chapter.

CHAPTER 1

Printed Enzymatic Current Sources

MATTI VALKIAINEN, SAARA TUURALA,
MARIA SMOLANDER AND OTTO-VILLE KAUKONIEMI

VTT Technical Research Centre of Finland, Biologinkuja 5, Espoo,
P.O. Box 1000, FI-02044 VTT, Finland

1.1 Introduction

Biofuel cells are devices capable of transforming chemical energy directly to electrical energy *via* electrochemical reactions involving enzymatic catalysis replacing precious metal catalysts. Operational principles are the same in biofuel cells and in conventional fuel cells, but the operating conditions, catalysts and materials, as well as fuels utilised differ considerably from conventional fuel cells.

In a microbial fuel cell the chemical energy is converted to electrical energy by the catalytic reaction of microorganisms, which produce their own enzyme catalysts. These microbial fuel cells have application areas such as wastewater treatment. Another group of biofuel cells comprises of enzymatic fuel cells that are equipped with purified enzymes which are further on dealt with in this chapter.

In an enzymatic biofuel cell various oxidising and reducing enzymes, *i.e.* oxido-reductases are applied as biocatalysts for the anodic or cathodic half-cell reactions. The electron transfer process in glucose oxidase (GOx) half-cell reaction using glucose as the fuel was first shown to take place by Yahiro *et al.*[1] Recently, the state-of-the-art of enzyme catalysed fuel cells were reviewed by Minteer *et al.*,[2] Davis and Higson[3] and Cooney *et al.*[4] The introduction of

RSC Energy and Environment Series No. 2
Innovations in Fuel Cell Technologies
Edited by Robert Steinberger-Wilckens and Werner Lehnert
© Royal Society of Chemistry 2010
Published by the Royal Society of Chemistry, www.rsc.org

enzymes enables the operation of the cell under mild conditions and the utilisation of various, renewable chemicals as fuels. Biofuel cells can be utilised in various applications, including miniaturised electronic devices, self-powered sensors and portable electronics. It is also anticipated that implanted biofuel cells could utilise body fluids, particularly blood, as the fuel source for the generation of electrical power, which may then be used to activate pacemakers, insulin pumps, prosthetic elements, or biosensing systems.

A power source integrated with printed electronics could have a remarkable market potential in several mass-marketed consumer products, *e.g.* as package integrated functionalities (sensors, displays, or entertaining features, *etc.*) or as part of diagnostic devices. One of the main requirements is that the power source should be biodegradable or possible to incinerate with normal household waste. This demand is not easily met by traditional battery technology. The material costs of the power source should be reasonable, should not significantly increase the price of the product, and the cells should also to be made from roll to roll in a cost effective way. As an alternative power source the miniaturised biological, enzyme catalysed fuel cell, has the potential to be developed to meet these demands.

Biofuel cell research started in the early 1960s and activity has increased greatly since year 2000. The cumulative number of publications up to early 2010 is about 1800 including 1500 SciTech publications and 300 patents. The research has been very lively at St Louis University, where Professor Shelley Minteer and her group have authored altogether more than 100 papers and patents. The patenting has been led by companies such as Sony, Toyota and Canon.

In this chapter the possibility to utilise biological catalysts, enzymes as the active components of a printed power sources, *i.e.* biofuel cells, will be discussed. As a background for the realisation of this type of innovative concept we will first describe, in detail, those biological fuel cells that are potentially applicable for series production, with special focus on their performance figures. Potential printing methods and existing applications of power sources will also be discussed generally, thereafter mass-producible applications involving the use of enzymes are discussed first generally and then focussing on the production of enzymatically active layers by printing. Finally, the concept of a printed biofuel cell is presented.

1.2 Enzyme Catalysts in Fuel Cells

An active and stable biocatalyst is necessary to realise an enzyme-based power source. Enzymes are biocatalysts consisting of amino acids with the exception that the active centre of the enzyme may contain metal ions or other non-metallic compound co-factors. The three-dimensional structure of the enzyme molecule determines its substrate specificity. Substrate refers here to the compound, which is modified during catalysis. The most suitable redox enzyme types for bioelectrodes are those having relatively tightly bound co-factors

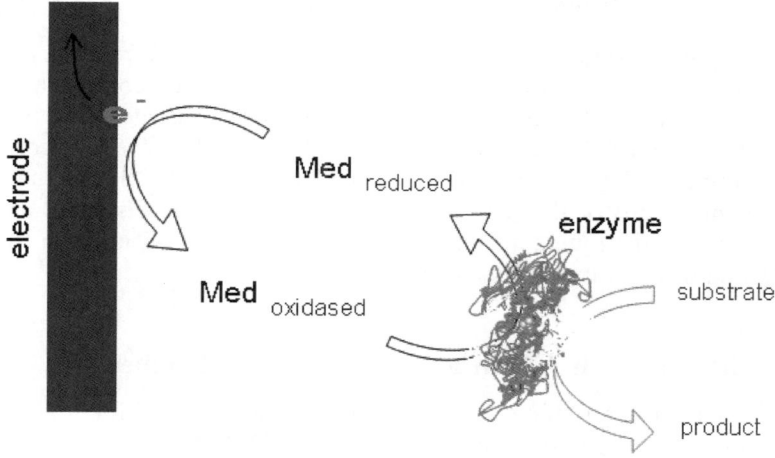

Figure 1.1   Schematic figure of the mediated electron transfer process.

(such as metal ions, pyrroloquinoline quinone (PQQ) and flavin adenine dinucleotide (FAD) or haem), which are needed to carry out the electron transfer within the enzyme.

The majority of enzymatic biofuel cells utilise mediated electron transfer (MET) type bioelectrocatalysis, which is applicable for many redox enzymes. Mediators are redox species with reversible electrochemistry that can transfer electrons between the co-enzyme/co-factor of an oxido-reductase enzyme and the electrode. These mediators can be either in the electrolyte solution or immobilised in the electrodes. Figure 1.1 illustrates the principle of mediated electron transfer on a bioelectrode. The MET type bioelectrocatalysis often offers current density advantage over the direct electron transfer (DET) type as long as the mediator concentration is sufficiently high; however, MET has also some disadvantages like thermodynamic loss and potential mediator leakage.[5] Recently, the studies on DET have become an object of growing interest as reviewed by Cooney *et al.*[4]

The indicator of the catalytic performance, enzyme activity, is related to the reaction rate. The SI unit used to express this is katal (kat) which expresses the amount of enzyme that is needed to transform 1 mol of substrate in 1 s (distinct from the activity unit, U, which expresses the amount of enzyme needed to transform 1 μmol in 1 min. 1 kat = 60 000 000 U). The unit is defined in a way that represents the maximum catalytic power of the enzyme in specified environmental conditions. When the amount of enzyme activity on the electrode is known the approximate theoretical limit for electron transfer and cell current caused by the characteristic enzyme activity can be calculated. For instance, in the case of two electrons liberated per reaction at the molecular level and enzyme activity of $1\,\text{nkat}\,\text{cm}^{-2}$ the theoretical limiting current of $190\,\mu\text{A}\,\text{cm}^{-2}$ is obtained.

1.3 Enzyme-based Microsystems for Power Production

In this chapter the state of the art of biofuel cells is examined from the point of view of the developments in the cell structure. Biofuel cell constructions are presented in three different categories: biofuel cells constructed in a liquid chamber, biofuel cells based on carbon fibre design and biofuel cell constructions suitable for large-scale production. Different biofuel cell structures and their potential construction or manufacturing methods are discussed and the performance of the different biofuel cell constructions is reviewed in the following chapters and in Tables 1.1 and 1.2.

1.3.1 Biofuel Cells Constructed in a Liquid Chamber

A liquid chamber or a bulk electrochemical cell is a basic tool for studying the performance of enzymatic electro-active layers. Disk electrodes are covered with a paste containing electro-catalytic agents. The size of the disk electrodes in the characterisation studies is usually a few square centimetres and a common practice is to add the component acting as a mediator into the buffer solution. In the studies considered below, the electrode surface material was typically gold or glassy carbon, but porous carbon has also been used.

In 1998 Willner *et al.*[6] presented a liquid chamber biofuel cell construction based on golden disk electrodes (*ca.* 0.2 cm^2). The open circuit voltage (OCV) was 0.31 V and the short circuit current density 114 µA cm^{-2}. The power output of the cell was 32 µW at ambient temperature, which corresponds to a power density of *ca.* 160 µW cm^{-2}. Glucose oxidase from *Aspergillus niger* was used as the anode enzyme on a roughened golden disk electrode modified with a cystamine monolayer and electron acceptors PQQ and FAD. Microperoxidase-11 (MP-11) was harnessed as the cathodic enzyme and covalently linked to a cystamine monolayer on a roughened golden electrode surface.

In Katz *et al.*[7] Willner and collaborators utilised the similar electrode structure in a membraneless liquid chamber. In this design the anode worked in an aqueous electrolyte and the cathode in a non-aqueous electrolyte (Dichloromethane), where cumene peroxide was used as the oxidant. The open circuit voltage for this design was 0.99 V. The area of the disk electrodes was 0.4 cm^2 and the maximum power output of the device was 540 µW, which corresponds to the maximum power density of 4300 µW cm^{-2}. The short circuit current density was 830 µA cm^{-2}.

When the suitability of a fuel cell concept for mass-production is considered, the affordable price of the electrode materials, biocatalysts and mediators are essential requirements. The study reported by Liu *et al.*[8] confirmed that glassy carbon and noble metal electrodes could be substituted with a porous carbon support. Here the enzymes used were glucose oxidase from *Aspergillus niger* and laccase from *Coriolus versicolor* and the cost-efficient mediators were ferrocene monocarboxylic acid (FeMCA) and 2,2'-azino-bis-(3-ethylbenzthiazoline-6-sulfonic acid) diammonium salt (ABTS). Carbon nanotubes and chitosan were mixed with both enzymes before they were spread onto porous

Table 1.1 Experimental set-ups for enzyme-catalysed fuel cells reported in literature.

Reference	Set-up	Anode (enzyme/mediator)	Cathode (enzyme/mediator)	Separator	Fuel (anode/cathode)
	Coated electrodes in a liquid chamber				
Willner et al.[6]	Au wire electrodes	GOx/PQQ-FAD	MP-11	glass frit	Glucose/H_2O_2
Katz et al.[7]	Au disc electrodes	GOx/PQQ-FAD	MP-11	liquid–liquid interface	Glucose/cumene peroxide
Liu et al.[8]	Porous carbon, CNTs	GOx/FeMCA	Laccase/ABTS	membrane	Glucose/O_2
Zhang et al.[9]	Graphite foil	MDH/TMPD	No enzyme/mediator	cation exchange membrane	Methanol/$KMnO_4$
	Carbon supported electrodes				
Chen et al.[11]	7 μm carbon fibre	GOx/redox-polymer	Laccase/redox-polymer	none	Glucose/O_2
Heller[10]	7 μm carbon fibre	GOx/redox-polymer	BOD/redox-polymer	none	Glucose/O_2
Mano et al.[13]	7 μm carbon fibre	GOx/redox-polymer	BOD/redox-polymer	none	Glucose/O_2
Sakai et al.[15]	Carbon fibre sheet	GDH + NADH + DI/VK$_3$	BOD/$K_3[Fe(CN)_6]$	cellophane	Glucose/O_2
Tokita et al.[14]	Carbon fibre sheet	GDH + NADH + DI/ANQ	BOD/$K_3[Fe(CN)_6]$	cellophane	Glucose/O_2
Akers et al.[18]	Carbon felt + methylene green	ADH/NADH	Pt	Nafion	Ethanol/O_2
Akers et al.[18]	Carbon felt + methylene green	ADH + AldDH/NADH	Pt	Nafion	Ethanol/O_2
Akers et al.[18]	Carbon felt + methylene green	ADH + FdDH + FDH/NADH	Pt	Nafion	Methanol/O_2

ABTS, 2,2′-azino-bis-(3-ethylbenzthiazoline-6-sulfonic acid) diammonium salt; ADH, alcohol dehydrogenase; AldDH, aldehyde dehydrogenase; ANQ, 2-amino-1,4-naphthoquinone; CNT, carbon nanotube; DI, diaphorase; FAD, flavin adenine dinucleotide; FDH, formate dehydrogenase; FdDH, formaldehyde dehydrogenase; FeMCA, ferrocene monocarboxylic acid; GDH, glucose dehydrogenase; GOx, glucose oxidase; MDH, methanol dehydrogenase; MP-11, microperoxidase-11; NADH, β-nicotinamide adenine dinucleotide disodium salt; PQQ, pyrroloquinoline quinone; TMPD, 2,2,4-trimethylpentane-1,3-diol; VK$_3$, vitamin K$_3$.

Table 1.2   Performance data of enzyme-catalysed fuel cells.

Reference	Area (mm^2)	OCV (V)	P_{MAX} (μW)	P_{MAX} $(\mu W\,cm^{-2})$	I_{MAX} $(\mu A\,cm^{-2})$	Conditions
Coated electrodes in a liquid chamber						
Willner et al.[6]	20	0.31	32	160	114	Exernal load 3.0 kΩ
Katz et al.[7]	12.6	0.99	520	4140	830	External load 0.4 kΩ
Liu et al.[8]	160	0.66	159.6	99.8	950	pH 4.0
		0.55	23.6	14.75	738	pH 5.0
		0.16	3.2	2.0	144	pH 7.0
Zhang et al.[9]	1600	1.4	4000	250	380	0.67 V
Carbon supported electrodes						
Chen et al.[11]	0.44	0.65	0.3	64	160	0.40 V, 23 °C, pH 5.0
			0.6	137	343	0.40 V, 37 °C, pH 5.0
Heller[10]	0.82	0.72	2.4	290	558	0.52 V, 25 °C, pH 7.2
			4.3	540	1040	0.52 V, 37 °C, pH 7.2
Mano et al.[13]	0.44	0.72	1.2	280	538	0.52 V, 25 °C, pH 7.2
			1.9	440	846	0.52 V, 37 °C, pH 7.2
Sakai et al.[15]	2040	0.80	29 600	1450	4830	0.30 V, pH 7.0
Tokita et al.[14]	–	0.80	–	3000	6000	0.50 V, pH 7.0
Akers et al.[18]	100	0.61	1160	1160	1900	20 °C, pH 7.15
Akers et al.[18]	100	0.82	2040	2040	2800	20 °C, pH 7.15
Akers et al.[18]	100	0.67	1550	1550	2900	20 °C, pH 7.15

carbon, and the area of the electrode was $7\,mm^2$. The important dependence of the cell performance on pH was studied. The highest power density of $99.8\,\mu W\,cm^{-2}$ was measured at pH 4.0, whereas it decreased to $14.75\,\mu W\,cm^{-2}$ at pH 5.0 and was only $2.0\,\mu W\,cm^{-2}$ at pH 7.0. In addition, the same behaviour appears in the open circuit voltage, which was 0.66 V at pH 4.0, 0.55 V at pH 5.0, and 0.16 V at pH 7.0. The short circuit current density was $950\,\mu A\,cm^{-2}$ at pH 4.0, $738\,\mu A\,cm^{-2}$ at pH 5.0, and $144\,\mu A\,cm^{-2}$ at pH 7.0.

Zhang et al.[9] studied a novel type of direct methanol biocatalytic fuel cell (DMBFC) based on enzymatic conversion of methanol by methanol dehydrogenase (MDH) from *Methylobacterium extorquens* at the anode. The terminal electron acceptor at the cathode was potassium permanganate.

Performance characteristics achieved were: open circuit voltage 1.4 V, power density $0.25\,mW\,cm^{-2}$ and current density $0.38\,mA\,cm^{-2}$ at the operating voltage of 0.67 V, and a continuous operation time of 2 weeks.

1.3.2 Miniature Membraneless Biofuel Cells

Miniature biofuel cells described in this chapter are comprised of two bio-electrocatalyst-coated carbon fibre electrodes inside a small test cell without a separator. This miniature cell structure differs from the liquid chamber construction where covered disk or wire (*i.e.* bigger) electrodes are mainly used and usually a separator is needed. The most remarkable features of these miniature biofuel cells are their small size and structural simplicity due to which they are potential power sources for medical applications, *e.g.* hypodermic implants. For example, a sensor–transmitter could be powered by a miniature biofuel cell that continuously produces a few microwatts, of which less than 1 µW will be consumed by the sensor; the transmitter will require most of the power. The transmitted information will be easily acquired outside the body with a small ultra-capacitor, that stores about 10 µJ, which is enough for a 1 ms burst of 10 mW (1–10 GHz) every 10 s.[10]

In 2001 Chen *et al.*[11] studied a cell where carbon fibres were placed in two $1\,mm \times 1\,mm$ grooves bored in a 3 cm long polycarbonate support. The gap between the grooves was 400 µm and the active area of the fibres was $0.44\,mm^2$. Glucose oxidase from *Aspergillus niger* and laccase from *Coriolus hirsutus* were used as the anode and cathode enzymes, respectively, and osmium complexes were used as mediators. A cross-linking agent poly(ethylene glycol) diglycidyl ether (PEGDGE) was used to immobilise the enzymes. The power density for this design was $64\,\mu W\,cm^{-2}$ at 23 °C and $137\,\mu W\,cm^{-2}$ at 37 °C. The true power output of the device was 280 nW and 600 nW, respectively.

In 2002 Tsujimura *et al.*[12] demonstrated that a membraneless biofuel cell is capable of operating in a physiological buffer solution, which enables biofuel cells to be used *e.g.* in living plants. In 2003, Heller[10] and Mano *et al.*[13] demonstrated a miniature glucose–O_2 biofuel cell implanted in a grape (Figure 1.2). The electrodes were 7 µm in diameter and 2 cm in length.[10,13] On the anode, glucose oxidase from *Aspergillus niger* was immobilised and electrically connected onto the carbon fibre. Respectively, on the cathode, bilirubin oxidase from *Trachyderma tsunodae* was immobilised and electrically connected onto the carbon fibre.

1.3.3 Biofuel Cell Constructions Suitable for Large-scale Production

The Sony Corporation has developed a biofuel cell that uses carbohydrates (sugars) as its fuel and enzymes as its catalyst. The first published passive-type test cells achieved a power output of 50 mW.[14] This output was high enough to power an MP-3 player with two speakers (Figure 1.3). The second passive-type

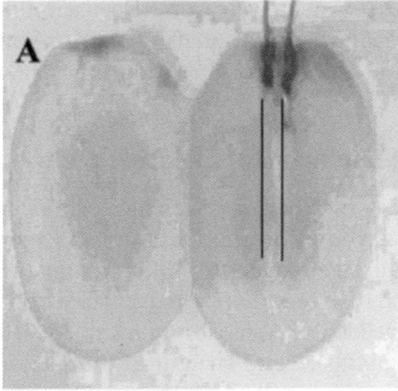

Figure 1.2 Two carbon fibre electrodes implanted into a grape. Reprinted with permission from Mano *et al.*[13] Copyright 2003 American Chemical Society.

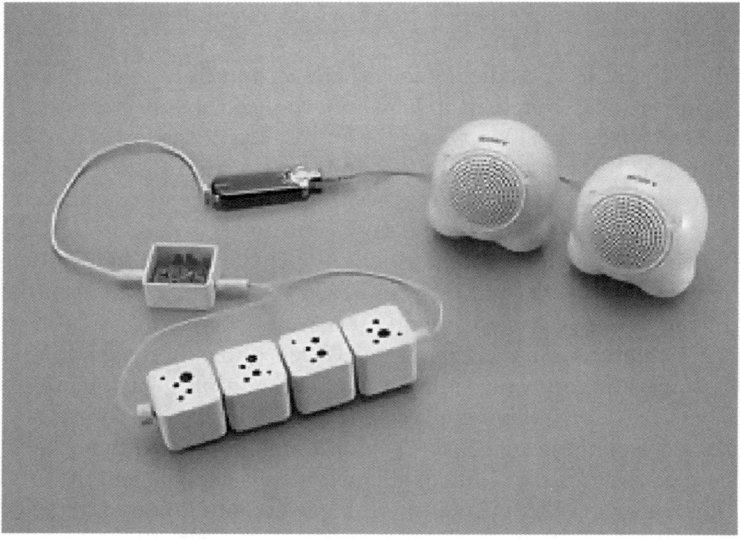

Figure 1.3 Sony's four biofuel cell units connected to an MP-3 player and two speakers, Tokita *et al.*[16]. Reproduced by permission of ECS - The Electrochemical Society.

test cells were capable of generating a power of over 100 mW, high enough to operate a radio-controlled car.[15]

Sony's biofuel cell is based on glucose dehydrogenase (GDH) from a *Bacillus* sp. and diaphorase (DI) from *Bacillus stearothermophilus* as the anode enzymes and methyl-1,4-naphthoquinone (vitamin K_3, VK_3) or 2-amino-1,4-naphthoquinone (ANQ) as the anode mediator. β-Nicotinamide adenine dinucleotide

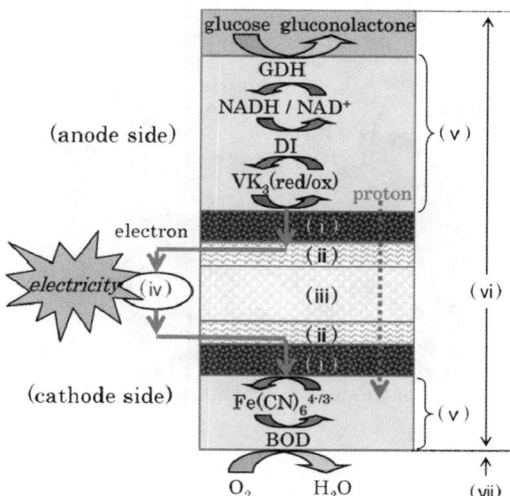

Figure 1.4 Reaction scheme of the Sony's biofuel cell. (i) Carbon fibre electrode, (ii) Ti mesh current collector, (iii) cellophane membrane, (iv) external circuit, (v) immobilised enzyme/mediator layer, (vi) electrolyte solution and buffer, (vii) air. Sakai *et al.*[15] – Reproduced by permission of The Royal Society of Chemistry.

disodium salt (NADH) is added as the soluble co-factor of the dehydrogenase. The cathodic enzyme and mediator are bilirubin oxidase (BOD) from *Myrothecium verucaria* and $K_3[Fe(CN)_6]$, respectively (Figure 1.4).[15,16] The performance of Sony's biofuel cells was achieved by applying different approaches described below.

In order to increase the maximum current density, enzymes and mediators were effectively entrapped on ozone treated carbon fibre (CF) electrodes retaining the enzymatic activity. CF sheets were used because CF has a higher surface area and a higher porosity, allowing an undisturbed transport of fuels.[15]

Sakai *et al.*[15] reported that the catalytic current density of the manufactured CF bioanode was lower than assumed at first, despite the fact that the evaluated surface area of the CF electrode was higher than previously. It was presumed that the low catalytic current density was due to insufficient proton diffusion in the CF bioanode. In order to solve this problem, the effects of buffer concentration in the glucose containing electrolyte solution (phosphate buffered saline, PBS) were examined. The current density was considerably improved in 1.0 M PBS compared to the previously used electrolyte of 0.1 M PBS. It was also noticed that the current density decreased when the buffer concentration was higher than 1.0 M, presumably due to enzymatic activities of GDH and DI.[15] On the other hand it was observed that the higher ionic strength of the buffer could beneficially suppress the pH drop during the operation of the CF bioanode.[15] As a conclusion, Sakai *et al.* proposed two rate determining steps for the reaction: the insufficient proton diffusion at low buffer concentrations

anode collector   cathode collector

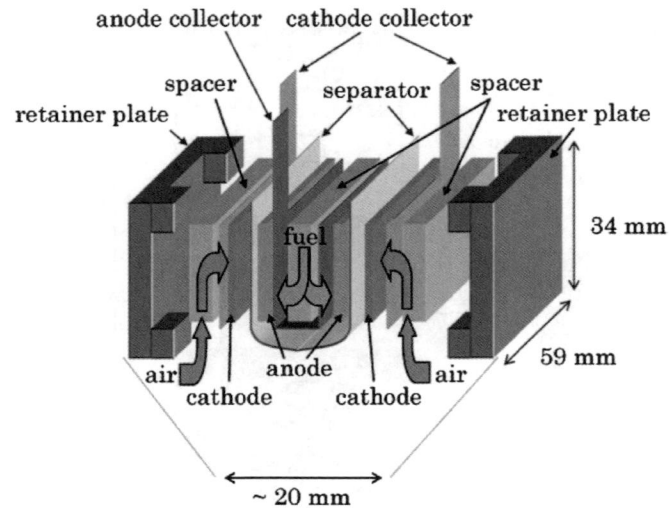

Figure 1.5   Schematic view of two biofuel cells connected in parallel. Sakai *et al.*[15] –
Reproduced by permission of The Royal Society of Chemistry.

(<1.0 M) and inactivation of the enzymatic activity at higher buffer concentrations (>1.0 M).[15] The concentration of the buffer was therefore optimised for the immobilised enzymes.

The structure of the cathode was also optimised for efficient oxygen
absorption. First, Sakai *et al.* used a 'sink type' cell where O_2 was supplied to
the cathode through the bulk solution phase. A decrease of the current density
was noticed due to insufficient O_2 supply. In order to obtain an efficient O_2
supplying system, a biocathode exposed directly to air was designed. The
resulting structure was an open-air type cell with a cathode structure in a
polymer electrolyte fuel cell or an air battery, because the concentration in air
and the diffusion coefficient in gas are much higher than those in aqueous
solution (Figure 1.5).[14,15]

With these optimisations Sony developed a biofuel cell with a maximum
power density which is 1.45 mW cm^{-2} with VK$_3$ as the anodic mediator and
3.0 mW cm^{-2} with ANQ as the anodic mediator.[15,16]

At St Louis University, Department of Chemistry, Professor Shelley Minteer's group is focusing their research on increasing the power density and
lifetime of biofuel cells. They have developed immobilisation techniques for
enzymes at the electrode surface based on chitosan and Nafion.[17] Fragile
enzymes stabilised and protected in tiny pore-like structures using this technique have increased operation time of months instead of days and increased
power and lifetime. High current density bioanodes are formed from poly
(methylene green) (an electrocatalyst for NADH) modified electrodes coated
with a layer of tetrabutylammonium bromide salt-treated Nafion with dehydrogenase enzymes immobilised within the layer. Ethanol/O_2 biofuel cells
employing these bioanodes have yielded power densities of 1.16 mW cm^{-2} with

a single-enzyme system (alcohol dehydrogenase) and $2.04\,\mathrm{mW\,cm^{-2}}$ with a double-enzyme system (alcohol dehydrogenase and aldehyde dehydrogenase) in the polymer layer.[18] The group has also applied this technique to a wide variety of fuels including carbohydrates, fatty acids and alcohols. Professor Shelley Minteer is also co-founder of a company, Akermin, located in St Louis, and developer of stabilised enzyme solutions for producing power, food, fuels and specialty chemical intermediates. Akermin, Inc., in conjunction with St Louis University, has achieved a record level of 3 years for enzyme stability in an energy-generating biofuel cell electrode.[19]

1.4 Printing Processes as Manufacturing Methods for Power Sources

Several printing techniques such us flexography, screen printing, gravure printing and inkjet printing offer interesting possibilities in the manufacture of thin power sources. These processes offer a wide variety of manufacturing methods that can be utilised in creating thin batteries in a cost efficient way. Table 1.3. contains some characteristics of different kind of printing methods and parameters of printed layers created by them.

As can be seen from Table 1.3, printing processes are true mass manufacturing methods that have really high production capacities. The other advantage of printing processes is that they enable patterning. This means that it is relatively easy to tailor the shape of the printed batteries to meet the requirements of different applications. In addition the capacity of the battery can be easily altered by changes in battery geometry.

Different printing methods are used for different purposes in publication printing. For example, gravure printing is typically used for printing high quality magazines; offset printing is used for printing newspapers; flexography for printing packages; inkjet for personifying printed products; and screen

Table 1.3 Typical industrial printing process characteristics for different printing methods.[a]

Method	Line speed $(m\ s^{-1})$	Machine width (m)	Calculated production capacity $(m^2\ h^{-1})$	Typical achievable printed layer thickness (μm)
Gravure printing	20	2,4	129 600	~2, typically less
Screen printing	2	1	7 200	100–300
Flexography	10	1,5	54 000	5
Ink jet (industrial)	1	1	3 600	1
Offset printing	15	1,5	81 000	~2

[a]From: *Handbook of Print Media*, ed. Helmut Kipphan, Springer, 2001, ISBN 3-540-67326-1.

printing for different kinds of specialty products such as posters or Braille writings. These different printing methods all state different kinds of demands for the printing substrate as well as for ink quality. For example, rotogravure printing is very sensitive for smoothness of the printing substrate whereas flexography can also be used for rougher substrates.

In rotogravure printing the printing ink is transferred to printing substrates by engraved cups in the metallic printing cylinder in a nip that is formed between the printing cylinder and a backup roll. With rotogravure it is possible to reach very good resolution and high printing quality and extremely high production capacity. Inks used in rotogravure have very low viscosity.

Flexography printing works, in principle, like a stamp. Printing ink is lifted with an engraved anilox roll to a patterned printing cylinder made from polymer materials. The ink is then transferred from the printing plate to the printing substrate. The amount of ink transferred is mainly controlled by the anilox roll parameters. The inks in flexography have higher viscosity compared to rotogravure inks.

Offset printing is another high output printing method. It has a rather complicated ink transfer process and it is therefore not widely used in the printing of active materials. Tailoring the active inks to be suitable for offset is also demanding. Offset printing is a lithographic printing method in which ink is fed to a separate ('offset') printing blanket through a impression cylinder. The printing image is formed in the impression cylinder by areas having different surface chemistries. Printing ink wets areas in the impression cylinder that have compatible surface energy with the ink and an image is formed on these areas. Offset printing is the most widely used printing method in publication printing because it has good quality, high production capacity and is relatively quickly prepared for starting the production process.

Screen printing is a method in which the ink is forced through a patterned wire to the printing substrate. It is relatively easy to adopt because it places virtually no demands for the quality of the printing substrate. It can also be used to create extremely thick printed layers, which is the reason that it is used in the printing of many different kinds of functional inks. Inks used in screen printing have high viscosity which is also beneficial for printing many functional inks because the dry matter content of the ink can be extremely high.

Printing can also be used in combination with other manufacturing techniques such as laminating and heat sealing. For example, printable battery anodes and cathodes could be printed and a separator membrane laminated between them. However, printing machines are usually unsuitable for printing power sources without some tailoring. Typically, all the necessary unit operations are not installed in one single machine. Additionally, printing of power sources requires somewhat different characteristics than publication printing. In publication printing, the appearance of the printed image is very important, whereas for printed batteries the structure of the printed layer is the most important aspect.

1.4.1 Types of Thin and Printable Power Sources

Thin and printable power sources can be divided, for example, into the following categories:[20]

- Printed batteries (*e.g.* lithium ion, zinc–alkaline)
- Micro fuel cells (direct methanol, direct ethanol, polymer electrolyte, bio)
- Micro energy sources (*e.g.* radio-frequency power, energy harvesting)
- Photovoltaics (solar cell)
- Capacitors

In principle, most of these devices could be manufactured by printing techniques. However, not all printing techniques are suitable for the different kinds of power sources because they require different kinds of structures that need to be manufactured by printing. For example, chemical batteries such as zinc–alkaline batteries require thick anode and cathode layers for the high energy content of the battery. Printed solar cells on the other hand require thin, pinhole free and very uniform anode and cathode layers for the cell to be functional at all. This means that different types of power sources need different printing methods for the optimum performance. For instance a zinc–alkaline battery should, preferably, be manufactured by a silk screen process, which creates thick layers, while solar cells are best prepared by rotogravure printing, which can be used to create extremely thin layers.

Printed thin film batteries are made by a couple of industrial companies: Power Paper and Enfucell. Their products are primary power sources based on zinc/manganese dioxide chemistry.[21,22]

1.5 Enzymes in Mass-production Applications

In addition to the use in biofuel cells, enzymes have been successfully and widely harnessed for analytical and well as other functional purposes. The interest towards mass production of enzyme-based functional products and materials has been one driving force for the development of methods enabling the incorporation of enzymes into or on flexible substrates like paper or plastics by printing or other reel-to-reel applicable processes. Striving towards disposable, single-use electrochemical devices, biosensors for biomedical, environmental, microbiological and food applications has also increased interest towards printing as a way to cost-efficient production of enzymatically active thin multilayer structures.[23–25]

Biosensors are defined as analytical tools consisting of a biological component specific to the analyte and a physical component which is able to transduce the biological signal to a physical one. For instance, enzymes, antibodies and cells can be used as the biological component of biosensors and the signal can also be detected in many ways, *e.g.* with amperometric, potentiometric, optical and calorimetric methods.

Several types of enzymatic biosensors have been developed, particularly for the detection of glucose. For many applications like point-of-care diagnostics and measurement of food quality, the availability of low cost, single-use, disposable sensor technology would be a considerable benefit. Production of the respective sensors using printing technologies would enable automated, reproducible fabrication processes. Basically, several different printing methods are, in principle, suitable for the production of bioelectrochemically active layers with high reproducibility from sensor to sensor, the possibility of mass production and long-term storage stability. Some examples are described in detail later in this chapter.

In addition to bioelectrochemical applications, *i.e.* biosensors and biofuel cells, printed enzymes can be applied as active components in true high-volume applications like packaging. In this application area, the utilisation of enzymes can be based on their substrate specificity or on the sensitivity of the reaction rate to external conditions like temperature or humidity. Time–temperature integrators (TTIs) are inexpensive intelligent labels that can show an easily measurable, time–temperature dependent change that reflects the full or partial temperature history of a food product to which it is attached.[26] Several TTI systems have been proposed; among them are a few concepts based on enzymatic reactions,[27–30] *e.g.* a system marketed as CheckPoint® TTI, (VITSAB A.B., Malmö, Sweden). This concept is not actually printed, but can be included into the package in such a way that the indicator cannot be removed or replaced at a later stage.[31]

Another example of enzyme-based mass-product is Bioett®, a time–temperature history-indicating system for transport packages that incorporates an enzyme-based device in a passive radio-frequency circuit. After activation the conductivity changes as a function of time and temperature due to the enzymatic reaction and can be monitored and linked to the temperature history of the product.[32]

Moreover, enzymatic reactions have been harnessed in humidity and moisture sensors and indicators proposed to be used for instance as layered disc or strip type systems for moisture indication in diapers.[33–35] Enzyme-based, printed humidity indicators were studied by Smolander and collaborators in an EU project SUSTAINPACK.[36]

Enzymes have also been used as active compounds in package-integrated quality indicators, for instance Cameron and Talasila[37] and Smyth *et al.*[38] have explored the potential of detecting the unacceptability of packaged, respiring products by measuring ethanol in the package headspace with the aid of the ethanol-oxidising enzyme, alcohol oxidase.

In addition to mass-produceable indicators and sensor systems, enzymes can be used to control undesirable microbial surface contamination of foods. For instance, hydrogen peroxide, a substance with antimicrobial activity, can be introduced into the package in the enzymatic reaction catalysed by glucose oxidase, which has been covalently immobilised onto amino or carboxyl plasma-activated bi-oriented polypropylene films *via* suitable coupling agents.[39] In addition to antimicrobial functions, other enzyme-based functionalities can also be added to packaging materials. Hotchkiss[40] presented an example of *in situ* processing of milk by removal of lactose by attaching lactase enzyme to PE whereas Lehtonen *et al.*[41] patented a packaging material which removes oxygen from a package. In this case the oxygen-removing layer comprises an enzyme (preferably glucose oxidase) in a liquid phase sandwiched between plastic films. The same oxygen scavenging ability of glucose oxidase was investigated by Nestorson *et al.*,[42] who either immobilised glucose oxidase on latex film surfaces or incorporated it directly into the paper coating colour.

1.6 Printing and Coating of Enzymes

In this chapter some examples are presented on the production of enzymatically active layers using the different printing methods.

1.6.1 Screen Printing

Screen printing has been widely used for the production of biocatalytic layers, *e.g.* in various amperometric biosensors.[43–46] Frequently, the fabrication of enzyme electrodes for biosensor applications is proposed as a two-step process including manufacturing of the basic sensor structure by screen printing potentially combined with other processes like lamination, and thereafter the application of the enzymatic catalyst dropwise at a precise location of the sensor surface. For instance, Trivedi[46] constructed a fructose sensor based on membrane-bound fructose dehydrogenase applied on the Pt sensor surface in a polymer matrix of polyethyleneimine (PEI) and poly(carbamoylsulfonate) (PCS) hydrogel. The enzyme electrode was found to be stable during several months of storage in buffer. Even when operated frequently, the sensor was able to function for about 25–35 days. This kind of sensor can not yet be considered truly mass-producible; however, it is a good starting point towards reel-to-reel production of enzyme-based biosensors.

However, the need to simplify the fabrication process of enzyme electrodes by reducing the number of layers was noticed over 10 years ago and some reports dealing with the actual incorporation of the enzymatic catalyst into the conductive ink have also been proposed. For instance. Koopal *et al.*[47] incorporated glucose oxidase into carbon ink containing polypyrrole and demonstrated the feasibility of the ink for screen printing. Galan-Vidal *et al.*[48] described a glucose oxidase-based enzyme electrode based on enzyme incorporated directly to a screen-printable graphite–epoxy composite. Interestingly, they report for example that cyclohexanone added to adjust the viscosity of the printed composition did not have an effect on the enzymatic activity of GOx. Good fabrication reproducibility is reported; however, the curing time (72 h, 40 °C) is not compatible with reel-to-reel production methods. Recently, Fang and collaborators proposed an enzyme electrodes for the detection of glucose, 3-β-hydroxybuturate or frucosyl valine, mixed into ink consisting of iridium–carbon particles, polyethyleneimine and hydroxyethyl cellulose in aqueous solution.[49–51]

Several stabilising substances including polylysine, PEG, dextran, lactitol, PEI, hydroxyethyl cellulose and polymers like Gafquat and Agrimer have been added to biocatalytic screen-printing thick-film pastes as stabilisers.[44,45,52–54] Shitanda *et al.*[55] incorporated glucose oxidase into polyamide microcapsules together with tetrathiafulvalene (TTF) before entrapping the capsulated biocatalysts into carbon based ink used for screen printed electrodes.

1.6.2 Inkjet Printing

Another printing method, inkjet printing deposition, is also a promising manufacturing tool for enzyme-based mass production due to its characteristics, *i.e.* relatively high production speed, low cost and the possibility for contactless printing. Different ink ejection technologies can be used in inkjet printing, the most commonly used being piezoelectric and thermal inkjet printing. Application of inkjet printing for biotechnical applications, *e.g.* for the development of biosensors, has been widely studied.[44,56–60] The crucial issue of inkjet printing lies in the engineering of inks, which can be reliably jetted due to their optimal viscosity and surface tension, but which simultaneously are compatible with the biocatalyst itself and do not have an inverse effect on the enzymatic activity.[59]

Setti *et al.*[57] used a commercial inkjet printer, for instance, to fabricate a two-layered enzyme electrode based on horseradish peroxidase (HRP). A conductive layer consisting of conductive polymer PEDOT/PSS was deposited first onto ITO-coated glass. Thereafter HRP was deposited in an aqueous ink containing EDTA as an antimicrobial agent and 10% glycerol as a wetting agent and stabiliser. As could be expected, based on the composition of the ink, the incorporation of HRP into the ink did not have a considerable effect on the enzyme activity. The enzymatic activity actually ejected from the printer was evaluated with the aid of the dye brilliant blue (E133) as a marker for the amount of deposited material. It was found that no considerable inactivation took place due to the printing procedure. It was presumed that glycerol, originally used as the wetting agent for the prevention of nozzle clogging, increased the enzyme's stability due to interaction between the protein and the polyol.

Di Risio *et al.*[59] formulated an inkjet ink containing HRP when studying the effect of ink additives, especially viscosity modifying substances like PEG, EG, glycerol, polyvinyl alcohol (PVA) and carboxymethyl cellulose (CMC) on the enzyme activity as well as on the jetting properties. It was found that some typical viscosity modifiers affected significantly HPR activity and that CMC was the best viscosity modifier in this case. Additionally, they studied the oxidative colour formation of ABTS catalysed by inkjet printed HRP and ended up in suggesting that the structure and chemistry of the printing substrate have a remarkable effect on the chromogenic reaction of a bioactive paper, better response being provided by substrates with less penetration and minimum spreading.[60]

Nishioka *et al.*[58] report an activity loss of peroxidase when printed using an ink jet with a piezo-ceramic actuator. The damage was found to decrease with slower compression rates and by the addition of trehalose and glucose to the ink. The stabilising effect of trehalose is proposed to be due to its extensive hydrogen bonding with proteins. In general the stability of the enzymes can be increased by chemical modification or by mixing them with various compounds such as polyvinyl alcohols, glycerol and synthetic polymers.[61] These compounds can be used to prevent the drying of the enzyme and to increase the shelf-life of enzymes.

Kouismi and Rochefort[62] proposed the encapsulation of laccase into polyethyleneimine microcapsules in order to prevent the leaking of the enzymes from the material and in order to create a stabilising environment for the biocatalyst.

1.7 Printed Biofuel Cells

As described above, biofuel cells can be utilised in various applications, including miniaturised electronic devices, self-powered sensors and portable electronics. In other applications the need of a low-cost, disposable power source is obvious. If realised using roll-to-roll manufacturing methods miniaturised biological fuel cells can be developed to meet these demands. These biocatalyst-based devices could be applicable in communicative packaging applications as disposable power sources for, *e.g.* active radio-frequency identification (RFID) tags, where a local power source enables longer reading distance than presently and more functions, such as memory and sensors.

In our own studies we have aimed at the realisation of cheap, disposable enzyme-based power sources. In the first step of the work we have focused on the construction of a printable enzymatic cathode based on a high redox potential fungal laccase from *Trametes hirsuta*[63] (Figure 1.6). Fungal laccases (*p*-diphenol dioxygen oxido-reductases, EC 1.10.3.2) are oxido-reductases which contain integrated redox centres in the form of four copper atoms and have, depending on the enzyme, a relatively high redox potential for oxidation of phenolic and other aromatic compounds. In addition to the potential applications in biofuel cells and biosensors,[11,13,64] it is notable, that laccases have a wide application potential in detergents, pulp bleaching, adhesives, fibre functionalisation, detoxification, denim bleaching, textile dye decolorisation and baking.[65,66]

Formulation of the electrochemically active enzyme-containing layer is a challenging task since most conductive inks are based on various solvents, which is contradictory to the fact that most enzymes need aqueous solutions for their stability and catalytic activity. In our studies the properties of the printed biocathode were optimised by varying the ink composition and printing substrate. Three principal ink types were studied as the basis for the

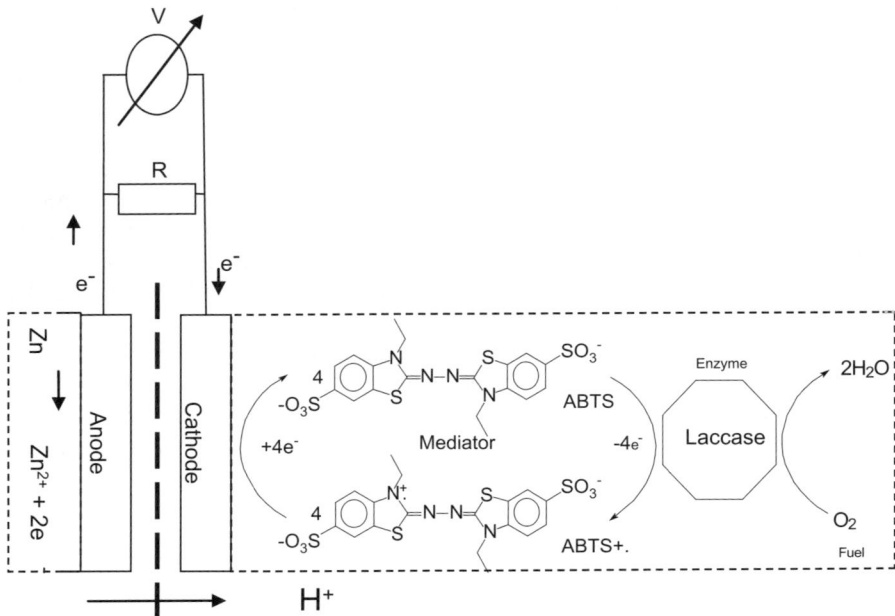

Figure 1.6 Schematic structure of the laccase: Zn half fuel cell. Reprinted, with permission, from Smolander et al.[63] © 2008 Elsevier.

enzyme-containing, conducting layers, two commercial inks from DuPont based on silver or carbon were further tailored with carbon nanotubes, additionally a third, experimental ink based on polyvinyl alcohol, carbon nanotubes and some additives was developed (Table 1.4). At this stage the coating was performed with a rod coater but basically the same inks are applicable to screen printing. The results showed that the enzymatic activity can be retained and maintained for months in different conductive inks, depending on the storage conditions (Table 1.5).

The biocathode was tested and optimised in a printed fuel cell construction using a zinc-based anode (a schematic structure of the cell is shown in Figure 1.7). Under optimised conditions, a fuel cell containing a laccase-based cathode maintained its capacity to generate power for several days.

The structure of the power source presented by us is manufactured by stacking different layers on top of each other. The cell contains a printed anode and cathode, printed current collectors, laminated separator membrane and outer shell that is manufactured from multilayered cardboard/plastic laminate. The cell can be sealed by heat sealing or gluing, for example.

Properties of the printed enzymatic fuel cell are primarily determined by the properties of anode and cathode inks. These can be tailored by for example suitable enzyme and mediator selection and optimising the structure of printed layer. The structure of printed layer influences for example the oxygen transfer

Table 1.4 Conductive ink formulations used in the printed enzyme-containing layers.

Name of the ink	Ink composition
Ink 1 (carbon-based ink)	Commercial carbon-based ink (DuPont BQ231) with carbon nanotubes (Hydrocell), mediator and enzyme mixed in the ink
Ink 2 (silver-based ink)	Commercial silver-based ink (DuPont BQ129) with carbon nanotubes (Hydrocell), mediator and enzyme mixed in the ink
Ink 3 (experimental ink)	Carbon nanotubes (Hydrocell), α-sorbitol, polyvinyl alcohol, binder (Sicpa 1100), mediator and enzyme mixed in ink binder

Reprinted, with permission, from Smolander *et al.*[63] © 2008 Elsevier.

Table 1.5 Storage stability of laccase-containing printed layers in different storage conditions.

	Storage condition				
Ink[a] and storage time	$-22\,°C$	$+4.5\,°C$	$+4.5\,°C$, RH 80%	$+4.5\,°C$, under N_2	$+22\,°C$
Ink 1[b]					
2 weeks	32	40	27	46	15
6 weeks	79	70	5	40	27
12 weeks	92	14	8	35	10
Ink 2					
2 weeks	145	41	42	50	19
6 weeks	218	83	34	60	40
12 weeks	166	89	20	46	4
Ink 3					
2 weeks	72	104	76	89	72
6 weeks	109	80	76	106	59
12 weeks	110	124	62	96	53

[a]See Table 1.1 for composition.
[b]Formula without carbon nanotubes.
The activity in the printed layers was compared to the initial activity of the particular printed layer monitored with oxygen consumption measurement (error of the measurement max. 10%).
The table is reprinted, with permission, from Smolander *et al.*[63] © 2008 Elsevier.

of the cell. The structure can be altered by selecting a suitable pigment/binder combination, for example.

The capacity of the printed enzymatic fuel cell is determined by the amount of anode and cathode pastes that are optimised from a functional point of view. Therefore, the preferred printing method in manufacturing of this kind of cells is screen printing that creates thick layers. From the application point of view, the cell voltage can be increased by the use of serial connections. Currently, the

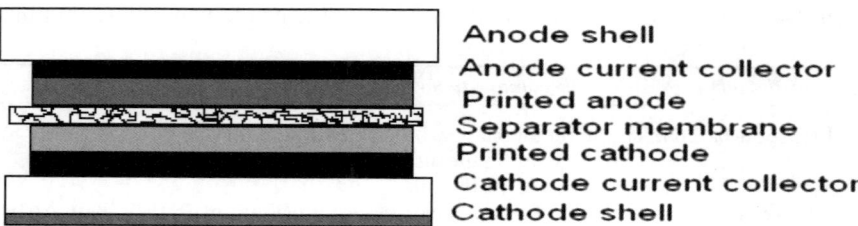

Figure 1.7   Generic structure of a printed enzymatic fuel cell. The cell consists of
anode and cathode cell materials, printed current collectors, a printed
anode and cathode, and a separator membrane.

steady state current output is in the range of 1 µW cm^{-2} allowing 10-fold power
peaks momentarily.[67] This performance level is compatible with, for example,
RFID tags with data logging functions. The higher peak current demand,
which can be some milliamperes depending on the application, duration of load
being typically in the range of milliseconds to a few tens of milliseconds, could
be met, however, by integrating a printed supercapacitor into the cell structure.[68] The main components needed in a double layer capacitor can be made
by printing techniques and as the structure of the biofuel cell current collector
components and the supercapacitor electrodes are partly similar, single stage
manufacturing and integration is feasible.

1.8   Conclusions

Enzymatic biofuel cells employ renewable catalysts and have several significant
properties such as the ability to operate optimally at temperatures between
room temperature and body temperature, also the chemical environment is
mild; several renewable fuels can be employed, including glucose.

Biofuel cells are a subject of increasingly intensive research. They still
encounter fundamental scientific and engineering challenges on the way from
the laboratory to the anticipated applications. The research in progress
addresses the weak issues, which include lifetime, power density and efficiency.

As to the application areas that are being considered for enzymatic biofuel
cells, the *in vivo* implantable power supplies would certainly have applications
from sensors to pacemakers based on glucose in the blood stream as fuel.

For *ex vivo* applications mostly alcohols and sugars are being considered, but
the fuel choice can be based on other alternatives, too. The application areas
considered are battery replacement in small portable devices such as music
players.

The feasibility of the concept for printed enzyme catalysed fuel cells has also
been demonstrated. These low-cost, enzymatic power sources would be disposable with normal package material or hospital waste. The application area
would be active RFID tags especially for measurement functions and

upholding memory in medical applications and packages in the distribution chain of food products. The potential application area in consumer products is also remarkable including entertainment features.

References

1. A. T. Yahiro, S. M. Lee and D. O. Kimble, *Biochim. Biophys. Acta. – Specialized Section on Biophysical Subjects*, 1964, **88**, 375.
2. S. D. Minteer, B. Y. Liaw and M. J. Cooney, *Curr. Opin. Biotechnol.*, 2007, **18**, 228.
3. F. Davis and S. P. J. Higson, *Biosens. Bioelectron.*, 2007, **22**, 1224.
4. M. J. Cooney, V. Svoboda C. Lau, G. Martina and S. D. Minteer, *Energy Environ. Sci.*, 2008, **1**, 320.
5. Y. Kamitaka, S. Tsujimura, N. Setoyama, T. Kajino and K. Kano, *Phys. Chem. Chem. Phys.*, 2007, **9**, 1793.
6. I. Willner, E. Katz and F. Patolsky, *J. Chem. Soc., Perkin Trans.*, 1998, **2**, 1817.
7. E. Katz, B. Filanofsky and I. Willner, *New. J. Chem.*, 1999, 481.
8. Y. Liu, M. Wang, F. Zhao, B. Liu and S. Dong, *Chem. Eur. J.*, 2005, **11**, 4970.
9. X.-C. Zhang, A. Ranta and A. Halme, *Biosens. Bioelectron.*, 2006, **21**(11), 2052.
10. A. Heller, *Phys. Chem. Chem. Phys.*, 2004, **6**, 209.
11. T. Chen, S. C. Barton, G. Binyamin, Z. Gao, Y. Zhang, H. Kim and A. Heller, *J. Am. Chem. Soc.*, 2001, **123**, 8630.
12. S. Tsujimura, K. Kano and T. Ikeda, *Electrochemistry (Tokyo)*, 2002, **70**, 940.
13. N. Mano, F. Mao and A. Heller, *J. Am. Chem. Soc.*, 2003, **125**, 6588.
14. Y. Tokita, T. Nakagawa, H. Sakai and T. Hatazawa, SONY's biofuel cell, Advanced Materials Laboratories, Sony Corporation 4-16-1, Okata, Atsugi-shi, Kanagawa 243-0021, Japan.
15. H. Sakai, T. Nakagawa, Y. Tokita, T. Hatazawa, T. Ikeda, S. Tsujimura and K. Kano, *Energy Environ. Sci.*, 2009, **2**, 133.
16. Y. Tokita, T. Nakagawa, H. Sakai, T. Sugiyama, R. Matsumoto and T. Hatazawa, *ECS Trans.*, 2008, **13**, 89.
17. M. J. Cooney, C. Lau, M. Windmeisser, B. Y. Liaw, T. L. Klotzbach and S. D. Minteer, *J. Mater. Chem.*, 2008, **18**, 667.
18. N. L. Akers, C. M. Moore and S. D. Minteer, *Electrochim. Acta*, 2005, **50**, 2521–2525.
19. http://www.akermin.com/ [Accessed 15 June 2009].
20. R. Gee, Portable Power, Explorer, SRI Consulting Business Intelligence (www.srici-bi.com), 2008.
21. Power Paper: http://www.powerpaper.com/index.php?categoryId = 33403, [Accessed 22 June 2009].

22. Enfucell: http://www.enfucell.com/products-and-technology, [Accessed 22 June 2009].
23. M. Albareda-Sirvent, A. Merkoçi and S. Alegret, *Sens. Actuators*, 2000, **69**, 153.
24. E. Crouch, D. Cowell, S. Hoskins, R. Pittson and J. Hart, *Anal. Biochem.*, 2005, **347**, 17.
25. M. Smolander, H. Boer, M. Valkiainen, R. Roozeman, M. Bergelin, J.-E. Eriksson, X.-C. Zhang, A. Koivula and L. Viikari, *Enzyme Microb. Technol.*, 2008, **43**, 93.
26. P. S. Taoukis and T. P. Labuza, *Food Sci.*, 1989, **54**, 783.
27. P. S. Taoukis and T. P. Labuza, in *Novel Food Packaging Techniques*, ed. R. Ahvenainen, Woodhead Publishing, Cambridge, and CRC Press, Boca Raton, FL, 2003, p.103.
28. S. Tsoka, P. S. Taoukis, P. Christakopoulos, D. Kekos and B. J Macris, *Food Biotechnol.*, 1998, **12**, 139.
29. H. Reichert, P. Simmerdinger and T. Bolle, Enzyme based Time Temperature Indicator, 2006, WO2006015961 A2.
30. Y. Sun, H. Cai, L. Zheng, F. Ren, L. Zhang and H. Zhang Hengtao, *Food Control*, 2008, **19**, 315.
31. H. Laursen, S. Jönsson, C. Ahlberg and A. Troedsson, Package and method of making the same, 1998, WO 98/38112.
32. J. Sjöholm and P.-E. Erlandsson, Biosensor and its use to indicate the status of a product, 2001, WO 01/25472.
33. B. Sahlberg, J. Sjöholm, E. Wehtje, and P.O. Erlandsson, Moisture sensor, 2003, WO 03/012419 A1.
34. N.B. Powell, Moisture indicator, US Patent 4 327 731, 1982.
35. T.G. Eakin, Wound dressing, US Patent 6 559 351, 2003.
36. H. Linna, Innovation and sustainable development in the fibre-based packaging value chain, Research and development activities in printed intelligence, VTT, 2008, pp. 32–33 (download: www.vtt.fi/files/download/scientific_reports/cpi_scientific_activities_2008.pdf)
37. A.C. Cameron and P.C. Talasila, *IFT Annual Meeting/Book of Abstracts*, 1995, p. 254.
38. A. B. Smyth, P. C. Talasila and A. C. Cameron, *Postharvest Biol. Technol.*, 1999, **15**, 127.
39. J. Vartiainen, M. Rättö and S. Paulussen, *Packaging Technol. Sci.*, 2005, **18**, 243.
40. J. H. Hotchkiss, *Active Additives 2005 Conference, 27–28 September 2005, Brussels*.
41. P. Lehtonen, P. Aaltonen, U. Karilainen, R. Jaakkola and S. Kymäläinen, A packaging material which removes oxygen from a package and a method of producing the material. WO 91/13556 (1991).
42. A. Nestorsson, K. G. Neoh, E. T. Kang, L. Järnström and A. Leufvén, *Packaging Technol. Sci.*, 2008, **21**, 193.
43. D. Ogonczyk, L. Tymeckia, I. Wyzkiewicz, R. Konckia and S. Glab, *Sens. Actuators*, 2005, **106**, 450.

44. A. L. Hart, A. P. F. Turner and D. Hopcroft, *Biosens. Bioelectron.*, 1996, **11**, 263.
45. A. L. Hart, H. Cox and D. Janssen, *Biosens. Bioelectron.*, 1996, **11**, 833.
46. U. B. Trivedi, D. Lakshminarayana, I. L. Kothari, P. B. Patel and C. J. Panchal, *Sens. Actuators B*, 2009, **136**, 45.
47. C. G. Koopal, A. A. C. M. Bos and R. J. M. Nolte, *Sens. Actuators B*, 1994, **18**, 166.
48. C. A. Galan-Vidal, J. Muñoz, C. Dominguez and S. Alegret, *Sens. Actuators B*1998, **52**, 257.
49. L. Fang, W. Li, Y. Zhou and C. C. Liu, *Sens. Actuators B*, 2008, **137**, 235.
50. L. Fang, S. H. Wang and C. C. Liu, *Sens. Actuators B*, 2008, **129**, 818.
51. J. Shen, L. Durik and C.-C. Liu, *Sens. Actuators B*, 2008, **125**, 106.
52. G. A. M. Mersal, M. Khodari and U. Bilitewski, *Biosens. Bioelectron.*, 2004, **20**, 305.
53. J. T Schumacher, I. Münch, T. Richer, I. Rohm and U. Bilitewski, *J. Mol. Catal. B: Enzym.*, 1999, **7**, 67.
54. A. L. Hart and W. A. Collier, *Sens. Actuators B*, 1998, **53**, 111.
55. I. Shitanda, M. Konya, M. Itagaki, K. Watanabe and Y. Asano, *Electrochemistry (Japan)*, 2008, **76**, 569.
56. L. Setti, A. Fraleoni-Morgera, B. Ballarin, A. Filippini, D. Frascaro and C. Piana, *Biosens. Bioelectron.*, 2005, **20**, 2019.
57. L. Setti, A. Fraleoni-Morgera, I. Mencarelli, A. Filippini, B. Ballarin and M. Di Biase, *Sens. Actuators B*, 2007, **126**, 252.
58. G. M. Nishioka, A. A Markey and K. Holloway, *J. Am. Chem. Soc.*, 2004, **126**, 16320.
59. S. DiRisio and N. Yan, *Macromol. Rapid Commun.*, 2007, **28**, 1934.
60. S. DiRisio and N. Yan, *J. Pulp Paper Sci.*, 2008, **34**, 203.
61. J. Kim, H. Jia and P. Wang, *Biotechnol. Adv.*, 2006, **24**, 296.
62. L. Kouisni and D. Rochefort, *J. Appl. Polymer Sci.*, 2009, **111**, 1.
63. M. Smolander, H. Boer, M. Valkiainen, R. Roozeman, M. Bergelin, J.-E. Eriksson, X.-C. Zhang, A. Koivula and L. Viikari, *Enzyme Microb. Technol.*, 2008, **43**, 93.
64. G. T. R. Palmore and H. H. Kim, *J. Electroanal. Chem.*, 1999, **464**, 110.
65. S. Riva, *Trends Biotechnol.*, 2006, **24**, 219.
66. C. S. Rodriguez and T. Herrera, *Biotechnol Adv.*, 2006, **24**, 500.
67. M. Smolander, H. Boer, M. Valkiainen, O.-V. Kaukoniemi, P. Saurus, R. Roozeman, M. Bergelin, J.-E. Eriksson, X.-C. Zhang, A. Koivula and L. Viikari, *213th ECS Meeting*, Phoenix, AZ May 18–May 22 2008, Electrochemical Society, 2008.
68. J. Keskinen, E. Sivonen, M. Bergelin, J.-E. Eriksson, M. Valkiainen, M. Smolander, A. Koivula and H. Boer, *ESSCAP'08, 3rd European Symposium on Supercapacitors and Applications*, Rome, 6–7 Nov 2008, paper ESS-001, European Symposium on Supercapacitors and Applications, 2008.

CHAPTER 2
Potential of Multilayer Ceramics for Micro Fuel Cells

MICHAEL STELTER

Fraunhofer Institute for Ceramic Technologies and Systems, Dresden, Germany

2.1 Challenges of Micro Fuel Cell System Development

Just like any fuel cell system, a micro fuel cell consists of major sub-systems, such as:

- Components for the electrochemically active part (membrane electrode assemblies (MEAs), gas diffusion layers (GDLs) or other contact materials, interconnects, electrical conductors)
- Mechanical integration (system package, gas and liquid tubing, fluid connectors, housing, environmental shielding)
- Balance of plant (fuel and oxidiser tanks and delivery systems, pumps, blowers, valves, sensors, controller, user interface)
- Electrical interface (direct current (DC)/DC converter or inverter, hybridisation).

In many of today's small fuel cell systems, all these components are shrunk versions of classic fuel cell system designs, i.e. miniaturised blowers, individual micro-tubings etc. are used. Much effort has been put into scaling down balance of plant (BoP) components. That included the application of completely new micro-fabrication techniques such as micro-wire eroding for metallic parts in pumps or fluid connectors. But still such micro fuel cell systems have

RSC Energy and Environment Series No. 2
Innovations in Fuel Cell Technologies
Edited by Robert Steinberger-Wilckens and Werner Lehnert
© Royal Society of Chemistry 2010
Published by the Royal Society of Chemistry, www.rsc.org

approximately the same parts count as large fuel cells. In some cases, this may hinder the commercial introduction of such systems, for various reasons.

2.1.1   Cost of Assembly

The cost of assembly is mainly a function of the parts count in a system, not a function of part size. Thus the cost of assembling and testing a 1 kW system is largely the same as assembling a 100 W system, provided both systems have the same technical concept. On the other hand, these fixed costs can not be allocated to a higher absolute system price in the case of the small system. Consequently, the sheer number of individual parts needs to be reduced in a micro system to compete cost wise with existing technologies.

2.1.2   Component Failures

The probability of component failures is a function of parts counts as well. The more individual components are assembled, the more potential failure modes may occur. Again, it is desirable to reduce the number of components in a system for that reason.

Just removing parts from a system in order to shrink or simplify it eventually leads to a loss of system functionality. In terms of micro fuel cells, this may lead to systems that have a reduced efficiency, reduced load-following capabilities or to fuel supply systems that, once activated, can not be stopped in a proper way anymore due to a lack of control actuators.

With multilayer ceramics, a technological platform is proposed that can be employed to provide mechanical integration functionality as well as active components to a fuel cell system, while at the same time the assembly effort is drastically reduced.

2.2   Introduction to Multilayer Ceramics

Multilayer ceramics can, in principle, be compared to multilayer printed circuit board (PCB) technology. Individual layers of unfired, so-called 'green', ceramics are modified to carry a specific functionality, such as being electrically conductive, and then these layers are laminated. Upon lamination, inter-layer connections are formed. After a final sintering step, the ceramic 'PCB' is ready for component assembly (Figure 2.1).

Multilayer ceramics has formed out of the so-called 'hybrid electronics', a technology based on planar ceramic substrates that were imprinted with electrical conductor paths. These ceramic parts could carry multiple silicon semiconductor chips at once. As aluminium oxide was used as the principal ceramic, the sintering temperature of the material was rather high (> 1500 °C). Thus, glass–alumina hybrid materials were developed that could be sintered at much lower sintering temperatures of *ca.* 850 °C. All functional features (vias, conductors *etc.*) were still imprinted on these new ceramic sheets when they were

Figure 2.1   Large size (200 × 200 mm) LTCC substrate with mass produced, highly repetitive ceramic micro-components. This multilayer substrate size is the industry standard.

still unfired ('green'). During the sintering step, these features were burnt in together with the main substrate material. This process is thus called 'co-firing'. Consequently, the new material class was called low-temperature co-fired ceramic (LTCC).

Today, LTCC is one of the main applications of multilayer ceramics. Because of its particularly good dielectric properties it was quickly adopted for radio-frequency (RF) electronic components. Almost any modern cell phone contains multiple LTCC-based parts. The mechanical and thermal robustness as well as the good thermal conductivity of LTCC on the other hand has led to wide-spread application in the automotive and aerospace industries. About 20% of all electronic control units (ECUs) in a modern passenger car are based on LTCC. Today, the manufacture of LTCC is clearly a mass production technology with a broad selection of commercially available raw materials and production equipment. See Figure 2.2 for a schematic of the LTCC process and Table 2.1 for typical properties of some ceramic and organic substrates.

2.3   Fuel Cell Relevant Subsystems and Geometries

2.3.1   Geometrical Shaping

To integrate a fuel cell system based on the LTCC platform, crucial geometric and electric features of the stack, fuel supply system and BoP systems need to be replicated in ceramic. The typical size of these structures in a fuel cell of

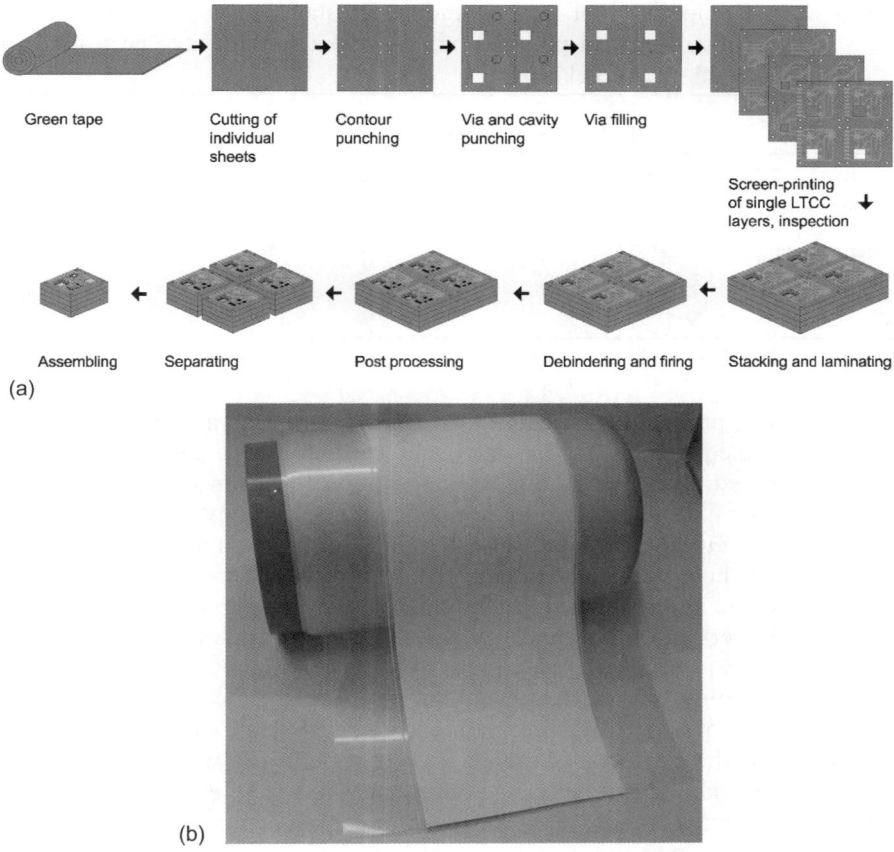

Figure 2.2   (a) The principal pathway of multilayer ceramic processing. The technological chain is clearly focused on mass production, using very productive techniques such as screen printing and multiple cuts. (b) Alumina green tape that can be manufactured to multilayer ceramic fuel cells. The white ceramic tape is still on its plastic transfer tape from tape casting.

the 1 W class ranges from 100 µm to several millimetres. Numerous structuring technologies were investigated by IKTS to generate three-dimensional (3D) elements that are necessary for this fuel cell functionality (see Figure 2.3 for an overview).

The most flexible technology to generate exact microstructures in LTCC green (unfired) laminates is milling with computerised numerically controlled (CNC) machinery. However, this technology should be kept confined to rapid prototyping of single pieces, as it is slow, sequential and causes excessive wear of delicate micro-cutter heads (see Figure 2.3a).

Green tape embossing is a method to imprint 3D structures in a green tape. A heated die carrying the negative structure is pressed into the thermoplastic tape, which is then irreversibly deformed. Upon firing, these structures sustain. This

Table 2.1   Comparison of several industrial standard multilayer ceramic
substrates with polymer-based substrates that have been used to
build micro fuel cells.

Property	Al_2O_3	BeO	AIN	LTCC	CCM	Kapton/ polyimide	FR4
Maximum process temperature (°C)	>1500	>1800	>1400	900	1000	>400	120
TEC (10^{-6} K^{-1}).	7.5	8.5	3.4	5.0–7.0	12.5	27	30
Therm. conductivity (W/mK)	20	230	150	4–6	25	1.16	0.2
Permittivity	9.5	7.0	10.0	3–5	5–6	3.5	5.0
Cost factor	1	50	<40	10	4	0.5	0.25

It becomes clear that LTCC provides the most well-balanced profile of properties and cost.

method is very productive, as complex structures can be generated at once. On the
other hand, the precision and the degrees of freedom for the design are limited.

A high speed, high precision machining option for green tapes is punching
(Figure 2.3b). With automated matrix punch tools, as many as 2500 holes per
second can be cut into the tape. A single punch tool can cut about 25 holes per
second, with hole diameters ranging from 60 μm to 4 mm. Certainly, only
through-holes can be punched, but complex shapes are possible.

A fourth, very flexible technology is laser ablation (Figure 2.3c). By using
ultraviolet (UV) lasers with several tens of watts continuous power and proper
3D beam shaping and deflection equipment, very precise structures can be
machined out of green and fired laminates at very high speeds. The lateral
resolution of these structures can be as precise as 5 μm, whereas large-scale
material removal can also be carried out in the range of several mm^3 s^{-1}. Laser
ablation structures do not necessarily need to consist of through-holes; almost
any shape can be written.

At IKTS, the two most productive structuring technologies – matrix punching
and high power laser ablation – have been combined in a completely new
machine concept (see Figure 2.4). The two tools work together in one traversial
set-up, controlled by optical fiducials on the individual substrates as measured
by a high density charge-coupled device (HD-CCD) machine vision system.
Thus, the coordinate systems are perfectly aligned, while laser and matrix punch
work together in parallel. The fuel cell structures can be generated at very high
speeds without the need for re-chucking the ceramic substrates for each individual step. The development of such mass production capable machinery is
considered to be crucial for ceramic micro fuel cell production, as fuel cells, by
their nature, require relatively large active areas with proper 3D structures.

2.3.2   Relevant Features of Fuel Cells

Based on the structure technologies described above – punching, laser ablation,
green tape embossing, and milling – a range of micro fuel cell components can
be manufactured.

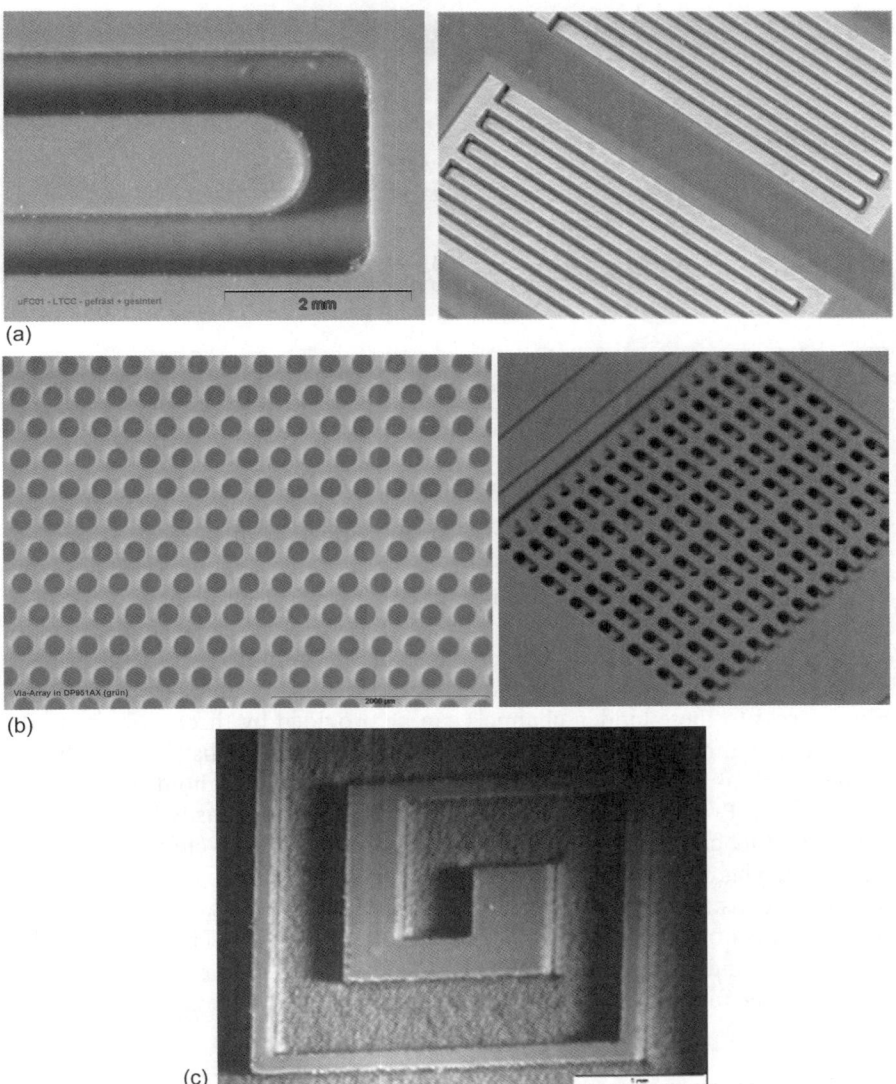

Figure 2.3 (a) CNC-milled PEM flowfield structures. Very high precision and aspect ratio, but long manufacturing times and wear of tools. (b) Array of high-speed micro-punched through-holes and flowfield geometry formed out of an arrangement of puched holes. Hole diameters can be as low as 50 μm and up to *ca.* 1.5 mm with a typical diameter of *ca.* 300 μm. (c) Flowfield structure manufactured by high-speed laser ablation in the green tape with subsequent firing. The precision and aspect ratio is at least as good as with CNC machining, whilst processing time and cost of tools are drastically reduced. CNC, computerised numerically controlled.

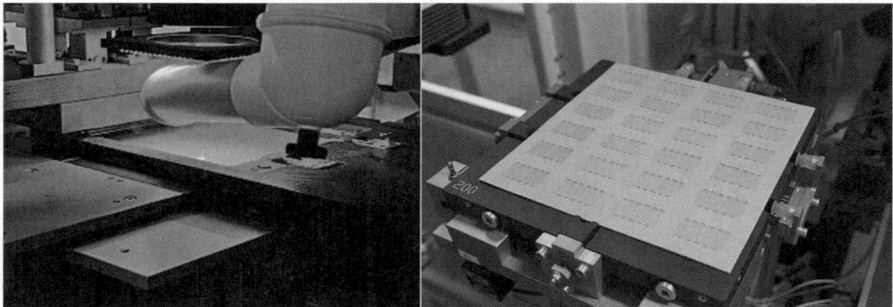

Figure 2.4   Combined laser ablation/matrix punching of flowfield geometries in LTCC (left) and finished large-scale (200 mm) substrate with multiple flowfield geometries.

2.3.2.1   Integrated Fluid Distribution Channels

Multilayer integrated fluid channels can be formed with laser ablation. The starting point is a LTCC or high-temperature co-fired ceramic (HTCC) laminate. By moving the laser in a hachure-like manner, trenches or cavities can be formed without puncturing the base material. Afterwards, a single sheet of ceramic tape is laminated on top of the structured multilayer. Upon firing, a hidden channel is formed in a homogeneous piece of ceramic (see Figure 2.5). Fluid access to these hidden channels can be provided by through-holes in the top layer. Channels formed by that method can be used to distribute hydrogen or reformat as well as liquid fuels and coolants throughout the ceramic microsystem. Proper routing provided, these channels can easily be mixed with the electrically conducting 3D microstructure in the form of current conductor paths and vias in the very same ceramic part. Thus, a massive increase of functional density can be achieved. For instance, the fuel or air can be delivered right to the centre of the flowfield and then spiral outwards. There is no need to provide fuel and oxidant to the edge of the flowfield anymore, as it is the case with all other fuel cell configurations.

2.3.2.2   Heterogeneous Microreactors

If a micro fuel cell needs a reformer, it can be manufactured in multilayer ceramics as well. Alumina is often used as the carrier for heterogeneous catalysts; thus it is beneficial to use an Al_2O_3 multilayer material directly. By placing punched holes in regular patterns in Al_2O_3 green tapes and subsequent lamination, a broad range of microreactor geometries can be manufactured: mixer geometries, parallel flow as well as meander flow monolith reactors all are possible. Catalyst layers can be applied prior to firing by using open holes in the laminate, before lamination by using methods like localised spray coating, or after firing by infiltration of the finished fired rector. A great advantage of multilayer microreactors is that additional functionality can easily be

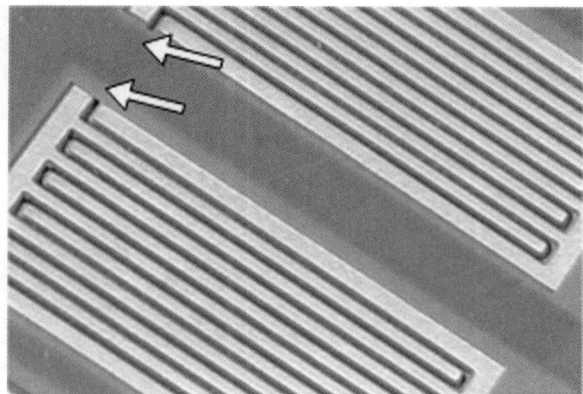

Figure 2.5 Hidden (buried) channel in a ceramic micro fuel cell connecting two flowfields to each other.

integrated into the reaction vessel by printing. Electric heaters or temperature sensors can be printed directly onto the tapes that form the reactor wall. Such functions not only add to the compactness of the reactor, but also the thermal coupling of the reaction zone to the bulk material of the reactor wall is greatly improved.

Even in the direct design of the flowfield new approaches can be implemented through the use of multilayer ceramics.

2.3.2.3 Flowfield Current Pick-up

A major flowfield design rule comes from the fact that LTCC is not conductive. Thus, to form a current pick-up, conductive layers have to be applied on top of the flowfield. An obvious solution is to use thin gold platings, manufactured by screen printing of gold particles containing glass pastes, electrochemical plating or by vapour deposition techniques (Figure 2.6). This technique is similar to gold platings that are used on PCB-based planar fuel cells. However, these platings still have a considerable thickness in the range of 10 μm, which considerably adds to the cost of the component. At IKTS, we thus have developed noble-metal free coatings that can be applied to the ceramic by screen printing or stencil printing. They have a high electronic conductivity, facilitate very good contact to the GDL but are considerably cheaper.

2.3.2.4 Flowfield Electrical Contact

An important issue when using surface conducting flowfields is the current transport away from the active area. As the bulk material is not conductive, the lateral conductivity of the metallisation becomes the critical limiting factor, and even more so the thinner the conductive layer is. A second design restriction is

Figure 2.6   Flowfield geometry without and with gold plating. Only the topmost layer
needs to be metallised. The electrical contact to the gold plating is carried
out with vias from one of the layers below the surface. Note that there is
no need to apply meander-like continuous flowfield structures as in a
conventional fuel cell.

that, in this case, only flowfield features can be manufactured that provide an
uninterrupted current path from the active area to the edge of the area, such as
meanders, but not stand-alone features such as dimples. A special design feature of multilayer ceramics can turn this disadvantage into an advantage:
through-holes filled with conductive material, so-called 'vias', can connect the
active area of the flowfield with deeper layers that provide paths for current
conduction. The usage of vias can greatly improve the degrees of freedom for
the flowfield design in ceramic fuel cells, as the designer is not restricted to any
design rules any more. The formation of vias in multilayer ceramics is a well
established technique that can be applied at very high speeds. As the flowfield
side of the vias is coated with the contact layer, no special electrochemical
requirements apply for the via fill material.

In general, the mechanical rigidity, stiffness and resistance against high
temperatures and swelling give multilayer ceramics a significant advantage over
other multilayer technologies that are used to build planar micro fuel cells, such
as PCB material. Highly functional planar stack configurations can be built for
polymer exchange membranes (PEMs) and direct methanol fuel cells (DMFCs)
that provide novel flowfield features, but still are only 1.5 mm thick. The main
advantage, however, is that with ceramics many other fuel cell functions can be
integrated right into the stack.

2.4   Examples

Several micro fuel cell systems and system components have been built using
multilayer ceramics, mainly based on the LTCC variety.

Figure 2.7 shows a very small system in the power range of a few 10 mW. The
anode side of the PEM system carries the electronics in the form of a DC/DC
converter and consists of two cells. Both cells are arranged on one single sheet

of MEA to save sealing space. Sealing and mechanical connection between the two shells is provided by an adhesive silicone that was applied using screen printing.

Figure 2.8 shows a larger system in a four-cell planar configuration. The dimensions are *ca.* 60×60 mm. Again, the anode side carries the more complex electronics. It is beneficial to use the anode side for the functional integration,

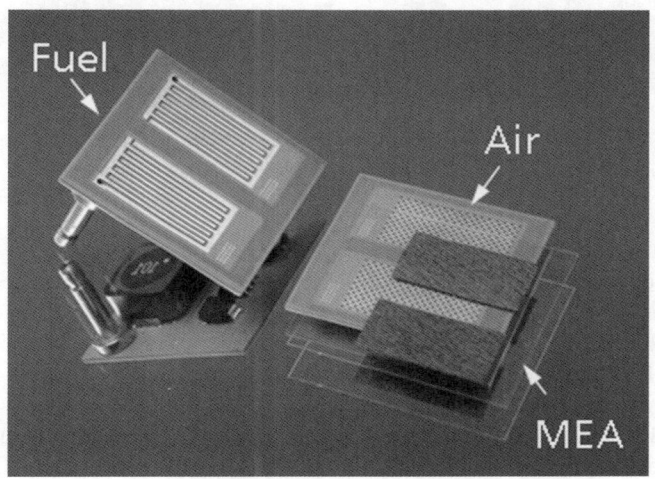

Figure 2.7 Very small (*ca.* 30 × 30 mm) two-cell micro fuel cell in LTCC.

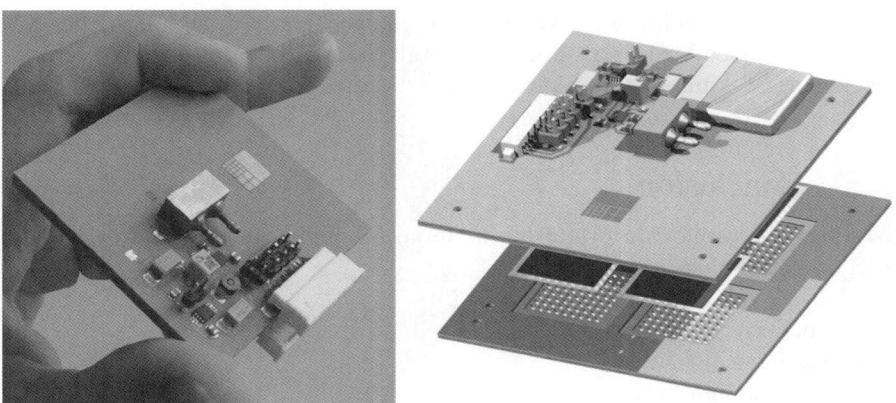

Figure 2.8 Larger ceramic fuel cell system (*ca.* 1.5 W), also in a co-planar configuration. The complex anode flow fields in this system were manufactured using a matrix of punch holes (see Figure 2.6). As the electronics is already quite elaborate and fully functional, the complete system is only 7 mm thick (2.4 mm for the ceramics), and was used in real-world field testing units.

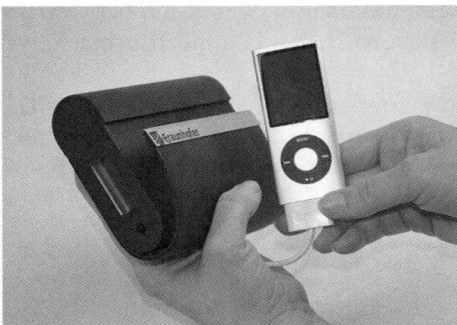 

Figure 2.9   Complete, fully functional fuel cell re-charger for portable electronics based on LTCC fuel cells. The internal mechanical set-up is very simple due to the functional integration inside the ceramic multilayer monolith.

because in a fuel cell with passive air flow (self-breathing mode) the anode ceramic incorporates far more complexity than the cathode. In the anode, all gas channels need to be routed, along with the current distribution and rather complex flowfields. In contrast, the cathode ceramic is rather simple, and because of the large open channels for unhindered air access it is also mechanically not as robust.

Figure 2.9 shows fully functional system prototypes based on ceramic multilayers that are used in field trials. The prospective application for this kind of system is a re-charger for portable electronics in use cases, where the next power plug is too far away. This may include business applications as well as leisure use, such as portable navigation devices (GPS) or telephone re-charges during outdoor hikes. Correspondingly, the systems are rated 2.5 W continuous power plus battery hybridisation. Again the high mechanical integration, the low number of assembled parts, and the overall mechanical robustness of ceramic micro fuel cells make these systems very well suited for the planned application.

2.5   Conclusion

Multilayer ceramic is a promising material for highly integrated novel micro fuel cells. Its main advantages are:

- Chemical inertness
- Mechanical rigidity
- Great design flexibility
- Ability to incorporate fuel cell functions as well as electric and mechanical features
- Availability of material and processing equipment for mass production.

To make use of this potential, however, a range of techniques needed to be developed that significantly go beyond the standard production processes for

LTCC in the electronics industry. Among these techniques that IKTS has developed are:

- High-precision, high-speed 3D structuring combining a UV laser with matrix punching
- Green tape structuring technologies
- Technology to form gas tight open and buried channels in the ceramic monolith
- Interface technologies to connect such channels to the outside world
- Electrochemically compatible coatings for flowfields.

From our research we estimate that ceramic integrated micro fuel cells are best suited for power ranges between *ca.* 50 mW to *ca.* 5 W. If the power output is smaller, the system complexity is so low that the advantages of functional integration of ceramic do not apply any more. In the case of greater power output, the active area that is necessary becomes so large that the cost of ceramic can not be over-compensated by advantages in system integration any more. A conventional stacked configuration of cells might then better suit the needs of such larger systems.

Part 2
High-Temperature Polymer Electrolyte Fuel Cells

Introduction

Lifetime is a critical issue with fuel cells. One method to reduce degradation (corrosion) in high-temperature fuel cells (*i.e.* solid oxide fuel cells (SOFCs)) is to lower the operating temperature from 800 °C to 600 °C since most deteriorating processes are temperature driven. On the other hand with low-temperature fuel cells (*i.e.* polymer electrolyte fuel cells (PEFCs)), an increase in temperature from 100 °C to 180 °C is desirable in order to simplify the fuel processing and system architecture. These interesting conflicting trends may, in effect, lead to a convergence of system concepts between low- and high-temperature fuel cells. Nevertheless, these respective temperature ranges pose considerable challenges for materials development. The following two chapters discuss the materials and system issues in HT-PEFC development and the prospects for fuel cell application created by the new temperature range.

Chapter 3, 'Trends in High-Temperature Polymer Electrolyte Fuel Cells', introduces the HT-PEFC which is based on phosphoric acid-doped polybenzimidazole-type membranes. The typical operating temperature is between 160 °C and 180 °C. In contrast to classical PEFCs water management is not a major issue for this fuel cell type. The CO tolerance amounts to 1–2% which makes a HT-PEFC particularly suitable for operation in combination with reformers. A drawback of this technology is that the temperature of a stack has to be at least 120 °C before current can be drawn when the stack is operated with reformate.

HT-PEFCs are close to commercialisation but the oxygen reduction reaction and the lifetime are issues which still have to be considered carefully.

Possible applications of HT-PEFC are described in Chapter 4, 'Large Auxiliary Power Units for Vessels and Airplanes'. Electricity requirements in mobile applications are increasing in nearly all forecasts for the future by electric devices for more comfort and a guaranteed energy supply during idling mode. Today, combustion engines and turbo jet engines are applied as auxiliary power units (APUs) on-board airplanes and ocean-going vessels. For logistical reasons, APUs should use the same fuel as the main engine. This will be kerosene or JET A-1 for airplanes and diesel or marine gas oil for vessels. The chapter will show the requirements for aeronautical and maritime applications and will give an overview on today's developments in fuel cell technology and fuel processing.

CHAPTER 3

Trends in High-Temperature Polymer Electrolyte Fuel Cells

WERNER LEHNERT, CHRISTOPH WANNEK AND
ROSWITHA ZEIS

Forschungszentrum Jülich GmbH, Institute of Energy Research,
IEF-3: Fuel Cells, 52425 Jülich, Germany

3.1 Introduction

The high-temperature polymer electrolyte fuel cell (HT-PEFC), based on polybenzimidazole-type membranes doped with phosphoric acid, has a typical operating temperature of 160 °C. As a result of the high temperature, it has a high CO tolerance, which means that it is particularly suitable for operation in combination with reformers. Compared to Nafion-based polymer electrolyte membrane fuel cells, water is not required for the ionic conductivity of the membrane; therefore the gases do not need to be humidified. Another advantage of HT-PEFC technology is that due to the high temperature difference between the stack and the ambient temperatures, cooling can be designed in a much more compact manner than is necessary in classical polymer electrolyte membrane fuel cell (PEMFC) systems.

However, for all these advantages, prevention measures must be in place to ensure that the acid is not discharged from the membrane. Operating conditions that could create liquid water in the cells should be avoided. Safe operation below 120 °C is possible only with pure hydrogen. Due to the decreasing CO tolerance with decreasing temperature, operation above 120 °C is necessary if the fuel contains typical CO concentrations of 1%. Furthermore, it was shown that the membrane resistance increases significantly at a

RSC Energy and Environment Series No. 2
Innovations in Fuel Cell Technologies
Edited by Robert Steinberger-Wilckens and Werner Lehnert
© Royal Society of Chemistry 2010
Published by the Royal Society of Chemistry, www.rsc.org

temperature below 60 °C.[1] Operation in low-temperature ranges should therefore be avoided. This has been taken into account by operating strategies and the specific areas of application.

Only a few years ago, intrinsic proton-conductive materials such as phosphonic acids or *N*-heterocycles were regarded as prospective alternative electrolytes for HT-PEFCs, but today it seems to be unlikely that these electrolyte systems may sufficiently be optimised to become technically attractive,[2] and research on ionic liquids is still in its infancy.[3] All-solid electrolytes such as mixed solid acids, undergoing a so-called superprotonic phase transition at elevated temperature (*e.g.* $CsHSO_4$),[4] as well as doped diphosphates (MP_2O_7 with M = Ce, Zr, or Sn, for example),[5] have shown promising results in very small laboratory cells but face serious scaling-up problems such as leak tightness in larger cells. As a consequence, they are not considered in this review focussing on near-term technical applications.

We will concentrate on the phosphoric-acid-doped polybenzimidazole membrane cells as these are closest to commercialisation. In this context the oxygen reduction reaction and operation behaviour of cells and stacks will be discussed.

3.2   The Oxygen Reduction Reaction

In principle, the anode and the cathode both contribute to the polarisation loss (over-potential) of a hydrogen–oxygen fuel cell. In practice, however, the kinetics of hydrogen oxidation reaction (HOR), which occurs at the anode, is much faster than the oxygen reduction reaction (ORR) on the cathode side. This disparity is quantified by the exchange current density. The values of the exchange current density for HOR are at least six orders of magnitude higher than those of the ORR in hydrogen–oxygen fuel cells.[6] The slow ORR kinetics is associated with its complex reaction mechanism, which involves several steps and significant molecular reorganisation. Despite the higher operating temperature of HT-PEFCs, the sluggish ORR in concentrated phosphoric acid is still a dominant factor that limits the fuel cell efficiency. Therefore understanding the reaction mechanism and kinetic rates of the ORR is fundamentally important.

3.2.1   Tafel Slope and Reaction Pathway

A great deal of basic research on this subject[7–17] was carried out from the 1970s until the early 1990s within the framework of the phosphoric acid fuel cell (PAFC) development. To obtain the reaction parameter and study the reaction pathway, the rotating disk electrode technique had often been utilised.[9,13,14,16] The well-defined mass transfer condition due to the rotation of the disk electrode is an obvious advantage of this technique. The challenge was to perform the measurements under conditions similar to those in real fuel cells; for example, at elevated temperature and relative low humilities. Yeager and co-workers[9] built a high-pressure high-temperature rotating ring disk set-up. Such a sophisticated measurement set-up was difficult for other research groups to rebuild. Bockris[11] overcame this problem by utilising a platinum micro-electrode and stepping the potential of the electrode from a value where no chemical change occurs to one where the oxygen reduction reaction takes place. The current versus time response of the rapid change of potential, was then recorded, from which the kinetic parameters of the reaction were obtained.

The concept of HT-PEFCs using phosphoric-acid-doped polybenzimidazole (PBI) as the electrolyte is relatively new. There are only a few publications[18-21] regarding the ORR kinetics in these HT-PEFCs phosphoric acid. Liu *et al.*[19] conducted a comprehensive study on this subject by using a micro band electrode which they specifically designed for their experiments.[22] This technique is similar to the potential stepping method described in the previous paragraph. Zecevic *et al.*[20] employed a platinum rotating disk covered with a thin film of PBI to investigate the oxygen reduction. Both studies concluded[19,20] that the presence of PBI at the interface of platinum and phosphoric acid had little or no effect on the kinetic parameters and reaction pathway of the ORR because the amorphous phase of H_3PO_4 functions as an electrolyte. Therefore under similar conditions $Pt/PBI-H_3PO_4$ and Pt/H_3PO_4 should be comparable in terms of ORR kinetics.

The ORR kinetic current density can be approximated as an exponential function of the over-potential, at least in the low-current regime. The over-potential appears as a straight line when plotted against the logarithm of the kinetic current density. The slope of this line is called the Tafel slope, typically in the range from 60 to 300 mV per decade. A smaller Tafel slope is desired because it corresponds to less polarisation loss. The Tafel slope is directly associated with the ORR reaction pathway. Therefore, measurement of the Tafel slope is of critical importance for the basic understanding of ORR mechanisms.

To our knowledge today, the experimental results regarding the Tafel slope for ORR on platinum in phosphoric acid are inconclusive. As a consequence, the reaction pathway is still not fully understood. In particular, the nature of the rate-limiting step is still under debate.

From dilute solutions measured at room temperature, the Tafel slope was found to be approximately 60 mV/decade and 120 mV/decade stepping for the low-current (high-potential) and high-current (low-potential) regimes, respectively.[17,20,23]

Possible explanations[23,24] for this phenomenon are all linked to the influence of oxide formation on the platinum surface. Damjanovic and Genshaw attributed this behaviour to the change of adsorption isotherm for the reaction intermediates. For a partial coverage of oxide on the platinum surface the adsorption is governed by the Temkin isotherm (Tafel slope ~ 60 mV/decade). On the oxide-free surface the Langmuir isotherm becomes predominant (Tafel slope ~ 120 mV/decade). Meanwhile, Tarasevich[23] claimed that the change in the surface coverage of the chemisorbed oxygen-containing species forming the anodic film is responsible for the change of the Tafel slopes at the various potential regimes. The cyclic voltammogram indicates an onset potential where the anodic film is formed. Typically, this onset potential coincides with the change of the Tafel slope.[13] A strong influence of an oxide film on the platinum surface on the ORR kinetics was also found in experiments[25,26] with specially prepared oxide-covered and oxide-free platinum electrodes. Damjanovic and Hudson[26] not only investigated the effect of the oxide film on the ORR reaction but also examined the pH dependence of the reduction. From this study they concluded that the reaction order regarding the ORR is 1/2 with respect to $[H_3O^+]$ in the high-current regime and that the order is 1 in the low-current regime. The unusual reaction order of 1/2 was accounted for by the barriers of the oxide layer and the Helmholtz layer. The rate-determining step therefore seems to be associated with protons as reactants.

However, Yeager and co-workers[15,27] concluded that the rate-determining step in concentrated phosphoric acid does not involve making or breaking any bonds involving protons and that proton transfer does not play an important role in the oxygen reduction kinetics. They reached this conclusion based on the absence of any isotopic effect[15] as they measured polarisation curves for the oxygen reduction in deuterated and undeuterated phosphoric acid. Furthermore, the reaction rates in concentrated acid solution and concentrated alkaline solution are within one order of magnitude whereas the proton activities differ by many orders of magnitude. According to Yeager et al.[9] the oxygen reduction proceeds through the dissociate adsorption of the oxygen molecule on the platinum surface, which occurs simultaneously with an electron transfer. From their polarisation curves they extracted a Tafel slope of $\sim 120\,mV/decade$ independently of temperature (Figure 3.1). Such temperature independence is difficult to reconcile with simple theoretical models, which predict that the Tafel slope is usually directly proportional to the absolute temperature assuming the transfer coefficient is temperature independent. In the case of a temperature-independent Tafel slope the transfer coefficient must be temperature dependent. This might be an effect of impurities as proposed by Scharifker et al.[28] Although impurities may certainly affect Tafel slopes, this seems to be an unlikely explanation. The cyclic voltammograms (CVs) presented in the study by Clouser et al.[9] showed no traces of impurities. The hydrogen adsorption–desorption region was not affected. The unusual phenomenon of the temperature dependence of the transfer coefficient made Yeager believe that the reaction mechanism in concentrated phosphoric acid is different, and that the rate controlling step is associated with oxygen adsorption instead of proton transfer.

The more recent study by Liu et al.[19] supports the analysis of Damjanovic regarding the rate-determining step. From their investigations regarding the effects of H_3PO_4 doping level and water content on oxygen reduction on Pt interfaced with PBI-H_3PO_4 at elevated temperatures (150 °C) and low relatives humidity they concluded that the proton transfer is rate determining. Based on their experimental results, they proposed the following reaction as the rate determining step: $O_2 + H^+ + e^- \rightarrow O_2H$. A Tafel slope of 95 mV/decade

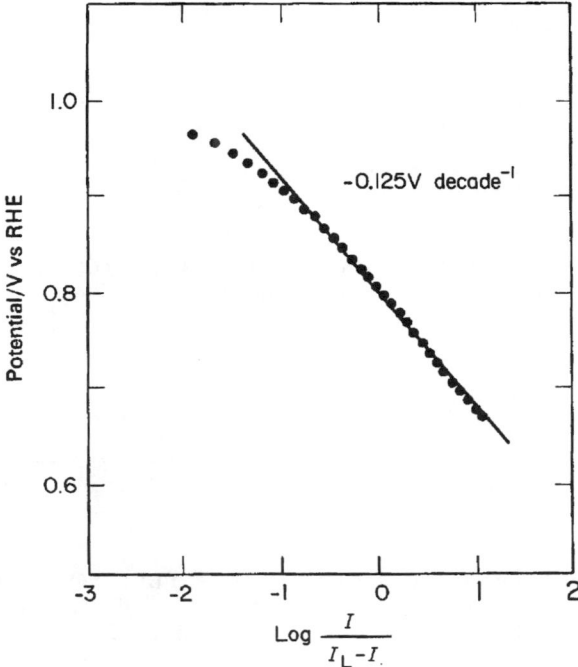

Figure 3.1 Tafel plot obtained for oxygen reduction on a platinum disc electrode in 100.3% H$_3$PO$_4$ at 206 °C and 1 atm oxygen pressure. Reprinted, with permission, from Clouser *et al.*[9] © 1993 Springer.

(Figure 3.2) indicates the rate-determining step is a one-electron transfer reaction. Furthermore, the reaction order was first with respect to the proton concentration. They observed an increase of the oxygen reduction rate with an increase of the water content which is proportional to the proton activity. By varying the oxygen saturation concentration in the polymer electrolyte, a reaction order of 1 was extracted, which suggests that the rate-determining step involves O$_2$ as reactant. In this experiment, the oxygen solubility increased with the H$_3$PO$_4$ doping level, which resulted in an increase in the oxygen reduction rate.

3.2.2 The Adsorption of Phosphoric Acid Molecules and Phosphate Anions on Platinum

The occupancy of reaction sites by adsorbed molecules and anions may impede the oxygen reduction. It has long been speculated that strong adsorption of phosphoric acid molecules or phosphate anions on the platinum catalysts is the main cause for the large polarisation losses in fuel cells utilising phosphoric acid as the electrolyte. There have been a number of experiments to study the

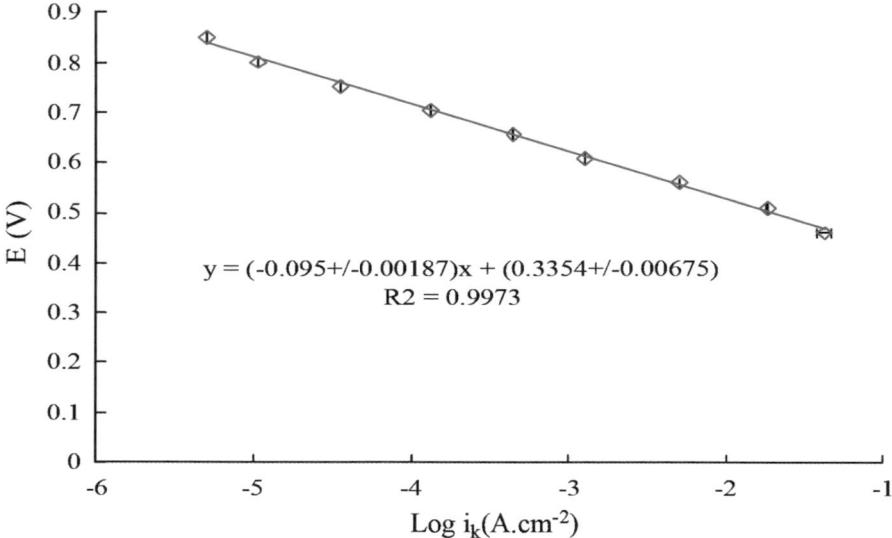

$$y = (-0.095 +/- 0.00187)x + (0.3354 +/- 0.00675)$$
$$R2 = 0.9973$$

Figure 3.2   Tafel plot obtained for ORR at the Pt/PBI-H$_3$PO$_4$ interface. The measurement was taken by a micro band electrode (O$_2$ purged; 150 °C; and 10% RH). Reprinted, with permission, from Liu *et al.*[19] © 2006 Elsevier Science.

adsorption and desorption of these species on a microscopic level including investigations employing radiotracers,[29] infrared spectroscopy,[30,31] and cyclic voltammetry.[32]

In 1978, Horanyi *et al.*[29] used the radiotracer technique to investigate the concentration and potential dependence of the adsorption of phosphoric acid species on platinised platinum electrodes. They varied the concentration of phosphoric acid from 0.1 mM to 10 mM in 1 M perchloric acid. Perchloric acid was used as the supporting electrolyte because the ClO$_4^-$ anion adsorbs very weakly and exhibits no significant effect on the adsorption of other ions and components. From the experiments, the authors concluded that the amount adsorbed onto platinum electrode first increases with the applied potential until it reaches a maximum at 800 mV, and then it decreases with further increase of the potential. In a certain potential range between 600 and 800 mV, the adsorption of the phosphoric acid species could be described by a Freundlich-type isotherm. Horanyi *et al.*[29] also studied the effect of different additives on the adsorption of H$_3$PO$_4$. They reported an increase of the adsorption in the presence of Cd^{2+} ions in the potential range where the electrosorption takes place. Similar cases of induced adsorption processes had previously been observed for Cl$^-$ and HSO$_4^-$ anions. Therefore, the authors suggested that it is the phosphate anions that adsorb onto the platinum electrode despite the fact that the main part of the phosphoric acid in the solution phase is in the non-dissociated form.

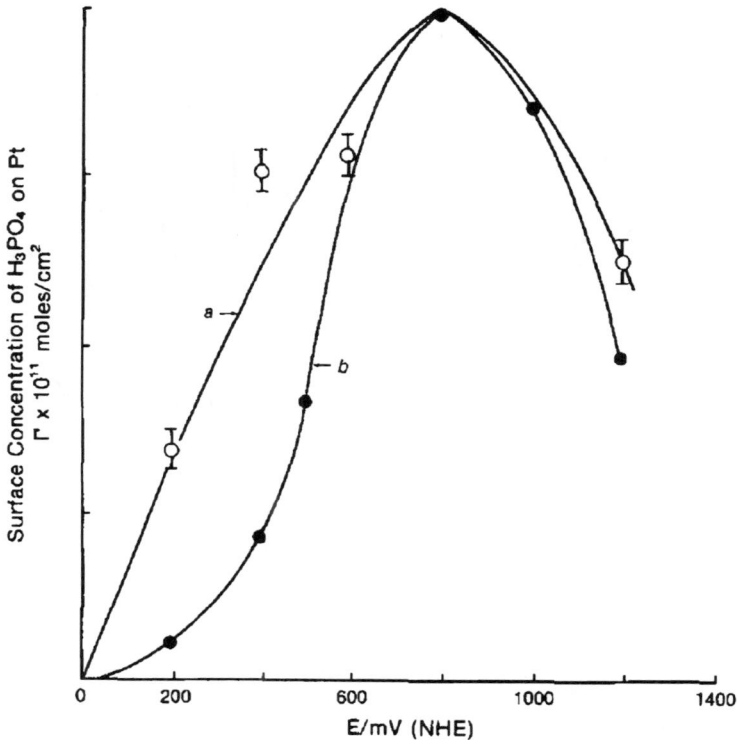

Figure 3.3 Surface concentration, Γ, of H_3PO_4 on Pt as a function of potential. (Curve a) Γ as normalised area of IR peak at $1074\,cm^{-1}$. The highest peak area is assumed to be the highest surface concentration, as obtained from (curve b), the radiotracer measurements. Reprinted, with permission, from Habib and Bockris.[30] © 1985 The Electrochemical Society.

In 1985 Habib and Bockris[30] made the first Fourier transform infrared (FT-IR) spectroscopic measurement to study the adsorption of H_3PO_4 on platinum and gold. The measurement was carried out in 1 M perchloric acid with various concentrations of H_3PO_4. The authors reported an FT-IR spectral peak at $1074\,cm^{-1}$ associated with a P–O stretch vibrational mode of H_3PO_4 molecules adsorbed on platinum surface. The IR absorption peak first increases with the applied potential, reaches a maximum, and then decreases with further increase of the potential. This agrees with the radiotracer experiment[29] as shown in Figure 3.3. From this result, the authors suggested that the adsorbed species is the H_3PO_4 molecule, which could be displaced by water or oxide on Pt at high potentials. If it were the $H_2PO_4^-$ anion, then the decrease of this peak with the increase of the potential would be difficult to justify.

In 1992 Nart and Iwasita[31] revisited the topic using FT-IR measurement techniques with much improved signal-to-noise ratio and spectral resolution. In particular, the authors measured the FT-IR spectra with both s and p

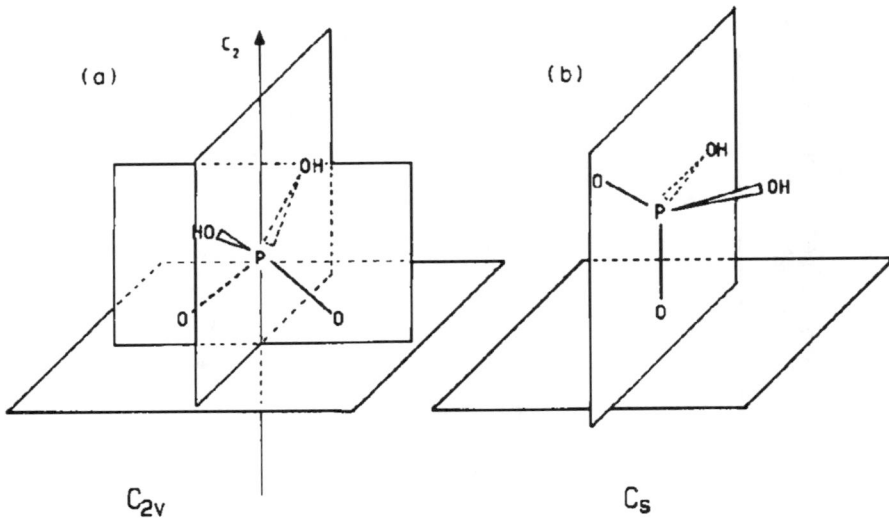

Figure 3.4   The $H_2PO_4^-$ ion under C_{2v} and C_s symmetries. These are the most probable orientations of $H_2PO_4^-$ at low and high positive potentials respectively. Reprinted, with permission, from Nart and Iwasita.[31] © 1992 Elsevier Science.

polarisations of light to exclude artifacts due to phosphoric acid consumption in the solution. The authors also replaced perchloric acid with hydrofluoric acid to avoid IR band interference due to perchlorate ion migration. From the new FT-IR measurement results, the authors reached quite different conclusions from those of Habib and Bockris.[30] First, by varying the pH value from 2.8 to 0.23, Nart and Iwasita[31] found that both $H_2PO_4^-$ anions and H_3PO_4 molecules could adsorb on platinum. At lower potentials, the undissociated H_3PO_4 molecule is likely to adsorb on platinum through the non-protonated oxygen atom under the C_{3v} symmetry, and only one IR absorption band at 1050 cm^{-1} associated with the P–O stretch vibration can be observed within the measurement spectral range. At higher potentials the adsorbed H_3PO_4 molecule dissociates to $H_2PO_4^-$ anion, and the onset of dissociation depends on the pH. The adsorbed $H_2PO_4^-$ anion undergoes a symmetry change as the potential further increases. At potentials below 900 mV the $H_2PO_4^-$ anion is doubly coordinated to the surface, presenting a C_{2v} symmetry. At higher potentials, the adsorption is singly coordinated with a lower C_s symmetry. This geometry change causes the IR absorption peak at 1000 cm^{-1}, which corresponds to the P–(OH)$_2$ stretch vibration, to decrease at higher potentials (Figure 3.4).

   Despite these research efforts, the exact nature of phosphate species adsorption on platinum has not been fully understood. Other experimental techniques have also been used to gain further insights into the phenomenon. However, new experiments tend to raise new questions and add new controversies to existing ones. In 2009, Mostany *et al.*[32] reported cyclic voltammetry studies of phosphate adsorption on single crystal Pt (111) electrode with various concentrations of NaH_2PO_4 in 0.1 M perchloric acid solution. A thermodynamic analysis of the data suggests that in the absence of co-adsorbed species, HPO_4^{2-} is the predominant adsorbed species although the concentration of HPO_4^{2-} is extremely low in the solution with a low pH value. The authors pointed out that the phosphate anion adsorption might have been complicated by the competitive co-adsorption of hydrogen, OH, alkaline cations and perhaps even perchlorate ions.

3.2.3 Enhanced ORR Activity: Platinum Alloy Catalysts and Alternative Electrolytes

3.2.3.1 Platinum Alloy Catalysts

Until now, carbon-supported Pt nanoparticles have been the most commonly used catalyst for oxygen reduction in low temperature fuel cells. The search for catalyst materials that are more active, less expensive, and more stable than Pt has led to the development of Pt alloys such as PtNi and PtCo. But the stability of these Pt alloy catalysts is questionable. This is partially because, under fuel cell operating conditions, transition metals such as nickel or cobalt are expected to form oxides or hydroxides and these are not very stable and tend to dissolve from the electrode surface.

Nevertheless, carbon-supported platinum transition metal alloy catalysts are often used in conventional low-temperature PEFCs, and there is strong experimental evidence that they perform better than carbon-supported Pt.[33,34] In the literature many possible explanations can be found for the enhanced catalytic activity of Pt alloys. Extensive reviews on this topic can be found in Antolini *et al.*[35] and Thompsett.[36] In brief, the enhanced catalytic activity of Pt alloys has been ascribed to various structural changes of the Pt caused by alloying, which may result in shortening of the inter-atomic Pt–Pt distance. The increase in catalytic activity has also been attributed to the exposure of a more active vicinal Pt(110) plane on dispersed platinum particles. The exposure of the more active plane is believed to take place during the heat treatments to induce alloy formation in the particles. Other researchers have suggested an enhancing mechanism of the ORR based on an increased d-electron vacancy of the thin Pt surface layer caused by underlying alloy. According to Colón-Mercado *et al.*[37] the improved stability of the Pt alloy catalysts is caused by the hindered mobility of the Pt on the carbon due to the presence of the alloy. The sintering effect of Pt is therefore suppressed.

Recently, superior performance of a PtCo catalyst has been reported for HT-PEFC. Rao *et al.*[38] prepared carbon-supported Pt-Co alloy nanoparticles of various Pt:Co atomic ratios (1:1, 2:1, 3:1 and 4:1). These catalysts were evaluated in HT-PEFC. Improved performance was observed for Pt:Co atomic ratios of 1:1 and 2:1. These HT-PEFCs, operating at 180 °C and 50 mA cm^{-2}, were stable during 50 h of test. However, longer term stability of such alloy catalysts is yet to be evaluated.

3.2.3.2 Alternative Electrolytes

Despite good physical and chemical properties, concentrated phosphoric acid is a rather poor fuel cell electrolyte. The slow ORR kinetics in concentrated phosphoric acid is associated with its weak acidity, strong phosphate species adsorption and low oxygen solubility. This has motivated researchers to search for alternative electrolytes or electrolyte additives. In the 1980s, Yeager and co-workers[39,40] conducted extensive research on fluoronated acids such as monofluorophosphoric acid [$FPO(OH)_2$] and trifluoromethane sulfonic acid (CF_3SO_3H or TFMSA). These studies also aimed at better understanding of the ORR mechanisms in phosphoric acid fuel cells. These fluoronated acids were initially anticipated to improve the ORR kinetics significantly through much increased proton concentration compared with phosphoric acid. Their experiments did show enhanced ORR kinetics and more than an order of magnitude higher diffusion limiting currents compared with concentrated phosphoric acid. However, the quantitative analysis indicates that the observed improvement is mainly due to the higher O_2 solubility in these fluoronated acids and the impact of proton concentration appears to be insignificant.

The use of these fluoronated acids in high-temperature fuel cells is not practical due to their high vapour pressure and their tendency to wet the Teflon-bonded gas-diffusion electrodes. Salts from similar fluoronated acids have been investigated as electrolyte additives,[41] which appear more practical. Examples include potassium perfluorohexanesulfonate ($C_6F_{13}SO_3K$), potassium nonafluorobutanesulfonate ($C_4F_9SO_3K$) (Figure 3.5), perfluorotributylamine [$(C_4F_9)_3N$], and, more recently,[42] ammonium trifluoromethanesulfonate ($CF_3SO_3NH_3$ or ATFMS). These electrolyte additives were all found to enhance the ORR kinetics in phosphoric acid fuel cells through increased O_2 solubility.

3.3 Membrane Polymers

Virtually all high-temperature PEFCs known today are based on phosphoric acid-loaded polymer membranes from the class of polybenzimidazoles (PBIs). The most widely known representative of this family of polymers is the well-developed poly(2,2'-*m*-phenylene-5,5'-bibenzimidazole) (*meta*-PBI, Figure 3.6) which is prepared by step-growth polymerisation from 3,3',4,4'-tetra-aminobiphenyl and diphenyl isophthalate. Due to its extraordinary high chemical and thermal stability, it was first commercialised by Celanese in fibrous form and was used in firefighter suits, high temperature protective gloves, *etc.* several years before its use in fuel cells was envisaged.

The use of H_3PO_4-doped *meta*-PBI as fuel cell membrane was first suggested by Wainright *et al.* in 1995.[43] In the following, intensive development of PBI-type membranes has been carried out both in academic and industrial

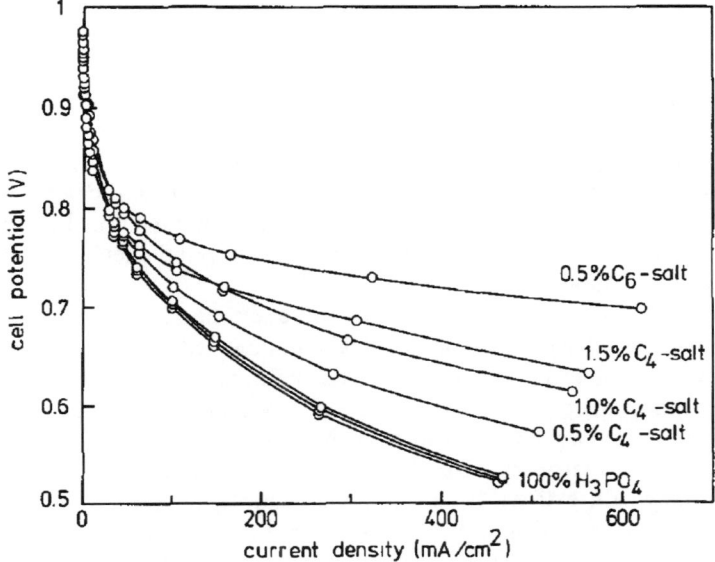

Figure 3.5 Polarisation curves for electrolyte modified by addition of C_6-salt ($C_6F_{13}SO_3K$) and C_4-salt ($C_4F_9SO_3K$) (fuel: H_2 oxidant, O_2; electrolyte, 100% H_3PO_4 and temperature 190 °C). Reprinted, with permission, from Gang *et al.*[41] © 1993 The Electrochemical Society.

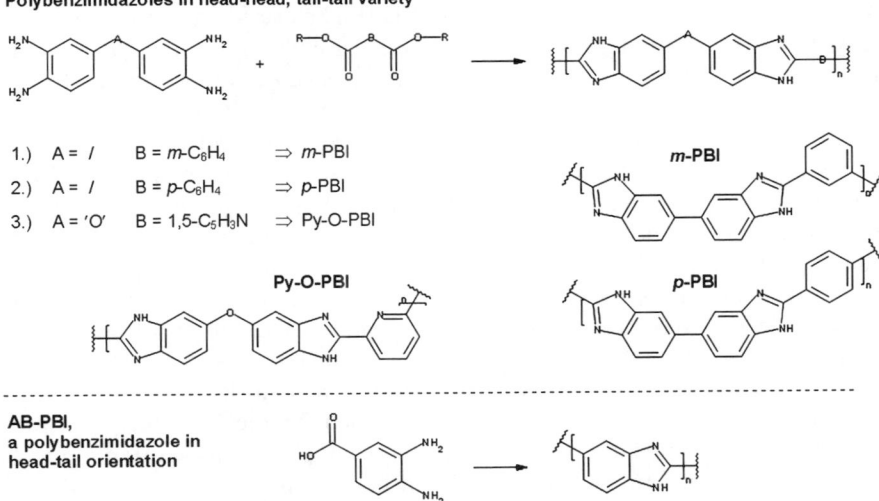

Figure 3.6 Structures of technically relevant polybenzimidazoles for fuel cell applications.

research. Among the different promising membrane materials are (1) PBI homopolymers produced from different (isomeric) tetramines and diacids, (2) polymers synthesised from a single monomer carrying both amino and acidic groups (*e.g.* AB-PBI from 3,4-diaminobenzoic acid) by head-tail poly-condensation, (3) sulfonated PBIs, (4) polybenzimidazoles with proton-con-ducting inorganic fillers such as heteropolyacids or zirconium phosphonates, (5) blends of polybenzimidazoles with other polymers, or (6) the replacement of the benzimidazole by a pyridine unit in the polymer chain. Two excellent reviews on all these types of HT-PEFC membranes and their specific advan-tages have been published only recently.[44,45]

All these polybenzimidazoles readily take up large quantities of liquid inorganic acids such as phosphoric acid when immersed in them. While the first two equivalents of H_3PO_4 being absorbed by *meta*-PBI, for example, do not lead to a significant proton conductivity as they undergo a classical acid–base reaction with the two basic nitrogen atoms in the imidazole rings, the con-ductivity increases strongly with all further ('free') acid molecules.

Membranes with extremely high polymer molecular weights and/or a care-fully adjusted degree of interchain linkage can absorb more than four times their own weight of phosphoric acid and still remain mechanically manageable. A somewhat lower 'doping level' of about six molecules H_3PO_4 per repeat unit of *meta*-PBI (= 65 wt% acid) represents a technically reasonable compromise between mechanical robustness and sufficient conductivity. An elegant way to produce robust acid-loaded membranes with extremely high doping levels is described in section 3.4.

The specific conductivity of such membranes increases strongly with tem-perature (curves b to d in Figure 3.7 represent measurements performed at 80, 140 and 200 °C), reaches values between 50 and 100 mS cm^{-1} at typical oper-ating temperatures of the HT-PEFC and is thus close to the conductivity of classical LT-PEFC materials such as Nafion®. Furthermore, the conductivity of acid-doped PBI is found to increase in presence of water vapour (for an explanation see below). Nevertheless, this dependence on humidity is much less severe than known for Nafion® (curve a in Figure 3.7), and even under com-pletely dry conditions the conductivity of acid-doped PBI is considerably high.

Depending on acid loading, temperature and water content of the membrane, the proton transfer occurs along different paths. For membranes with a high acid doping level and H_2O content (*e.g.* due to water formed during the fuel cell reaction), a Grotthus mechanism involving proton transfer along acid and water molecules has been proven.[47] The higher conductivity of water-con-taining membranes can been attributed to the fact that water molecules are (rotationally) more mobile than the much larger phosphoric acid molecules and that thus linear O–H–O hydrogen bonds, needed for the Grotthus transport, are more easily formed at higher concentrations of H_2O. This assumption is, in principle, confirmed by Daletou *et al.*,[48] although these authors stress that at temperatures above 120 °C the reversible dehydration of phosphoric acid according to the reaction $2H_3PO_4 \rightarrow H_4P_2O_7 + H_2O(g)$ sets in and that accordingly the reasoning above should refer to the differences in forming

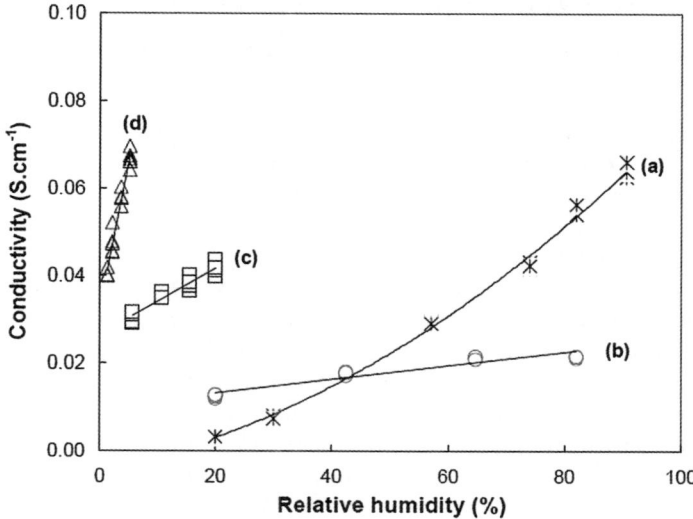

Figure 3.7 Specific proton conductivity of Nafion® at 50 °C (a) and acid-doped PBI at 80 °C (b), 140 °C (c) and 200 °C (d). Reprinted, with permission, from He *et al.*[46] © 2003 Elsevier.

hydrogen bonds between the relatively mobile orthophosphoric acid (H_3PO_4) and the more bulky pyrophosphoric acid ($H_4P_2O_7$) molecules. Time-dependent impedance measurements have shown that the cell resistance of an HT-PEFC decreases within a few minutes when the operating condition is changed from open circuit voltage to a given current density due to the water produced by the fuel cell reaction while the reverse process (dehydration of the cell) takes significantly more time (up to a few hours).[49]

On the one hand, polybenzimidazoles easily take up large amounts of H_3PO_4, but on the other hand at least the 'free' phosphoric acid which is not being bound to the polymer backbone by the acid–base reaction can similarly easily be washed out again. As a consequence, the presence of liquid water – for example, formed by condensation of the water produced by the fuel cell reaction itself – has strictly to be avoided in a HT-PEFC which imposes some precautions to be taken during start-up and shut-down of a cell to prevent acid leaching and thus loss of membrane conductivity and performance decrease. It seems that in industrial R&D this problem is mainly approached on an engineering level, whereas PBI membranes comprising an interpenetrating network of non-soluble polyvinylphosphonic acid (instead of free phosphoric acid) has only been proposed for application in direct methanol fuel cells operating around 100 °C.[50]

Although a large variety of polybenzimidazole-type materials has been evaluated, today *meta*-PBI is still the most often used membrane polymer in HT-PEFCs.

3.4 Catalyst and Diffusion Layer Development and Membrane Electrode Assembly Manufacture

While most attention during the last years was concentrated on the membrane development, research on optimised electrodes and diffusion layers is evolving only recently. This is due to the fact that, on the one hand, these functional layers must be adapted to the membrane material which has to be developed first, and on the other hand the materials used for classical phosphoric acid fuel cells work reasonably well also in HT-PEFCs.

Pure platinum is still the catalyst being most widely used for HT-PEFCs, both at the anode and the cathode. Nevertheless, different platinum alloys have been proven to show higher catalytic activity and thus to improve the performance, especially for the oxygen reduction reaction (*e.g.* Pt_3Ni, see section 3.2.3.1), but these novel catalysts tend to degrade much more quickly than platinum alone which may conflict with a broad technical use of these materials. Although the high operating temperature is believed to accelerate the electrode kinetics, the catalyst loading in HT-PEFCs ($\sim 1\,mg\,Pt\,cm^{-2}$ per electrode) is significantly higher than in classical LT-PEFCs operating around $80\,^\circ C$ (0.05–$0.5\,mg\,Pt\,cm^{-2}$). This is ascribed to the massive blocking of active catalyst sites by phosphate adsorption (*cf.* section 3.2.2). As mass transport losses at moderate current densities are more important in HT- than in LT-PEFCs,[18] platinum black as catalyst is often chosen for the fabrication of membrane electrode assemblies (MEAs) that must deliver extremely high power densities while carbon-supported Pt may be advantageous in terms of durability. This can be ascribed to two facts: catalyst sintering is slower in carbon-supported catalysts due to the higher platinum interparticle distance, and the rather thick layers can buffer some phosphoric acid to compensate acid losses from the membrane during long-term operation.

The careful design of the catalyst layers is a key issue in the development of any type of fuel cell. As somewhat contradictory requirements have to be met (*e.g.* intimate contact of the catalyst with the phosphoric acid electrolyte, easy access of reactants and discharge of product water at the cathode, good electrical contact with the current collectors), the micro-structures of the catalyst layers necessarily represent a compromise between these demands. Due to its chemical robustness poly(tetrafluoroethylene) (PTFE) has a long history in being used as a binder between the individual catalyst particles (either in form of carbon particles with a diameter of 30–100 µm being loaded with platinum or as nanometre-sized Pt black) in phosphoric acid fuel cells. By adjusting the volumetric ratio between the hydrophilic catalyst and the hydrophobic polymer, the degree of wetting the catalyst with phosphoric acid and the availability of channels for gas transport can be balanced. In typical HT-PEFC catalyst layers, there is an intimate mixing of catalyst and PTFE on the sub-micrometre scale including the formation of very fine PTFE fibres with diameters down to a few nanometres (Figure 3.8).

The incorporation of the membrane polymer as ionomer into the electrodes is, in principle, an auspicious option to fix phosphoric acid in the catalyst layers

Figure 3.8 SEM image of the microstructure of a catalyst layer based on 20% Pt/C and PTFE. Reprinted, with permission, from Wannek *et al.*[51] © 2009 Elsevier.

and thus prevent it from leaching out of the MEA. But due to the excellent film-forming capacity of the polybenzimidazoles it is, in fact, extremely difficult to effectively realise this approach without causing detrimental effects to the cell performance; for example, by enclosing and thus deactivating catalyst particles or by lowering the gas permeability of the catalyst layers. Nevertheless, as the vapour pressure of phosphoric acid is low at the operating temperatures of the cell, the need to introduce ionomer into the electrodes to assure good proton conductivity in the catalyst layers is very much smaller in HT-PEFCs than in the case of low temperature PEM fuel cells where Nafion® is clearly needed in the catalyst layers.

The use of additives such as potassium perfluorohexanesulfonate ($C_6F_{13}SO_3K$), for example, has been reported to enhance the oxygen reduction rate in the phosphoric acid fuel cell,[41] which might be due to increased oxygen solubility and diffusion in the cathode catalyst layer (*cf.* section 3.2.3.2). To our knowledge, this issue has not yet been re-invested systematically for HT-PEFCs.

As there is a re-distribution of phosphoric acid in the MEA and because some acid is needed in the catalyst layers to provide good proton conductivity, the gas diffusion electrodes should at least be slightly impregnated with phosphoric acid when the MEA assembly is done based on acid-doped membranes.[52] Too high acid loadings of the electrodes during fuel cell operation lead to a significant decrease in performance.[52–54]

Carbon fiber materials serve both as electrically conductive electrode substrates and (together with PTFE) the main constituency of highly porous gas diffusion layers (GDLs, see below). Woven GDLs have the advantage of being more flexible and thus easier to be coated with the catalyst layer in a roll-to-roll process. Non-woven materials (also referred to as carbon paper or felt) penetrate less into the fuel channels of an operating cell as they are stiffer. Similarly to the catalyst layers, the hydrophobicity of all these GDLs must be tuned by wet-proofing according to the needs of the other cell components and the envisaged operating conditions. In classical PAFCs, the GDLs also served as an electrolyte reservoir. Today, in HT-PEFCs where acid-loaded polymer membranes can be used, this is no longer necessary. Instead, so-called microporous layers can be inserted between the porous carbon substrate and the catalyst layer. Typically these layers consist of carbon ($> 50\,\text{wt\%}$), PTFE and optionally some additives. The microporous layers are designed on the one hand to be more dense and wet-proofed than the other layers so that they can act as barrier layer preventing liquid phosphoric acid from leaching out of the electrode, and on the other hand to be still porous enough to allow easy access of hydrogen and air to the reactive sites and removal of product water vapour from the cathode to the gas channels.

Several pathways for introducing the phosphoric acid needed for the proton conductivity of the membrane have been developed and described. The classical approach consists of doping free-standing sheets of the PBI-type membrane in concentrated phosphoric acid. As described beforehand, in this method, membranes with acid contents of up to 70–80 wt% can be obtained. However, the mechanical properties of these membranes may become critically poor, making the reproducible production of MEAs for cells with a large active area challenging. A second, very elegant way of fabricating this type of electrolyte yielding PBI membranes with an even higher content of phosphoric acid and thus showing very high conductivity is characterised by solution polymerisation of the monomers in polyphosphoric acid (PPA) followed by membrane formation as a consequence of controlled hydrolysis of the polyphosphoric acid to orthophosphoric acid[57] (Figure 3.9). However, this procedure is not applicable for those polybenzimidazole type polymers which are soluble in phosphoric acid. A third pathway consists of preparing a 'dry' catalyst-coated PBI membrane and contacting it with gas diffusion layers (GDLs) which are loaded with appropriate amounts of H_3PO_4 during MEA assembly. But in this type of MEA it appears to be difficult or perhaps even impossible to use microporous layers as the phosphoric acid would have to cross this barrier during the cell start-up process. A fourth pathway, the acid impregnation of the MEA *via* the catalyst layers, was identified at a very early point and revisited only lately.[51] The latter two procedures are attractive as all components during cell assembly are easy to handle. It has been proven that the redistribution of the phosphoric acid inside the MEA is a quick process yielding significant membrane conductivity within less than 10 min and a dynamic equilibrium after 24–72 h[55] so that the resulting cell performance can be independent of the acid introducing strategy.

As PBI-type polymers do not exhibit glass transition temperatures below $\sim 400\,^{\circ}\text{C}$, a hot-pressing step at temperatures between 100 and 200 °C does not necessarily boost the intimate contact between the membrane and the

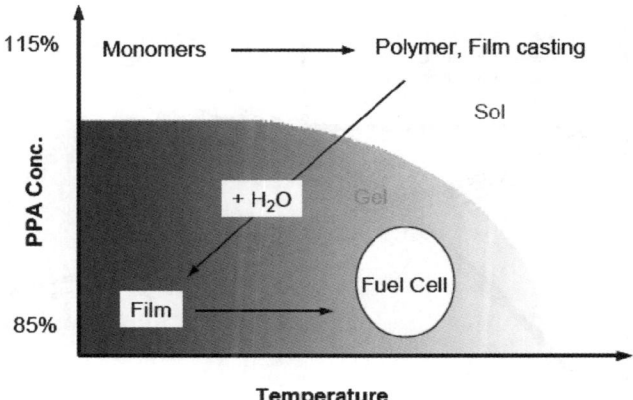

Figure 3.9 State diagram of the PPA sol–gel process. Reprinted with permission from Xiao *et al.*[57] © 2005 American Chemical Society.

electrodes from the electrochemical point of view. Nevertheless, it is usually performed to laminate the different components to yield a mechanically robust MEA. Different companies are currently engaged in the scaling-up of their HT-PEFC MEA production with BASF Fuel Cell certainly being at the head of the development having opened a new production facility in which advanced production and automation technologies are used to fabricate ready-for-use MEAs.[56]

3.5 Fuel Cell Performance and Durability

In a similar manner to other energy conversion techniques and especially all other types of fuel cells, the performance of HT-PEFCs depends strongly on the operating conditions. Typical operating temperatures are between 100 and 200 °C as on the one hand, the formation of liquid water has to be prevented to avoid leaching of the electrolyte, and on the other hand the durability of membranes, catalysts and other cell components becomes a critical issue at temperatures higher than 200 °C. As will be shown in more detail below, both pure hydrogen and reformate gas containing up to 3% carbon monoxide can be used as fuel. For all technical applications, the oxygen from ambient air is used at the cathode as the increase in cell performance obtained with pure oxygen does not overcompensate the expenses for the air fractionation. For the same reason, HT-PEFCs are most commonly operated at ambient pressure.

In the following, some general trends concerning the fuel cell performance at different operating conditions are shown by means of experiments conducted at Forschungszentrum Jülich. As the cell components used for these measurements were still in the developmental state at the time of conducting these experiments, the absolute performances displayed on Figures 3.10 to 3.12 are

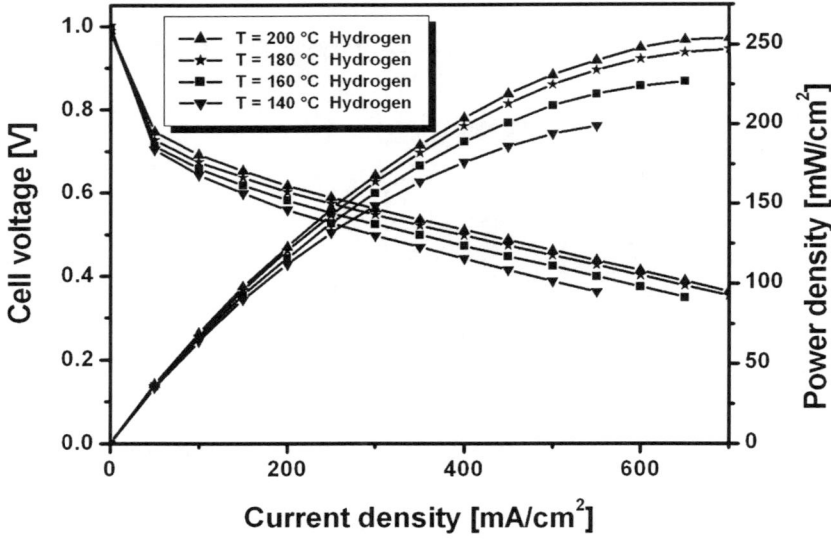

Figure 3.10   Cell performance of HT-PEFCs operated at different temperatures using hydrogen and air at ambient pressure as reactants. Reprinted, with permission, from Wannek *et al.*[59] © 2008 The Electrochemical Society.

somewhat lower than what can be achieved with state-of-the-art MEAs. Where it is of particular importance, the best data from other developers is also given in the text. Nevertheless, the features explained below hold true for the operation of any HT-PEFC with phosphoric acid-impregnated poly-benzimidazole-type membrane.

Normally, all developers of HT-PEFCs recommend their cells to be subjected to a break-in period of 24–100 h during which the cells are operated at a constant temperature (*e.g.* 160 °C) and moderate current densities with infrequent, if any, load changes. During this conditioning process, the cell voltage usually increases slowly but steadily (by 10–30 mV). It is believed that this time is needed before a steady-state distribution of phosphoric acid between the membrane, the catalyst layers and the gas diffusion layers is established. This is corroborated by the observation that an MEA assembled from acid-loaded gas diffusion electrodes and an acid-free membrane shows good performance at low and moderate current densities only 5 h after start-up, but a significant down-bending of the polarisation curve at high current densities.[55] These two facts can be ascribed to a quick activation of the electrode kinetics and a high hindrance of the mass transport in the electrodes – as some pathways for the reactants are still blocked with phosphoric acid – due to the uncompleted redistribution of H_3PO_4 within the MEA. Various attempts to develop an accelerated break-in procedure are currently being investigated,[58] as in view of future technical applications start-up times of several hours are to be avoided.

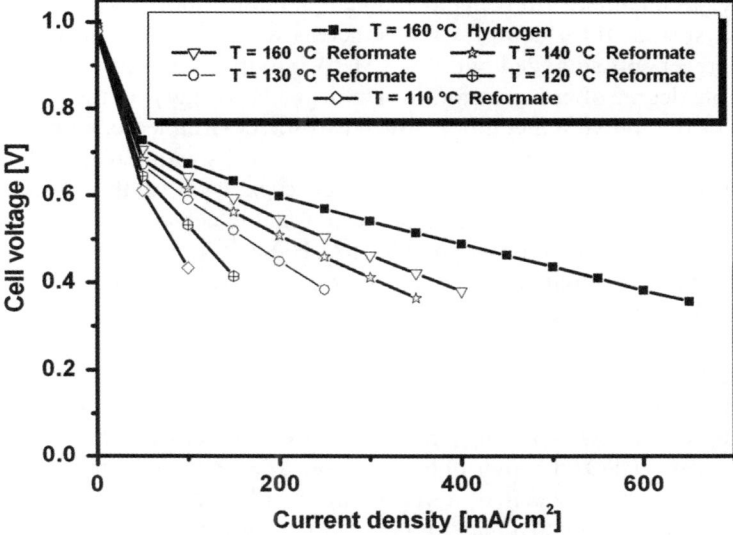

Figure 3.11 Polarisation curves obtained with reformate as fuel containing 1% CO and air at different temperatures. Reprinted, with permission, from Wannek *et al.*[59] © 2008 The Electrochemical Society.

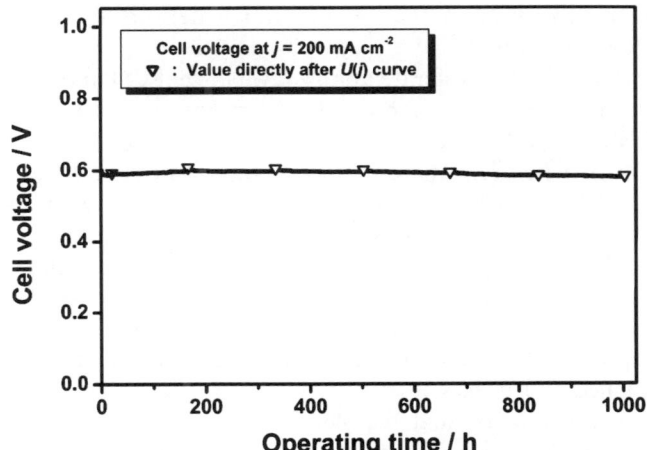

Figure 3.12 Cell voltage of a HT-PEFC operated under nearly steady-state conditions at 160 °C with H_2 and air. Reprinted, with permission, from Wannek *et al.*[51] © 2009 Elsevier.

To decrease all contact resistances between the different functional layers, the MEAs are usually compressed in the cell. But over-compression has strictly to be avoided as the pore volume available in the electrodes decreases during compression until the catalyst layers becomes nearly soaked with acid and with very few gas channels still being available for the transport for the reactants to the catalyst sites. If the compression is increased even further, the electrolyte is simply irreversibly squeezed out of the MEA into the flow fields. Accordingly, a reasonable degree of compression, which depends strongly on the porosity and rigidity of the individual components, has to be determined experimentally.

When carefully designed electrodes are used in cells with a reasonable degree of compression, relatively low stoichiometric excesses of the reactants (the so-called λ factors) are needed to achieve a stable cell operation. With pure hydrogen or reformate (containing 40–70% H_2) anodic λ values of 1.2 and 1.5 are technically reasonable even for large stacks (corresponding to fuel utilisations of 83% and 67%, respectively). The cathodic λ is usually fixed to 2 when ambient air is used.

The cell performance of HT-PEFCs increases steadily with temperature between 100 and $\sim$180 °C especially due to enhanced kinetics of the electrode reactions (Figure 3.10). At higher temperatures the dehydration of orthophosphoric acid H_3PO_4 to pyrophosphoric acid $H_4P_2O_7$ – already described above – and the associated decrease in membrane conductivity becomes dominant, which leads to a maximum of the cell performance around 200 °C. It is well known that carbon monoxide 'poisons' classical PEFCs by adsorbing on the catalyst and thus by blocking catalytically active sites. As a consequence, when LT-PEFCs are operated on hydrogen from reformed natural gas or liquid fuels, the CO content has to be lowered by a secondary gas treatment, the so-called 'selective oxidation step' which serves to convert carbon monoxide into CO_2 with oxygen from the air and which delivers fuel gases with CO contents below 100 ppm.

By contrast, HT-PEFCs tolerate up to a few per cent CO in the fuel gas as the higher operating temperature favours the desorption of carbon monoxide from the Pt catalyst. Naturally, the temperature dependence of the cell performance is much stronger when reformate is used as fuel compared to the use of pure hydrogen as can be seen from the example in Figure 3.11. When fuel gas from reformed middle distillates such as diesel with a comparably low hydrogen content is used ($\sim$40% hydrogen, $\sim$1% CO), at 180 °C more than 80% of the performance of a cell running on pure hydrogen can be obtained. However, this 'CO tolerance' decreases to 50–60% at 140 °C, amounts to less than 40% at 120 °C and approaches zero when the operating temperature of classical PEFCs (80 °C) is reached.[59]

Similarly to phosphoric acid fuel cells which can have excellent lifetimes, HT-PEFC-MEAs have already been shown to be able to be operated for 5000–20 000 h with small voltage drops under nearly steady-state conditions.[60,61] However, frequent and heavy load changes, very high temperatures and especially shut-downs are detrimental for the durability. Different degradation phenomena may be responsible for the performance loss of HT-PEFCs.

Among the phenomena responsible for membrane degradation are most notably mechanical stress caused by temperature cycling or swelling/shrinking of the membrane at load changes (causing creep, tearing and crack formation) and acid loss by evaporation or leaching due to the presence of liquid water.

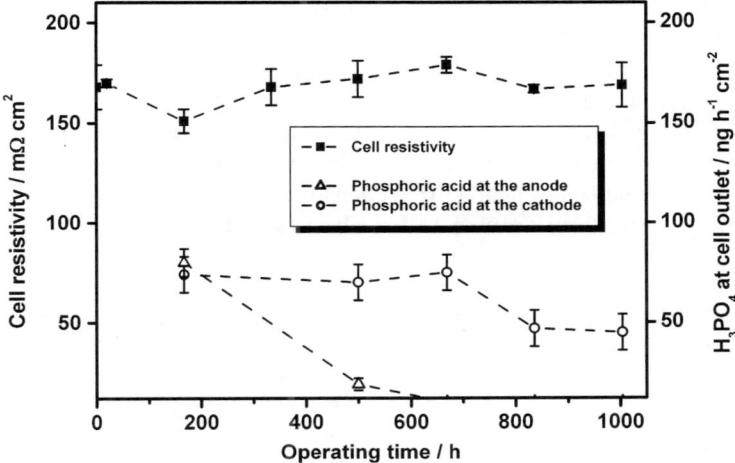

Figure 3.13 Cell resistance and acid loss from anode and cathode of a typical HT-PEFC at $T = 160\,°C$. Reprinted, with permission, from Wannek et al.[63] © 2008 Wiley-VCH.

The most critical part of the membrane prone to mechanical failure is the outer part of the MEA between the electrodes and the gaskets. To circumvent these problems several actions can be taken, including the use of a membrane polymer of very high average molecular weight, the introduction of internal cross-links, the preparation of blends of PBI with a second type of polymer, to limit the acid uptake and/or to strengthen the edges of the MEAs by using so-called 20–50 μm thin sub-gaskets that partially overlap with the electrode.

It has been shown that acid loss from the cell under normal operating conditions is very small (in the range of $1–100\,ng\,cm^{-2}\,h^{-1}$,[62,63] cf. Figures 3.12 and 3.13). As a consequence, this acid loss does not lead to a performance-limiting decrease of the membrane conductivity within tens of thousands of hours.

The main factors for electrode degradation are presumably the loss of active catalyst area by either platinum corrosion or Ostwald ripening, corrosion of the carbon catalyst support and the poisoning of catalytically active sites by gas impurities such as SO_2 or H_2S. Carbon corrosion is assumed to play a major role in cell degradation if the cathode is exposed regularly to high potentials. As a consequence, relatively stable carbon supports such as graphitised carbons or multi-walled carbon nanotubes are currently being developed. On the engineering level, all operating protocols of HT-PEFCs take care to avoid open circuit operation, for example by using an external shunt. While the adsorption of carbon monoxide on the platinum catalyst is completely reversible, it appears that the cell voltage does not fully recover after poisoning a HT-PEFC with sulfur in low concentrations (1–10 ppm).

Under optimal operating conditions, that is when the frequency of thermal and potential cycling is low and when reactants free of poisoning impurities are used, the loss of active catalyst area, especially in the cathode, is the main cause for the decrease in cell performance.

Summing up the performance and durability issues outlined above, it can be stated that HT-PEFCs are preferably operated nearly constantly at 160 to 180 °C with hydrogen or reformate and air at ambient pressure. At 600–650 mV, which represents a reasonable compromise between cell voltage (to achieve high efficiency) and current density (to minimise the cell area needed to deliver a given power output), today's best HT-PEFCs already deliver about 300 mW cm^{-2} in H$_2$/air cells (160 °C, ambient pressure) and about 200 mW cm^{-2} with reformate as fuel and can operate for around 20 000 h. Further improvements are expected especially from the use of new catalyst materials and electrode designs.

3.6 Stacks

In order to reach the desired voltage and power level of a HT-PEFC several cells have to be stacked together. Besides the membrane–electrode assembly the bipolar plates (BiPs) are major components of a stack. On the one hand they contribute significantly to cost, volume and weight of a stack. On the other hand they are important for the mechanical stability of stacks. They have to uniformly distribute fuel gas and air inside the cells, separate the individual cells in the stack, remove heat from the active area, prevent leakage of gas and coolant and carry away the electric current. Bipolar plates in a HT-PEFC have to fulfil these tasks under highly corrosive conditions at temperatures up to 200 °C.

Materials which are suitable for BiPs should have an electrical plate resistance lower than 0.01 Ω cm^2. The thermal conductivity should be higher than 10 W mK^{-1}, the gas permeability lower than 10^{-4} cm^3 s^{-1} cm^{-2}. A compressive strength of at least 4.2 N cm^{-2} is required.[64,65]

Materials which are discussed for BiPs in HT-PEFCs can be classified in (coated and non-coated) metals and carbon–polymer composites. Metals are potential candidates as BiP material because they can be manufactured very thin (<0.1 mm) and are easily stamped into the desired shape of the flowfield. However, due to corrosion effects at the typical operating conditions of HT-PEFCs the required long-term stability has not been reached yet, and consequently most activities are focused on carbon–polymer composite materials. These materials are available on the market and sold by several companies. The advantage of composite plates is their high corrosion resistance. Furthermore, the composite material can be moulded into nearly any shape which is a well known industrial process suitable for mass production. Detailed discussion of the advantages and disadvantages of the different BiP materials for low-temperature fuel cells can be found in the literature.[66,67]

To our knowledge all HT-PEFC stacks publicly available today are based on composite graphite material. Sartorius developed oil-cooled stacks in the 2 kW power range but they ceased their activities only recently. At the moment

Serenergy seems to be the only company offering HT-PEFC stacks. They developed air-cooled stacks in the power range of 900 W to 3 kW. Several research institutes are also developing HT-stacks either cooled with thermo oil, with air cooling, or externally cooled.[68]

In the following we will provide results of stacks developed at the Forschungszentrum Jülich. The objective of our stack development is to design, construct and test HT-PEFC stacks in the kilowatt power range. Development focuses specifically on applications in the area of on-board power supply, as the HT-PEFC is particularly suitable for such applications when combined with a reformer. The HT-PEFC becomes even more relevant when the fuel that is already in the vehicle can be used. As both diesel and kerosene will continue to be used in the future, the stacks are designed for reformate gas typically containing 33% hydrogen and 1% carbon monoxide in wet reformate. Due to the operating temperature of about 160 °C, stacks are capable of withstanding the high CO content in the reformate gas (*cf.* section 3.5). Stacks are designed for operation at ambient pressure and, because of the use of the PBI membrane doped with phosphoric acid, they require no external humidification, which leads to a simplified system layout.

Within the framework of stack development, different stack designs were tested. For the 5 kW power class, a concept was developed where the electrochemically active areas are strictly separated from the oil cooling areas. The advantage of this design is that single cells can be replaced without any coolant seeping into the electrochemically active areas. Furthermore, leakages in the area where the coolant is distributed through externally located manifolds can be detected immediately by an optical inspection and even in the case of an oil leakage the electrochemical active area remains unaffected by the oil which would destroy the MEA.

The bipolar plate unit is therefore composed of three components, as can be seen in Figure 3.14: two reactant bipolar plates and the separate cooling plate unit. The bipolar plates are composed of graphitic materials with milled flow fields, while the separate cooling cells are fabricated from metal. The stack has an active cell area of 320 cm^2. Around 60 cells are required when operating the stack with diesel reformate in order to achieve a power of 5 kW.

With regard to the development of another stack design based on purely graphitic materials, the primary emphasis was on simple fabrication technique and cost effectiveness. This stack is also supplied with reformate, which is fed into the cells with an active area of nearly 200 cm^2 (180 mm × 110 mm) *via* the manifolds, which are integrated into the bipolar plates. The cooling channels are also integrated into the bipolar plates and are located on the back side of the cathode. The bipolar plate unit composed of reactant distribution and cooling therefore consists of two parts: the anodic and cathodic bipolar plates. This design allows both air and conventional thermal oil to be used as coolants (Figure 3.15).

The stacks were tested with pure hydrogen and with synthetic reformate as fuel and air as oxidant. Commercial MEAs from the company BASF Fuel Cells were used. The stacks were tested up to 1000 h. A stoichiometry of λ = 2 for fuel

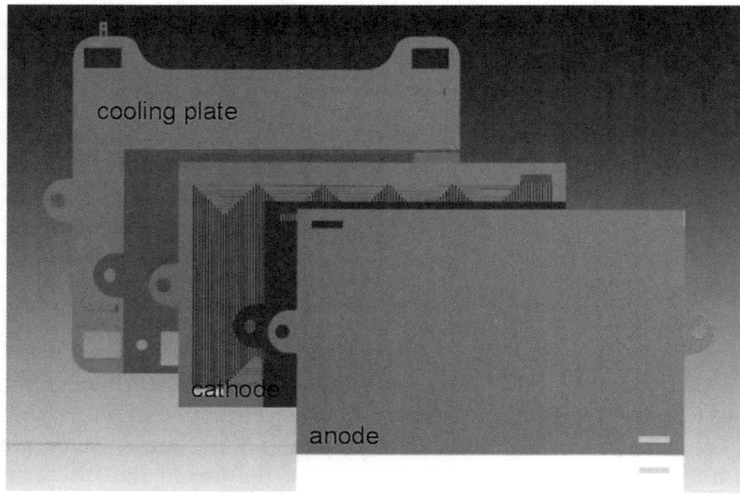

Figure 3.14   Repeating unit of a stack with separation of oil cooling and the electrochemical active region. Anode and cathode are machined into graphite composite material, the cooling plate is metallic.[69]

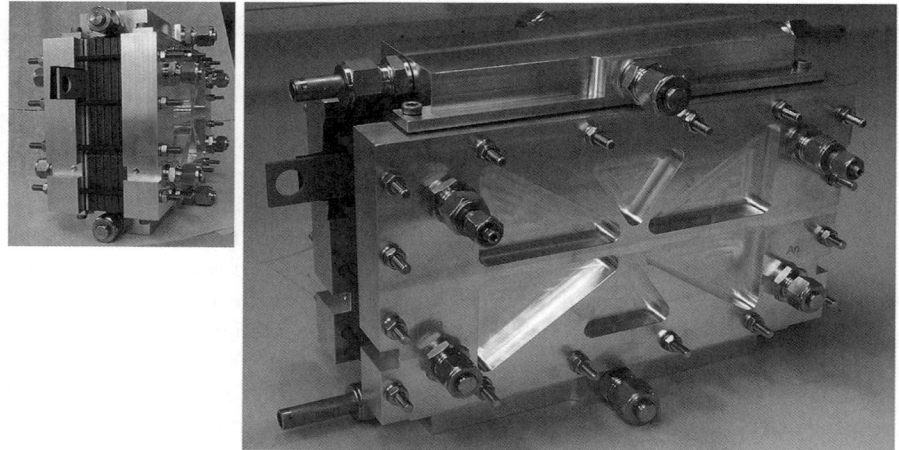

Figure 3.15   Five-cell HT-PEFC stack, $200\,cm^2$ active area and graphitic bipolar plates.

and air was chosen. The temperature was $160\,°C$, the current density was chosen to be $450\,mA\,cm^{-2}$ when the stack was operated with hydrogen and $250\,mA\,cm^{-2}$ when synthetic diesel reformate was used. The measurements were interrupted every day to record a polarisation curve. Typical polarisation

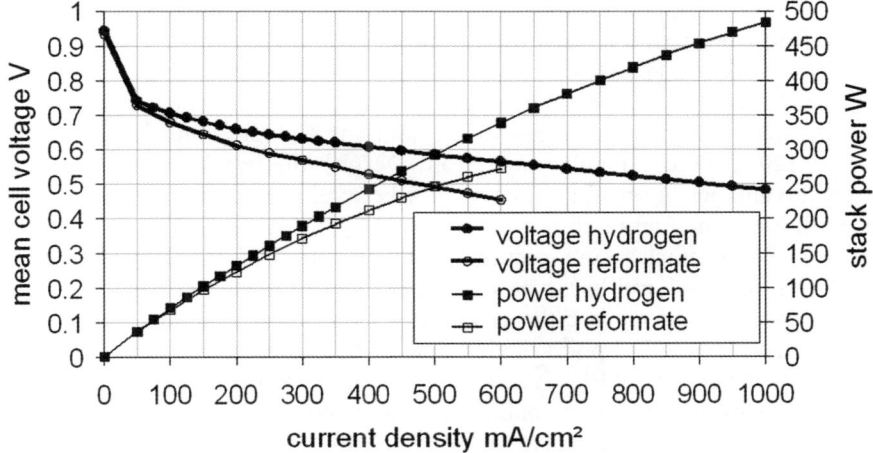

Figure 3.16   Mean cell voltage and stack power versus current density for a five-cell stack after 460 h of operation, operating conditions: 160 °C, hydrogen and synthetic reformate, λ: 2/2.

curves for the stack performance at ambient pressure using either pure hydrogen or synthetic diesel reformate as fuel and air as oxidant after 450 h of operation are shown in Figure 3.16 (the total time of operation was 1000 h). At 0.6 V the current density was 410 mA cm^{-2} for operation with hydrogen and 226 mA cm^{-2} when using synthetic diesel reformate. The decrease in current density can be attributed to the reduced hydrogen content and the CO content of 1% in the fuel. The average degradation rate during the 1000 h test was determined to amount to 60 μV h^{-1}.[69]

3.7   Perspectives

For a variety of applications with specific boundary conditions, high temperature PEFCs may represent the most auspicious option to produce electric current with high efficiency when compared to conventional internal combustion engines or other fuel cell types.

In comparison to classical PEFCs, a HT-PEFC with an operation temperature of 160–180 °C has a much higher tolerance for carbon monoxide. Furthermore, due to the high operation temperature and the phosphoric acid-based electrolyte water management is not a major issue for this type of fuel cell. Based on these facts the system layout will be much simpler: a fine purification of the fuel, *e.g.* by selective oxidation or methanation is not necessary. Additionally, there is no need for external humidifiers. Consequently, the number of system components can be reduced and the control engineering is simpler. Another advantage of the high operating temperature is the increased temperature difference between the stack and the ambient air which leads to a reduced size of the necessary radiator.

A drawback of a HT-PEFC is the missing cold-start capability. The stack has to be preheated before current can be drawn. In particular, when the stack is to be operated with reformate gas which contains CO the temperature should exceed 120 °C before regular operation. Despite this weakness a HT-PEFC offers advantages in several applications where reformate from diesel, jet fuel, or methane is the fuel of choice.

Commercial vehicles, construction machines, ships and aircraft will continue to be run on diesel and kerosene in the long term. Substantial energy savings can be made by increasing the efficiency of on-board generation of electric power. From a practical point of view and in the interests of the end user, these auxiliary power units (APUs) must run on the fuel that is already available on board. This means that these 'middle distillates' must be reformed. If the high-temperature polymer electrolyte fuel cell (HT-PEFC) is combined with an appropriate reforming technology on board, it becomes possible to efficiently generate electricity, even when the main engine is not running.

Although the HT-PEFC is characterised by a simple system design, the volumetric and gravimetric stack performance has to be increased to about 60% or 80% of that of a classical Nafion-based low-temperature PEFC before it can compete successfully with it. The most important issues that have to be addressed to reach this goal consist in a significant improvement of the kinetics of the oxygen reduction reaction (to minimise the adverse effect of phosphoric acid), an advanced catalyst layer design (to maximise catalyst utilisation and ensure optimal acid distribution within the MEA) and the design of corrosion-resistant thin bipolar plates (to reduce size and weight of the stack). Furthermore, while HT-PEFCs have already demonstrated extraordinarily good lifetimes of up to 20 000 h (at least on the cell level), the overall cost of a HT-PEFC system has to be drastically reduced in a later stage before a broad market introduction is possible.

References

1. A. Heinzel, G. Bandlamudi and W. Lehnert, High-temperature PEMFC, in *Encyclopedia of Electrochemical Power Sources*, ed. J. Garche, C. Dyer, P. Moseley, Z. Ogumi, D. Rand and B. Scrosati, Elsevier, Amsterdam, 2009, **vol. 2**, pp. 951–957.
2. M. Schuster, T. Rager, A. Noda, K. D. Kreuer and J. Maier, About the choice of the protogenic group on PEM separator materials for intermediate temperature, low humidity operation: a critical comparison of sulfonic acid, phosphonic acid and imidazole functionalized model compounds, *Fuel Cells*, 2005, **5**(3), 355–365.
3. M. Armand, F. Endres, D. R. MacFarlane, H. Ohno and B. Scrosati, Ionic-liquid materials for the electrochemical challenges of the future, *Nat. Mater.*, 2009, **8**(8), 621–629.

4. T. Uda and S. M. Haile, Thin-membrane solid-acid fuel cell, *Electrochem. Solid-State Lett.*, 2005, **8**(5), A245–A246.
5. K. Genzaki, P. Heo, M. Sano and T. Hibino, Proton conductivity and solid acidity of Mg-, In-, and Al-doped SnP_2O_7, *J. Electrochem. Soc.*, 2009, **156**(7), B806–B810.
6. M. B. Cutlip, S. C. Yang and P. Stonehart, Simulation and optimization of porous gas-diffusion electrodes used in hydrogen oxygen phosphoric-acid fuel-cells. 2. development of a detailed anode model, *Electrochim. Acta*, 1991, **36**(3–4), 547–553.
7. H. R. Kunz and G. A. Gruver, Catalytic activity of platinum supported on carbon for electrochemical oxygen reduction in phosphoric acid, *J. Electrochem. Soc.*, 1975, **122**(10), 1279–1287.
8. A. Parthasarathy, S. Srinivasan and J. Appleby, Temperature-dependence of the electrode kinetics of oxygen reduction at the platinum Nafion® interface – a microelectrode investigation, *J. Electrochem. Soc.*, 1992, **139**(9), 2530–2537.
9. S. J. Clouser, J. C. Huang and E. Yeager, Temperature-dependence of the tafel slope for oxygen reduction on platinum in concentrated phosphoric-acid, *J. Appl. Electrochem.*, 1993, **23**(6), 597–605.
10. L. D. Burke and J. A. Morrissey, Hydrous oxide formation on platinum in phosphoric-acid solution, *J. Electrochem. Soc.*, 1994, **141**(9), 2361–2368.
11. B. R. Scharifker, P. Zelenay and J. O. Bockris, The kinetics of oxygen reduction in molten phosphoric-acid at high-temperatures, *J. Electrochem. Soc.*, 1987, **134**(11), 2714–2725.
12. D. R. Lawson, L. D. Whiteley, C. R. Martin, M. N. Szentirmay and J. L. Song, Oxygen reduction at Nafion® film-coated platinum-electrodes – transport and kinetics, *J. Electrochem. Soc.*, 1988, **135**(9), 2247–2253.
13. J. C. Huang, R. K. Sen and E. Yeager, Oxygen reduction on platinum in 85% orthophosphoric acid, *J. Electrochem. Soc.*, 1979, **126**(5), 786–792.
14. A. J. Appleby and A. Borucka, Oxygen dissolution and evolution on platinum in 85 percent orthophosphoric acid at elevated temperatures, *J. Electrochem. Soc.*, 1969, **116**(9), 1212.
15. M. M. Ghoneim, S. Clouser and E. Yeager, Oxygen reduction kinetics in deuterated phosphoric-acid, *J. Electrochem. Soc.*, 1985, **132**(5), 1160–1162.
16. J. T. Glass, G. L. Cahen and G. E. Stoner, The effect of phosphoric acid concentration on electrocatalysis, *J. Electrochem. Soc.*, 1989, **136**(3), 656–660.
17. D. R. Desena, E. R. Gonzalez and E. A. Ticianelli, Effect of phosphoric acid concentration on the oxygen reduction and hydrogen oxidation reactions at a gas-diffusion electrode, *Electrochim. Acta*, 1992, **37**(10), 1855–1858.
18. T. J. Schmidt and J. Baurmeister, Properties of high-temperature PEFC Celtec®-P 1000 MEAs in start/stop operation mode, *J. Power Sources*, 2008, **176**(2), 428–434.
19. Z. Y. Liu, J. S. Wainright, M. H. Litt and R. F. Savinell, Study of the oxygen reduction reaction (ORR) at Pt interfaced with phosphoric acid

doped polybenzimidazole at elevated temperature and low relative humidity, *Electrochim. Acta*, 2006, **51**(19), 3914–3923.

20. S. K. Zecevic, C. R. Martin and S. Srinivasan, Kinetics of O_2 reduction on a Pt electrode covered with a thin film of solid polymer electrolyte, *J. Electrochem. Soc.*, 1997, **144**(9), 2973–2982.

21. T. J. Schmidt, Durability and degradation in high-temperature polymer electrolyte fuel cells, *ECS Trans.*, 2006, **1**(8), 19–31.

22. Z. Y. Liu, J. S. Wainright and R. F. Savinell. High-temperature polymer electrolytes for PEM fuel cells: study of the oxygen reduction reaction (ORR) at a Pt-polymer electrolyte interface, in *18th International Symposium on Chemical Reaction Engineering*, 2004, Chicago: Pergamon–Elsevier Science, 2004, pp. 4833–4838.

23. M. R. Tarasevich, Investigation of the consecutive stages in the reduction of O and H_2O_2. Pt. 9. Mechanism of the reduction of O on the Pt metals, *Elektrokhimiya*, 1973, **9**(5), 599–605.

24. A. Damjanov and M. A. Genshaw, Dependence of kinetics of O_2 dissolution at Pt on conditions for adsorption of reaction intermediates, *Electrochim. Acta*, 1970, **15**(7), 1281.

25. A. Damjanovic and J. O. M. Bockris, The rate constants for oxygen dissolution on bare and oxide-covered platinum, *Electrochim. Acta*, 1966, **11**(3), 376–377.

26. A. Damjanovic and P. G. Hudson, On the kinetics and mechanism of O_2 reduction at oxide film covered Pt electrodes. 1. Effect of oxide film thickness on kinetics, *J. Electrochem. Soc.*, 1988, **135**(9), 2269–2273.

27. S. Clouser, O_2 reduction in acidic electrolytes, PhD thesis, Case Western Reserve, 1980.

28. B. R. Scharifker, P. Zelenay and J. O. Bockris, The kinetics of oxygen reduction in molten phosphoric-acid at high-temperatures, *J. Electrochem. Soc.*, 1987, **134**(11), 2714–2725.

29. G. Horanyi, E. M. Rizmayer and G. Inzelt, Radiotracer study of adsorption of phosphoric-acid on platinized platinum-electrodes in presence of different ions and oxalic-acid, *J. Electroanal. Chem.*, 1978, **93**(3), 183–194.

30. M. A. Habib and J. O. Bockris, Adsorption at the solid-solution interface – An FTIR study of phosphoric-acid on platinum and gold, *J. Electrochem. Soc.*, 1985, **132**(1), 108–114.

31. F. C. Nart and T. Iwasita, On the adsorption of $H_2PO_4^-$ and H_3PO_4 on platinum – an *in situ* FTIR study, *Electrochim. Acta*, 1992, **37**(3), 385–391.

32. J. Mostany, P. Martínez, V. Climent, E. Herrero and J. M. Feliu, Thermodynamic studies of phosphate adsorption on Pt(111) electrode surfaces in perchloric acid solution, *Electrochim. Acta*, 2009, **54**(24), 5836–5843.

33. S. Mukerjee and S. Srinivasan, Enhanced electrocatalysis of oxygen reduction on platinum alloys in proton exchange membrane fuel cells, *J. Electroanal. Chem.*, 1993, **357**(1–2), 201–224.

34. U. A. Paulus, A. Wokaun, G. G. Scherer, T. J. Schmidt, V. Stamenkovic, V. Radmilovic, N. M. Markovic and P. N. Ross, Oxygen reduction on

carbon-supported Pt-Ni and Pt-Co alloy catalysts, *J. Phys. Chem. B*, 2002, **106**(16), 4181–4191.

35. E. Antolini, J. R. C. Salgado and E. R. Gonzalez, The stability of Pt-M (M = first row transition metal) alloy catalysts and its effect on the activity in low temperature fuel cells: A literature review and tests on a Pt-Co catalyst, *J. Power Sources*, 2006, **160**(2), 957–968.

36. D. Thompsett, Pt alloys as oxygen reduction catalysts, in *Handbook of Fuel Cells*, ed. L. A. Vielstich, W. Gasteiger H. A., Chichester, John Wiley, 2003, pp. 467–480.

37. H. R. Colón-Mercado, H. Kim and B. N. Popov, Durability study of Pt$_3$Ni catalysts as cathode in PEM fuel cells, *Electrochem. Commun.*, 2004, **6**(8), 795–799.

38. C. V. Rao, J. Parrondoa, S. L. Ghattya and B. Rambabu, High temperature polymer electrolyte membrane fuel cell performance of Pt$_x$Co$_y$/C cathodes, *J. Power Sources*, 2010, **195**(11), 3425–3430.

39. M. A. Enayetullah, E. J. M. Osullivan and E. B. Yeager, Oxygen reduction on platinum in mixtures of phosphoric and trifluoromethane sulfonic-acids, *J. Appl. Electrochem.*, 1988, **18**(5), 763–767.

40. M. A. Enayetullah, The role of electrolyte in O$_2$ electroreduction catalysis. PhD thesis, Case Western Reserve, 1986.

41. X. Gang, H. A. Hjuler, C. Olsen, R. W. Berg and N. J. Bjerrum, Electrolyte additives for phosphoric acid fuel cells, *J. Electrochem. Soc.*, 1993, **140**(4), 896–902.

42. S. G. Hong, K. Kwon, M. J. Lee and D. Y. Yoo, Performance enhancement of phosphoric acid-based proton exchange membrane fuel cells by using ammonium trifluoromethanesulfonate, *Electrochem. Commun.*, 2009, **11**(6), 1124–1126.

43. J. S. Wainright, J. T. Wang, D. Weng, R. F. Savinell and M. Litt, Acid-doped polybenzimidazoles: a new polymer electrolyte, *J. Electrochem. Soc.*, 1995, **142**(7), L121–L123.

44. J. Mader, L. Xiao, T. J. Schmidt and B. C. Benicewicz, Polybenzimidazole/acid complexes as high-temperature membranes, *Adv. Polym. Sci.*, 2008, **216**, 63–124.

45. Q. Li, J. O. Jensen, R. F. Savinell and N. J. Bjerrum, High temperature proton exchange membranes based on polybenzimidazoles for fuel cells, *Prog. Polym. Sci.*, 2009, **34**(5), 449–477.

46. R. He, Q. Li, G. Xiao and H. J. Bjerrum, Proton conductivity of phosphoric acid doped polybenzimidazole and its composites with inorganic proton conductors, *J. Membr. Sci.*, 2003, **226**(1–2), 69–184.

47. Y.-L. Ma, J. S. Wainright, M. H. Litt and R. F. Savinell, Conductivity of PBI membranes for high-temperature polymer electrolyte fuel cells, *J. Electrochem. Soc.*, 2004, **151**(1), A8–A16.

48. M. K. Daletou, J. K. Kallitsis, G. Voyiatzis and S. G. Neophytides, The interaction of water vapors with H$_3$PO$_4$ imbibed electrolyte based on PBI/polysulfone copolymer blends, *J. Membr. Sci.*, 2009, **326**(1), 76–83.

49. K. Wippermann, C. Wannek, H.-F. Oetjen, J. Mergel and W. Lehnert, Cell resistances of poly(2,5-benzimidazole)-based high temperature polymer membrane fuel cell membrane electrode assemblies: time dependence and influence of the operating parameters, *J. Power Sources*, 2010, **195**(9), 2806–2809.
50. L. Gubler, D. Kramer, J. Belack, Ö. Ünsal, T. J. Schmidt and G. G. Scherer, Celtec-V – a polybenzimidazole-based membrane for the direct methanol fuel cell, *J. Electrochem. Soc.*, 2007, **154**(9), B981–B987.
51. C. Wannek, W. Lehnert and J. Mergel, Membrane electrode assemblies for high-temperature polymer electrolyte fuel cells based on poly(2,5-benzimidazole) membranes with phosphoric acid impregnation *via* the catalyst layers, *J. Power Sources*, 2009, **192**(2), 258–266.
52. K. Kwon, T. Y. Kim, D. Y. Yoo, S. G. Hong and J. O. Park, Maximization of high-temperature proton exchange membrane fuel cell performance with the optimum distribution of phosphoric acid, *J. Power Sources*, 2009, **188**(2), 463–467.
53. J. Lobato, P. Canizares, M. A. Rodrigo, J. J. Linares and F. J. Pinar, Study of the influence of the amount of PBI-H_3PO_4 in the catalytic layer of a high temperature PEMFC, *Int. J. Hydrogen Energy*, 2010, **35**(3), 1347–1355.
54. S. Matar, A. Higier and H. Liu, The effects of excess phosphoric acid in a polybenzimidazole-based high temperature proton exchange membrane fuel cell, *J. Power Sources*, 2010, **195**(1), 181–184.
55. C. Wannek, I. Konradi, J. Mergel and W. Lehnert, Redistribution of phosphoric acid in membrane electrode assemblies for high-temperature polymer electrolyte fuel cells, *Int. J. Hydrogen Energy*, 2009, **34**(23), 9479–9485.
56. BASF Fuel Cell, press release P-09-231 (2009-05-06); available online: http://www.basf.com/group/pressrelease/P-09-231 [accessed 28 August 2009].
57. L. Xiao, H. Zhang, E. Scanlon, L. S. Ramanathan, E.-W. Choe, D. Rogers, T. Apple and B. C. Benicewicz, High-temperature polybenzimidazole fuel cell membranes via a sol-gel process, *Chem. Mater.*, 2005, **17**(21), 5328–5333.
58. T. Tingelöf and J. K. Ihonen, A rapid break-in procedure for PBI fuel cells, *Int. J. Hydrogen Energy*, 2009, **34**(15), 6452–6456.
59. C. Wannek, H. Dohle, J. Mergel and D. Stolten, Novel VHT-PEFC MEAs based on ABPBI membranes for APU applications, *ECS Trans.*, 2008, **12**(1), 29–39.
60. J. O. Park, T. Y. Kim, Y. Aihara, M. Takezawa and S. W. Choi, Catalyst degradation and cell voltage decay in high temperature PEMFC, in *Proceedings of the 217th ECS meeting*, 25–30 April 2010, Vancouver, Canada, Abstract #273.
61. Y. Oono, T. Fukuda, A. Sounai and M. Hori, Influence of operating temperature on cell performance and endurance of high temperature proton exchange membrane fuel cells, *J. Power Sources*, 2010, **195**(4), 1007–1014.
62. S. Yu, L. Xiao and B. Benicewicz, Durability studies of PBI-based high temperature PEMFCs, *Fuel Cells*, 2008, **8**(3–4), 165–174.

63. C. Wannek, B. Kohnen, H.-F. Oetjen, H. Lippert and J. Mergel, Durability of ABPBI-based MEAs for high temperature PEMFCs at different operating conditions, *Fuel Cells*, 2008, **8**(2), 87–95.

64. V. Mehta and J. S. Cooper, Review and analysis of PEM fuel cell design and manufacturing, *J. Power Sources*, 2003, **114**, 32–53.

65. J. Huang, D. G. Baird and J. E. McGrath, Development of fuel cell bipolar plates from graphite filled wet-lay thermoplastic composite materials, *J. Power Sources*, 2005, **150**, 110–119.

66. A. Hermanna, T. Chaudhuria and P. Spagnolb, Bipolar plates for PEM fuel cells – A review, *Int. J. Hydrogen Energy*, 2005, **30**, 1297–1302.

67. H. Tawfika, Y. Hunga and D. Mahajan, Metal bipolar plates for PEM fuel cell – A review, *J. Power Sources*, 2007, **163**, 755–767.

68. J. Scholta, M. Messerschmidt, L. Jorissen and Ch. Hartnig, Externally cooled high temperature polymer electrolyte membrane fuel cell stack, *J. Power Sources*, 2009, **190**, 83–85.

69. A. Bendzulla, Von der Komponente zum Stack: Entwicklung und Auslegung von HT-PEFC-Stacks der 5 kW Klasse, PhD thesis, RWTH Aachen, 2010.

CHAPTER 4

Large Auxiliary Power Units for Vessels and Airplanes

RALF PETERS[a] AND ANDREAS WESTENBERGER[b]

[a] Forschungszentrum Jülich GmbH, Institute of Energy Research,
IEF-3: Fuel Cells, 52425 Jülich, Germany; [b] Airbus Deutschland GmbH,
Hamburg, Germany

4.1 Introduction

Electricity requirements in mobile applications are increasing in nearly all forecasts for the future. Reasons for this development are the use of electric devices for increased convenience and a guaranteed energy supply during idling mode. Fuel cells are envisaged by the aviation industry as an environmentally friendly and highly efficient energy conversion system. The ideal energy carrier for fuel cells is hydrogen. The availability of hydrogen is a prerequisite for the use of fuel cells in mobile and stationary applications. However, an infra-structure for hydrogen as a future energy carrier does not yet exist. It is therefore essential that hydrogen be produced from readily available energy carriers. For stationary applications, natural gas and heating oil are suitable sources, while for mobile applications, gasoline, kerosene and diesel are options. Currently, the energy carriers listed above are mainly produced from crude oil as a fossil primary energy carrier. In the long term, some of the liquid energy carriers needed today will be produced from biomass. Following a single-fuel strategy, the same energy carrier should be used for traction and for electric power generation. Currently, some international development strategies are based on hydrogen as an energy carrier, which requires an additional storage system.

RSC Energy and Environment Series No. 2
Innovations in Fuel Cell Technologies
Edited by Robert Steinberger-Wilckens and Werner Lehnert
© Royal Society of Chemistry 2010
Published by the Royal Society of Chemistry, www.rsc.org

Auxiliary power units (APUs) have been considered for trucks,[1-4] passenger cars[5-7] and for military applications, *e.g.* for tactical vehicles[8] or as a propulsion system for submarines during strategic missions.[9] Lutsey *et al.*[10] showed that most APUs used for land transport require an electric power of about 2–5 kW$_e$. Possible applications for these APUs include luxury passenger vehicles, law enforcement vehicles, recreational vehicles, and line-haul heavy-duty vehicles. Refrigeration trucks and pick-up trucks, which are mainly used by farmers, craftsmen and contractors, need somewhat more electrical power, in the range of 10–30 kW$_e$. Larger APU systems with 35–75 kW$_e$ are only required for utility trucks such as cement mixers.

Targets for APU systems of this kind have been defined by the U.S. Department of Energy (DOE).[11] The data are summarised in Table 4.1 and will be discussed in connection with the targets for passenger cars, airplanes and stationary applications such as combined heat and power generation (CHP). As can be seen, CHP systems require as much electric power as aircraft. Efficiency should be high (*i.e.* 35–50%) for almost all applications. Additionally, combined heat and power generation systems require a total energy utilisation of up to 90%. Cost targets are most challenging for automotive systems; durability is critical for stationary systems.

Targets for aeronautic and maritime systems have not been defined in all aspects as has been done for stationary CHP systems, APUs and fuel cell drive systems for cars. It is obvious that APUs for aircraft have to meet specific requirements. Some of them correspond closely to other applications. The power density target of aircraft APUs is in the same order of magnitude as that of fuel cell drive systems for passenger cars based on hydrogen as fuel.

It can be expected that the durability of aeronautical and maritime APUs must withstand even longer operation periods than stationary systems. In addition, the load profile of APUs is often not well known and is therefore approximated using hypothetical consumer data such as for trucks.[12] Due to the lack of information on large APUs, a more detailed analysis would be worthwhile. This chapter presents some ideas and solutions for the introduction of fuel cell systems in this challenging and interesting application area.

4.2 Motivation

Figure 4.1 shows the principle parameters which determine the results of a technical evaluation for large APU systems. The main effects are the operation conditions of the application, the choice of fuel and available fuel cell technologies.

The use of fuel cell systems as APUs for airplanes was mentioned for the first time by Seidel *et al.*[13] The most important reason for introducing fuel cells is their improved efficiency. Commercial APUs in ground operation have an efficiency of 20%, it is likely that 40% will be reached with fuel cells. Higher efficiencies lead to lower kerosene consumption and lower fuel costs, which results in a higher overall efficiency on the aircraft mission level. An A330-300

Table 4.1 Targets for different APU applications and for stationary systems based on natural gas as an energy carrier.

Target	Aircraft[14,15b,16-18,22,23]	Passenger car APU[5]	DOE,[21] FC drive system (H_2), 2015	DOE,[21] stationary systems, 2011	DOE,[11] APU 2015/2020	DOE,[11] CHP 2015/2020
Electric power	100–400 kW	10 kW	80 kW	5–250 kW	1–10 kW	1–10 kW
Efficiency	40%	<35%	50%	40%	35%/40%	42.5%/45%
CHP energy efficiency	–	–	–	80%	–	87.5%/90%
Costs for 50 000/100 000 units	€1500 kW^{-1} @ 500 units	€40 kW^{-1}	\$30 kW^{-1}	\$750 kW^{-1}	\$600 kW^{-1}	€450 kW^{-1}
Durability	20 000/40 000 h^a	5000 h	5000 h	40 000 h	15 000/20 000 h	40 000/60 000 h
Degradation with cycling per 1000 h operation	–	–	–	–	1.3%/1%	0.5%/0.3%
Power density	750 W L^{-1}	333 W L^{-1}	650 W L^{-1}	–	35–40 W L^{-1}	–
Mass specific power	0.5–1 kW kg^{-1}	250 kW kg^{-1}	650 kW kg^{-1}	–	40–45 kW kg^{-1}	–
System availability	–	–	–	–	98%/99%	98%/99%
Cold start	^{b}	<1 s	5 s @ 20 °C; 30 s @ –20 °C	<30 min	10/5 min	30/20 min
Transient response (10% to 90% rated power)	–	<1 s	1 s	<3 min	3/2 min	3/2 min
Load range	–	1:50	–40 °C	–35 °C to 40 °C	–	–

^{a}Durability of aircraft APU critically depends on the fuel cell function and the maintenance plan.
^{b}Depending on the location of installation – inside or outside the pressurised cabin area.
CHP, combined heat and power generation.

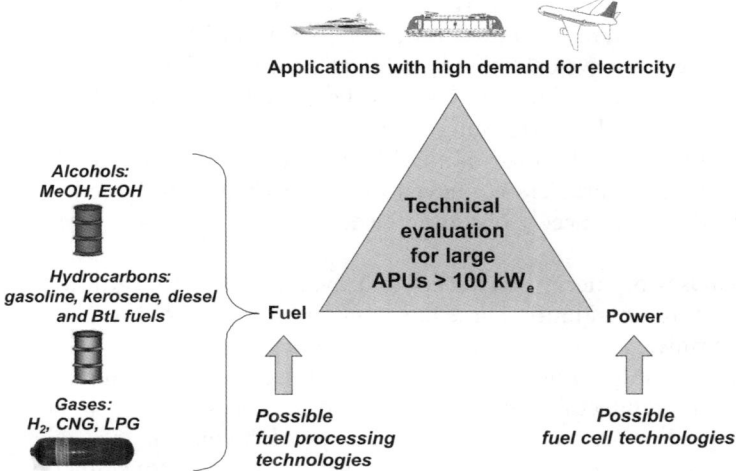

Figure 4.1 Main factors for the technical evaluation of large APUs.

type aircraft with a take-off gross weight of 230 t requires on average 100 000 L kerosene for a 10 000 km mission.[14] About 3–5% of the consumed fuel is required by on-board systems.[15a] During flight, the main engines produce electricity with an efficiency of 40–45%. Daggett *et al.*[16] analysed such systems with a 400 kW solid oxide fuel cell (SOFC) and a 40 kW gas turbine. Daggett *et al.* assumed a stack power density of $1 \, kW_e \, L^{-1}$ and an overall system power density of $0.5 \, kW_e \, L^{-1}$ for their calculations. For ground power generation, they assumed that fuel use would be reduced by 75% by increasing efficiency from 15% to 60%. It was therefore planned to achieve extremely high efficiencies, for example 75% in 2015[16] with hybrid systems based on SOFC technology and turbines,[12–14] and to reduce fuel consumption to 40% of today's level.

Such high efficiencies could only be calculated for systems with combinations of SOFC and coupled steam and gas turbine processes in the $10–100 \, MW_e$ electric power range for natural gas as energy carrier.[19] Additionally, this design requires high-temperature and high-pressure operation of SOFC and the availability of microturbines. Even the challenging targets of the US DOE are set to an efficiency of 60% for stationary SOFCs[20] and 40% for polymer electrolyte fuel cells (PEFCs), respectively.[21] Different SOFC system designs for aeronautical applications with efficiencies between 32% and 65% have been published in various reports[13–16] and will be discussed in more detail in sections 4.3.1.1 and 4.5.1.2.

Considering the present development status of the various fuel cell systems, partial replacement of the conventional main engine power generator capacity by a fuel cell (FC) system may not be obvious at first glance. Compared to the traditional electrical power generating system, fuel cell technology is far too heavy. Gas turbines, such as APUs in the tail cone or main engine, driving the generator in flight, have a power ratio of about $0.2\,kg\,kW^{-1}$, while today's most advanced fuel cell systems based on hydrogen reach about $1\,kg\,kW^{-1}$.[24] Consequently, fuel cell technology may not be considered as a direct replacement for traditional APUs performing the same functions.

Before finding the best use of novel fuel cell technology on board an aircraft, the operational characteristics have to be analysed and evaluated. To this end, the following tasks need to be addressed:

- Analyses of energy consumers and their characteristics during a mission on board an aircraft. These have only been carried out to a certain extent by Srinivasan *et al.*[17] and Gummalla *et al.*[18]
- How are the technologies and their architecture applied today? Could the power architecture and a 'power and a process manager' be modified, adapted or introduced into a future system layout?
- Do fuel cells, as electrochemical converters, have various functions? In the case of a multifunctional role, the needs of different consumers have to be understood.
- The options of different fuel cell technologies need to be understood.
- The hypothesis of the technology development status of conventional and 'alternative' technology need to be assumed.

If they fulfil these requirements, fuel cell technologies have potential benefits if they could take over more functions in commercial passenger aircraft.

The second reason for introducing fuel cell technology in aircraft is the ultra-low emissions of these systems. Airplane emissions account for 3% of total emissions worldwide and might increase to as much as 4%.[16] Typical jet engines emit about 40 ppmv CO, 300 ppm NO_x and 14 ppmv unburnt hydrocarbons (UHC). It can be expected that these emissions increase at rated power. Similar effects were measured for internal combustion engines during different idling modes of long-haul heavy-duty trucks.[25] Measured emission data for APU ground operation and for complete missions are published in Srinivasan *et al.*[13] and Gummulla *et al.*[18] A commercial aircraft APU emits about 5.3 kg NO_x, 6.2 kg CO and 0.4 kg HCs per flight[17] and $1.9\,t\,NO_x\,year^{-1}$, $2.4\,t\,CO\,year^{-1}$ and $0.2\,t\,UHCs\,year^{-1}$.[18] The yearly CO_2 emissions for APU ground operation amount to 1000 t CO_2 per aircraft.[17] These emissions can be summarised together with an estimate of the overall air carrier population in the United States of 6840 turbo jets[26] as 6.8 million t CO_2 and 13 000 t NO_x. Figures for long-haul heavy-duty trucks are in the same order of magnitude for CO_2 and one order of magnitude higher for NO_x (10.9 Mio. t CO_2 and 190 000 t NO_x).[27] Fuel cell-based APUs in aircraft would eliminate NO_x emissions and cut CO_2 emissions by half. The specific CO_2 production calculated from these data is $636\,g_{CO_2}\,kWh^{-1}$, which is only slightly higher than the US electricity grid's rate of $605\,g_{CO_2}\,kWh^{-1}$ in 2007.[28] Specific CO_2 emissions can be reduced further by using liquid biofuels or hydrogen as fuel.

If maritime applications are considered, a wide range of electric power requirements can be found, *i.e.* 300 W for yachts[29] and several megawatts for ocean-going ships.[30] APUs for tourist craft, leisure craft, offshore support vessels, research and survey vessels, fast ferries, ferries, passenger cruise vessels, coastal cargo vessels and international cargo vessels were discussed by Bensaid *et al.*[31] The preferred fuel cell types are PEFCs[29,32] and molten carbonate fuel cells (MCFCs).[31,33] The fuels used for hydrogen production range from hydrogen[34] and liquid petroleum gas (LPG)[29] to maritime diesel[30] and NATO F76 diesel.[32,35] Additionally, direct methanol fuel cells (DMFCs) with 50 W_e power will be developed by Smart Fuel Cells for sailboats.[36] The wide application range in sea transport causes a broad variation of fuels, fuel cell types and power classes. Driving forces for maritime applications are lower life-cycle costs, reduced NO_x emissions, low vibrations and low noise levels. Generally, maritime transport accounts for about 3% of global petroleum consumption but contributes 14% of total NO_x and 16% of total SO_x emissions.[31] Maritime applications can also benefit from synergy effects of fuel cell technology. Specchia *et al.*[33] proposed the production of sanitary water and the use of sensible heat for heating purposes on board.

In 1999, Karin *et al.*[37] presented a report on the feasibility of using fuel cells to provide shipboard electric power by ship service generators (SSGs) – corresponding to APUs in this chapter – to the U.S. Coast Guard. Most vessels require one or two SSGs in the classes < 500 kW and 500–1500 kW. In view of a possible market entry, smaller systems (< 500 kW) could be envisaged for fishing vessels, general cargo vessels, off-shore vessels and hoppers. All other ship types require more electric power, which can be provided using a modular design. In regard to the figures provided by Karin *et al.*,[37] such a design is typical of SSGs on board LPG/liquid natural gas (LNG) tankers, refrigeration cargo ships *etc.* Due to the publication date of the study and the slower progress made in fuel cell technology development, fuel cells in the maritime sector will be introduced with some delay. Considering the required APU power and the envisaged market segments for the United States, the focus could be placed on fishing and general cargo vessels.

Ten years later, in 2009, Lebutsch *et al.*[34] described ways of applying fuel cells in shipboard power generation. They analysed the applicability of fuel cells for megayachts, pleasure boats and sightseeing boats. Pleasure boats require about 2 kW_e electric power which could be delivered more easily by a DMFC. Sightseeing boats require an electric power of about 16 kW_e for propulsion purposes. For these boats, a PEFC with hydrogen as fuel was considered. Lebutsch *et al.*[34] proposed a high-temperature polymer electrolyte fuel cell (HT-PEFC) system combined with methanol reforming for a 500 kW_e megayacht APU. The power class of these boats corresponds very well to those of general cargo vessels and fishing vessels.[37] The impact of using APUs for megayachts on CO_2 reduction was assumed to be about 30 000 tons year^{-1} in Europe. This is one order of magnitude lower than for truck APU (600 000 tons year^{-1}) in Europe and one order of magnitude higher than for pleasure boats and sightseeing boats (4000 tons year^{-1}).[34] It is obvious from this study that the reduction potential of replacing truck APUs is ten times higher in the U.S. than in Europe. Nevertheless, the passenger car market in Europe has the largest potential for emission reductions: 270 million tons year^{-1} focused on propulsion systems. According to the data given in an APU case study for megayachts,[38] a comparison results in specific CO_2 emissions of $360\,kg\,CO_2\,h^{-1}$ for an HT-PEFC system combined with methanol steam reforming, $228\,kg\,CO_2\,h^{-1}$ for an SOFC – also with methanol as energy carrier – and $425\,kg\,CO_2\,h^{-1}$ for a conventional diesel engine.

Corbett and Winebrake[39] described the emissions of a container ship travelling between Hong Kong and Los Angeles. The vessel was equipped with a $55\,MW_e$ engine. The distance between Hong Kong and Los Angeles amounts to $16\,852\,km$. The trip takes 17 days at sea and 3 days in port. Therefore, the machine operates for $480\,h$. Corbett and Winebrake divided the trip by engine modes: 1.25% idling, 1.75% manoeuvring, 5% precautionary zone operations, 7% slow cruise mode and 85% full cruise mode. The emissions of the main engines amount to $15\,000\,t$ CO_2 per trip and 30–$225\,t$ SO_x per trip. Unfortunately, APU emissions were not considered. These case studies will be evaluated further in the next sections.

4.3   Conditions for Auxiliary Power Unit Operation

4.3.1   Aeronautical Applications

System application on board an aircraft has to follow certain rules and regulations. Such regulations are mainly enforced by the individual ministries of transport of each nation, while the national aviation authorities are responsible for their application. Basically, the rules and regulations require a certain safety standard for commercial means of transport. In the case of a new technology, where no existing regulations apply, a rule-making process has to be launched. The intention is to make air transport as safe as possible and as profitable as possible. Malfunctions with an impact either on safety or on the mission are unacceptable. Cooperation with partners in other sectors – mobile or stationary – always shows that very specific environmental conditions, operational conditions and the desired reliability need to be taken into consideration. Furthermore, there are dedicated rules for equipment qualification and accordingly the adaptation or creation of standards not applicable or even no existing standards, as already mentioned above. The institutions responsible for the applicable safety standards are usually the national ministries of transport which delegate these tasks to the respective airworthiness authorities. Other organisations like the Society of Automotive Engineers (SAE) may set up working groups in order to define mutually agreed safety standards in advance.

From a technical point of view, an aircraft as the 'location' for systems must meet special demands. First of all, aircraft have two very different sets of environmental conditions. There is the pressurised cabin area and the unpressurised area outside the cabin. Both areas need to be analysed carefully before starting the work on system layout. Operational aspects deserve special attention. The most important aspect is the effect of the handling and turn-around procedures and time. It has to be understood that the aircraft earns money in the air and not on the ground. Consequently, it would hardly be acceptable for a new technology to extend the time for ground procedures. For unconventional fuels, acceptability and the availability of airport infrastructure must be ensured.

The requirements on fuel cell systems will be discussed in more detail for aeronautical APU systems by analysing certain studies on system design.[13–16] All the authors generally discuss SOFC systems in combination with gas tur-bines or the main engine generator in a hybrid system. Mak and Meier[22] stu-died a regional jet with 90 passengers based on a 'more electric aircraft' (MEA) architecture for a distance of 2869 km (1549 nautical miles (n.m.)). Gummalla *et al.*[18] analysed a 162-passenger short-range aircraft sized for a distance of nearly 6000 km (3200 n.m., long range). Dollmeyer *et al.* considered a long-range aircraft for up to 7400 km (4000 n.m.).[23] Srinivasan *et al.*[17] discussed different missions of a 302-passenger aircraft designed for long-range missions of 14260 km (7700 n.m.).

4.3.1.1 Energy-consumer Analyses for On-ground Operation of an Aircraft Auxiliary Power Unit

Currently, aircraft APUs work only in ground operation. Srinivasan *et al.*[17] and Gummalla *et al.*[18] reported on typical APU electrical loads summarised in Figure 4.2. Gummalla *et al.* assumed a 300 kW SOFC system and a peak power

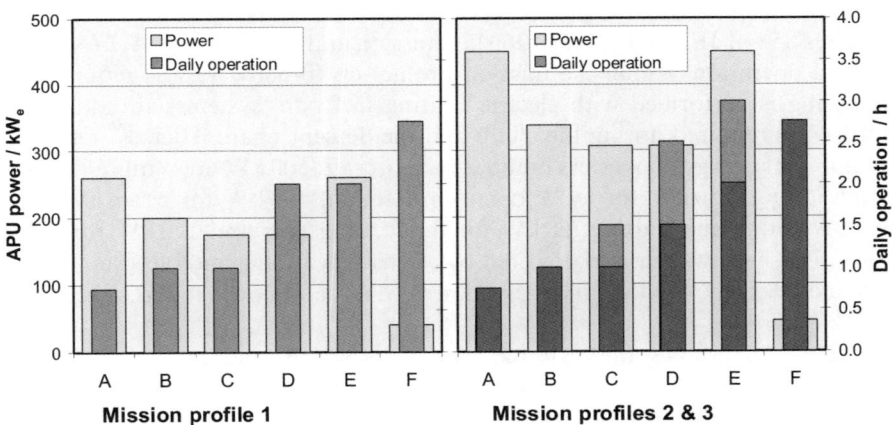

Figure 4.2 APU power required during ground operation for different mission pro-files, profile 1[18] and profiles 2 and 3.[17] Definitions: A, cabin pull down; B, cool cabin sustain; C, load phase of passengers; D, unload phase of pas-sengers; E, service phase between two flights; F, flight overnight maintenance.

of 260 kW for APU ground operation. For four flights per day, 3373 kWh must be delivered by the APU on ground. SOFC system efficiencies amount to between 53% and 68% at cruise conditions and between 39% and 48% on the ground.[18] The lowest efficiency corresponds to an architecture pressurised to only 1 atm, other system architectures work at higher pressures delivering higher efficiencies. Srinivasan *et al.*[17] defined the baseline APU electrical loads as 450 kW for cabin pull-down, 338 kW for sustaining cabin temperature, 450 kW during the passenger load and 306 kW during the passenger unload phase. Later on, electric power of about 306 kW is required during the service phase between two flights and 45 kW for the flight overnight maintenance. Srinivasan assumed five and three flights per day for two short-range missions of 926 km and 1852 km, resulting in a total energy of 1550 kWh day^{-1} and 2464 kWh day^{-1}, respectively. Mak and Meier consider different SOFC-based APU architectures with ambient and cabin air supply.[22] The APU was in operation for the main engine start (MES) with an electric power between 116 kW$_e$ and 185 kW$_e$ and for on-ground operation between 86.5 kW$_e$ and 107.5 kW$_e$. It must be noted that the duration of on-ground operation is limited to 10 min per flight for start-up, warm-up and tax-in and 10 min for tax-out and shut-down, respectively, while in[17,18] the time period for APU operation between two flights amounts to 75 min without tax-in and tax-out. Efficiency for on-ground operation amounts to 32–36% depending on system architecture.[22]

4.3.1.2   Energy-consumer Analyses during a Long-range Mission of an Aircraft

The baseline plane for 305 passengers defined in the study by Srinivasan *et al.*[17] requires 85 t of JET A-1 for a 14 260 km mission and 0.85 t of JET A-1 for APU ground operation. Along the mission profile, up to 550 kW$_e$ was required. If anti-icing is performed with electric heating in future systems, an additional 400 kW$_e$ is required during the climb and the descent phase. Hiebel[40] reported on different energy converters on board an aircraft: 500 kW are required for air conditioning, 250 kW for ice and rain protection, 100 kW for cabin systems, 300 kW for engine starting, 50 kW for the landing gear and 150 kW for flight controls. These systems are powered by electric, hydraulic and bleed-air power provided largely by the main engines (1 MW) and, on ground, by APUs (550 kW).

Figure 4.3 shows the required electric power during ground operation (profile 2, see Fig 4.2) and during a long-range mission taking the data from[17] into account. The mission profile was supplemented with the APU power requirements on ground at the departure airport, *i.e.* 0–146 min, and by the consumption at the arrival airport, *i.e.* 1110–1155 min. Additionally, the return flight is indicated. During a mission, the jet engines have to deliver 500–600 kW$_e$ as electric power. Srinivasan *et al.*[17] calculated an additional demand of about 400 kW$_e$ if anti-icing were to be realised by heater blankets. Anti-icing

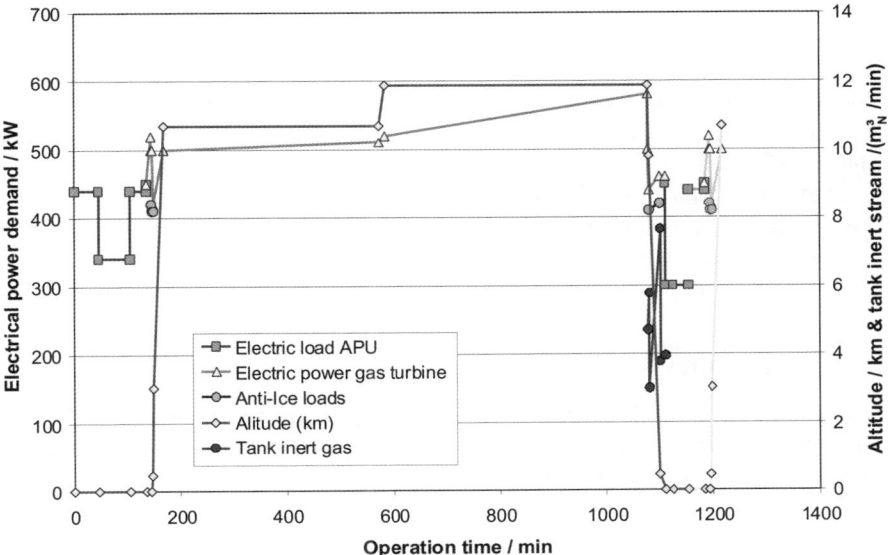

Figure 4.3 Required electric power during ground operation (profile 2, see Figure 4.2) and during a long-range mission.[17]

measures are necessary during climb and descent. These phases affect rather short periods in the mission profile, *i.e.* 145–150 min and 1080–1110 min, respectively. A tank inerting system is required to guarantee an oxygen content of less than 12% in the atmosphere above the liquid kerosene. Two effects increase the amount of gas in the tank: during the mission, kerosene is consumed by the jet engine and during descent, ambient pressure and temperature rise with decreasing altitude. If there are no counter-measures, ambient air will be drawn into the kerosene tanks. Tank inerting is required mainly during descent, *i.e.* in the period 1077–1110 min. The values between 3 and $8\,m^3_N$ min^{-1} shown in Figure 4.3 are based on an empty 182 000 L tank which was chosen in accordance with the long distance of 14 260 km in Srinivasan *et al.*[17]

Gummalla *et al.*[18] assumed a maximum electric demand of $360\,kW_e$ with anti-ice loads at the beginning of the initial climb and with $260\,kW_e$ during cruise. Mak and Meier[22] considered a power range between 84 and $164\,kW_e$ depending on the system architecture. Dollmeyer *et al.*[23] performed their analysis for a constant power output of $700\,kW_e$.

Fuel cell-based APUs for aircraft can be used in a multitude of ways with additional benefits compared to conventional solutions. Besides electricity production, fuel cell-based APUs can produce water and tail gases for tank inerting.[14,41] Water tanks in use today provide between 200 L for an Airbus aircraft type A320[42] to a maximum of 2200 L for an A380.[43] Tank inerting systems which make use of fuel cell tail gases require a maximum inert gas flow of $7.7\,m^3_N\,min^{-1}$ for a tank volume of 182 000 L. This tank size was assumed with respect to that of commercial aircraft for long-range missions. Such a flow rate corresponds to a fuel cell system with an electrical power of about $170\,kW_e$. Considering a fuel cell system with a total efficiency of 40%, a maximum water production rate of $67\,cL\,kW_e^{-1}\,h^{-1}$ can be obtained from the tail gases of the fuel cell and the catalytic burner. A more detailed study[44] calculates recovery rates of 30–$45\,cL\,kW_e^{-1}\,h^{-1}$ for a PEFC system with hydrogen as fuel and 19–$25\,cL\,kW_e^{-1}\,h^{-1}$ for an HT-PEFC system with kerosene fuel processing. The calculation method is based on a reasonable heat exchanger design and takes weather data all over the world into account.

4.3.2   Maritime Applications

Advantages of fuel cells for maritime applications are low emissions, no odour nuisance, no vibration, no pollution, and reduced noise. APUs should be reliable and simple. Additional requirements for fuel cell-based APUs are:

- No limited usage
- No additional taxes and fees, *i.e.* saving fees for the operation of diesel engines in port
- Lower operation costs
- Low investment in infrastructure.

Ansaldo Fuel Cells and Polytecnico di Torino investigated the modelling of APU systems based on MCFCs for maritime applications.[31,33] The envisaged hybrid system was initially based on a $500\,kW_e$ MCFC and a $100\,kW_e$ gas turbine. They selected cruise ferries, Ro-Ro vessels and Ro-Pax vessels, as objects of their study. Ro-Ro vessels carry containers, trucks and cars, while Ro-Pax vessels also provide passenger accommodation. All vessel types can use electricity and water from the fuel cell system. APUs should deliver between 2900 and $5000\,kW_e$ and between 1250 and $3330\,kg\,water\,h^{-1}$. The largest amount of water is required by cruise ferries, the largest amount of power by Ro-Ro vessels.[31] The electricity requirements are even higher for military applications, *i.e.* $2500\,kW_e$.[32]

In the report by Karin *et al.*[37] of 1999, one or two APUs in the power class from 500 to $1500\,kW_e$ were installed on a container ship. Compared to the propulsion power of $55\,MW_e$ published by Corbett and Winbrake[39] in 2008, the value of 10–$20\,MW_e$ indicated by Karin *et al.*[37] seems slightly underestimated. Due to the time which has passed between the two studies, an APU power of $1500\,kW_e$ seems to be more realistic. If the APU is operated in the harbour for 72 h, the maximum requirement is $108\,MW_eh$.

Lebutsch *et al.*[34] excluded commercial vessel types from their analyses due to their high power requirements. They targeted megayachts with $400\,kW_e$ required power as a market. Megayachts spend more time in the harbour than at sea and are not operated for 24 h. For harbours and coastal regions, especially, fuel cell systems could supply ships with electric energy. Machine data for the luxury yacht *Alysia* are given in.[38] The two main engines have a total power of 4 MW. Three diesel engines are installed for electricity generation, each with a capacity of 500 kW$_e$. Additionally, a sixth diesel engine delivers 400 kW for auxiliaries. The megayacht is designed for 36 guests. It is equipped with water tanks with a capacity of about 110 m^3 and diesel fuel tanks with a capacity of 234 m^3. The study assumed that one of the three generators could be replaced by a 500 kW$_e$ fuel cell to deliver the base load, while the conventional generators should be responsible for the dynamic load. Assuming an operation time of 3000 h year^{-1}, 1500 MW$_e$h is provided by fuel cells. This is in the same order of magnitude as the APU of a container ship for 18 harbour stops (2000 MW$_e$h) on the route Los Angeles to Hong Kong.[39]

Some special conditions for the fuel cell system and the balance of plant components should be considered:

- The system needs rigid housing and vibration dampers.
- Yachts operate worldwide in all types of climatic conditions. In general, fuel cells should not be stored below 0 °C. Although this problem has already been solved by the automobile industry, good water management is required for the stack and the periphery to avoid freezing inside the system.
- The salt content in the air requires good filtration to avoid negative effects for the flow fields and for the periphery components.

4.4 Fuel Cell Technologies

4.4.1 Fuel Cell Types and their Applicability for Large Auxiliary Power Unit Systems

Fuel cells are normally used as electrical generators providing power to consumers of electricity. This is the case for applications in nearly all sectors like surface vehicles or stationary applications. When analysing aircraft and ships, it is very obvious that both belong to the mobile sector, but the characteristics of the on-board systems are very similar to stationary ones. This fact may influence the selection of the fuel cell technology. Depending on the concept of use, different technologies and architectures may be of interest. Another aspect is the time horizon until the planned market entry of the product. Short-term solutions clearly require technologies which already show a certain maturity level today, while long-term solutions leave more room for development and adaptations until they go into service.

Tables 4.2 and 4.3 indicate the main characteristics of the most popular fuel cell technologies and their applications or intended applications. This overview helps to take initial decisions on the level of proposed system layouts. It is important to note that the different characteristics of the various fuel cell types affect the choice of fuel. The development of PEFC systems with on-board fuel

Table 4.2 Main characteristics of low-temperature fuel cell technologies and their applications or intended applications.

Fuel cell type	Electrolyte	Operating temperature/efficiency	Applications	Advantages	Disadvantages
Proton exchange membrane (PEFC)	Solid organic polymer polyper-fluorosulfonic acid (proton-conducting polymer)	60–100 °C/35–50% w/o and up to 60% with cogeneration	• Electric utility • Portable power • Transportation	• Solid electrolyte reduces corrosion and challenges for material handling • Operation close to ambient temperature • Quick start-up • High power density	• Low temperature requires expensive precious catalysts • High sensitivity to fuel impurities (CO poisoning) • Needs sufficient water management to prevent clogging
Alkaline (AFC)	Aqueous solution of potassium hydroxide soaked in a matrix	90–100 °C	• Military • Space (for decades)	• Fast response • Cathode reaction faster in alkaline electrolyte, i.e. high performance	• Limited durability • Expensive removal of CO_2 from fuel and air streams required
Phosphoric acid (PAFC)	Liquid phosphoric acid soaked in a synthetic fleece material (H_3PO_4)	150–200 °C/35–50% w/o and up to 85% with co-generation (CHP)	• Electric utility	• Up to 85% efficiency in cogeneration of electricity and heat (CHP) • Can use impure H_2 as fuel (tolerates >1% CO)	• Requires platinum catalyst • Low current and power density • Large size/weight • Low efficiency without heat extraction

	Electrolyte/membrane	Temperature	Advantages	Applications	Disadvantages
HT-PEFC	Polybenzimidazole membrane doped with phosphoric acid	120–200 °C	• High reliability • Considerable experience available • Tolerates up to 2% CO at 200 °C or up to 20 ppm H_2S and 1% CO and 10 ppm H_2S at 160 °C • No complex moistening in comparison to PEFC	• Transportation	• Loss of phosphoric acid • Limited durability • Requires platinum catalyst • Low current and power density • Lower efficiency due to higher cathode overpotential • Discharge of phosphoric acid (less than PAFC)
Direct methanol fuel cell (DMFC)	See PEFC	60–80 °C/20–30%	• Use of alcohol/methanol as fuel • Minor storage problems and less complex infrastructure for small systems	• Battery replacement for portables (*e.g.* phones, laptops)	• Lower performance than PEM cells • Methanol hazardous in operation • No relevance for hydrogen economy: CO_2 emissions high

Table 4.3 Main characteristics of high-temperature fuel cell technologies and their applications or intended applications.

Fuel cell type	Electrolyte	Operating temperature/ efficiency	Applications	Advantages	Disadvantages
Molten carbonate (MCFC)	Liquid solution of lithium, sodium, and/ or potassium carbonates soaked in a matrix (K-Li-Na-CO$_3$)	600–1000 °C/45–60% w/o cogeneration 85% with co-generation	• Electric utility	• High efficiency • Fuel flexibility (internal reforming) • Use of a variety of catalysts • Use of high-temperature heat (high-pressure steam) • No noble metals required • Better stack cooling by internal reforming	• High temperature increases corrosion and breakdown of cell components • Corrosive liquid • Requires expensive alloys • Low power density • CO$_2$ needed in the cathode feed air • Internal reforming is only possible for natural gas
Solid oxide (SOFC)	Solid zirconium oxide to which a small amount of yttria is added (Y-Zr$_2$O$_3$)	600–1000 °C/45–55% (operation at 1 atm) 50–60% for pressurised system up to 85% as CHP	• Electric utility	• High efficiency due to high cell voltages • Fuel flexibility (internal reforming) • Use of high-temperature heat (high-pressure steam) • Use of a variety of catalysts • No noble metals required • Better stack cooling by internal reforming	• High temperature increases breakdown of cell components • Sealing problems lead to leaky stacks • Low thermal cycling capability • Low redox capability • Slow start-up • Internal reforming is only possible for natural gas, not for higher hydrocarbons • High air ratios lead to high peripheral losses due to air compression – especially for diesel operation, internal reforming impossible

processing for transport applications was stopped by the go/no-go decision of the US DOE in 2004.[45] Presently, only Nuevera Fuel Cells and Renault are developing hybrid fuel cell systems for passenger cars with on-board reforming.[46]

Most automotive companies are focusing on hydrogen as a fuel. This leads to the following mismatch: some of the stacks are not applicable for reformate operation due to the usage of pure Pt as an anode material instead of Pt-Ru alloys.

Some applications are also based on H_2/O_2-PEFC systems for submarines[9] and passenger cars such as the Hy-Light car, a concept vehicle developed by the Paul Scherrer Institute (PSI) and Michelin.[47]

Alkaline fuel cells (AFCs) are limited to special military applications today. Direct methanol fuel cells (DMFCs) can only be used for small systems, *i.e.* up to a maximum of $2\,kW_e$, due to their low efficiency caused by methanol permeation and the subsequent oxidation on the cathode side. Other direct alcohol fuel cells – namely direct ethanol fuel cells – are still being researched electrochemically. Phosphoric acid fuel cells were developed as stationary combined heat and power systems. A new innovative type is HT-PEFC with a polybenzimidazole (PBI) membrane doped with phosphoric acid. It has the electrochemical advantages and in part the disadvantages of PAFCs, but it is optimised for transport applications and makes material handling easier and safer. Unfortunately, stack power is presently limited to some kilowatts, but rapid development is possible.

High-temperature fuel cells such as SOFCs and MCFCs were mainly developed as CHP systems. A few SOFC developers worldwide like Elring–Klinger and Delphi work on APUs for trucks and other heavy-duty vehicles. SOFCs and MCFCs are most effective with internal reforming, which is only possible with natural gas as an energy carrier.

4.4.1.1 Focus on Aeronautical Systems

The analysis of the technology readiness of different fuel cell technologies leads to the conclusion that the quite advanced PEFC technology could be proposed. It should be highlighted that at this stage, the different options of H_2 and O_2 supply should be taken into account for short-term application of fuel cells in aircraft. Nevertheless, the input could also be hydrogen and ambient air. After the chemical reaction, the output is electrical energy and thermal energy as well as pure water and exhaust gas, *i.e.* the air entering the fuel cell minus the oxygen which reacts with the H_2 at the cathode side. PEFCs are promising for the following reasons:

- Maturity
- Dynamic behaviour
- Operating temperatures which could allow its operation inside the pressurised cabin area

- Good efficiency for electrical power generation
- By-products like thermal energy, process water and low O_2 exhaust air may be utilised further.

Negative aspects are:

- It runs best on very pure H_2. The effort of fuel processing for PEFC is extremely large, which makes on-board reforming of kerosene unattractive.
- The low operating temperature may lead to cooling problems at maximum power operation on the ground in hot climates.
- Most membrane materials demand a completely humidified fuel and air stream, *i.e.* $\sigma = 100\%$. Additional humidification devices such as evaporators or moistening membranes are therefore required.

To mitigate those drawbacks, the fuel cell can be supplied with H_2 from an on-board storage system. In order to decrease the cooling problem, high-temperature PEFCs could be developed which operate at 120–180 °C instead of 50–85 °C. Taking all this into consideration, two options can be proposed for further studies. Option 1: an emergency electrical power system; option 2: a multi-functional fuel cell system based on a large APU.

Considering option 1, it is intended to provide a power source that is independent of the kerosene fuel system. Such a device is important in case of failure; that is, if the combustion engines are inoperative due to contaminated fuel or defective fuel supply. Approval directive CS-25 stipulates that an emergency system must be able to provide power until the aircraft has altered its route and landed safely despite complete power loss of the main engines. The usual technology currently consists of a ram air turbine (RAT) which deploys automatically or manually and supplies power using ram air. An alternative technology could be a simple fuel cell architecture supplied by GH_2 and GO_2 which takes over this function. The justification for this short-term application of fuel cells is that RATs are 'sleeping' systems, which means that a malfunction between the functional tests remains undetected. As a consequence, the RAT is tested frequently, which causes maintenance costs. Other aerodynamic aspects may also be taken into account. In any case, the alternative technology must perform better than the existing one, *e.g.* by causing lower maintenance costs and being of higher aerodynamic quality. Maintenance efforts could be reduced by so-called build-in testing using an easy, fast and cheap virtual function test.

Option 2 is a long-term application of fuel cell technologies. A multi-functional fuel cell system based on a large APU must be particularly suitable to produce electricity, but also the by-products heat, water and inert gas. The energy carrier has not yet been determined, but the main requirements as discussed in Sections 4.2 and 4.3 must be fulfilled by any system. With respect to the main characteristics listed in Tables 4.2 and 4.3, PEFC, HT-PEFC and SOFC fuel cells can initially be taken into consideration. The advantage of

PEFC stacks is still their availability for the $100\ kW_e$ power class. In most cases, PEFC stacks are limited to hydrogen as fuel. HT-PEFC technology allows the use of liquid hydrocarbons as fuel but both fuel cell stacks and fuel processing units must be scaled up to several tens of kilowatts. Power density must then be increased. Although SOFC stacks have a long research and development history, they suffer from the limited availability of stacks in the $100\ kW_e$ power classes. The tubular concept of Siemens–Westinghouse was developed for a $250\ kW_e$ stationary CHP system, but planar concepts presently reach only an electrical power of several kilowatts. In addition, these systems are unable to deliver tail gas with a specified quality for tank inerting due to the high air ratios.[44]

4.4.1.2 Focus on Maritime Systems

As large APUs for vessels, MCFCs, SOFCs and PEFCs are applicable. Currently, only MCFCs are available as large stationary systems in the power range from 250 to $500\ kW_e$. PEFCs were predominantly developed for automotive applications with hydrogen as the energy carrier and for distributed power generation based on natural gas. Tubular SOFC concepts for CHP application were also developed in the power range between $150\ kW_e$ and $250\ kW_e$ by Siemens–Westinghouse and Mitsubishi Heavy Industries for natural gas as an energy carrier. Presently, planar SOFC concepts and HT-PEFC stacks with PBI membranes are still under development and must be scaled up from $5\ kW_e$ at present to $25–50\ kW_e$ units in the future. One advantage of fuel cell technology is its modularity, but for large systems, a minimum stack size of $50–100\ kW_e$ should be envisaged. Therefore, the development of large APUs for maritime applications is focused on MCFC technology with different energy carriers. In 2009, CFC Solutions demonstrated a $320\ kW_e$ MCFC system on board a supply vessel for oil platforms off the Norwegian coast with LNG as the energy carrier.[48,49] Ongoing projects are investigating MCFC technology for passenger ships and yachts and PEFCs for ferries.[50]

4.4.2   Fuels for Large Auxiliary Power Units

The availability of hydrogen is a prerequisite for the use of fuel cells in mobile and stationary applications. At present, however, an infrastructure for hydrogen as a future energy carrier does not exist. Demonstration projects which have been set up during the last decade use liquefied hydrogen (LH2). The lower power density of compressed hydrogen (GH2) compared to liquefied hydrogen considerably reduces the effort for road transport of LH2. A typical trailer for compressed gaseous hydrogen (293 K and 200 bar) can deliver a net capacity of $5145\ m^3_N\ H_2$ per freight and a liquid hydrogen truck $38\ 809\ m^3_N\ H_2$ per freight.[51] In the long term, hydrogen will be distributed by pipelines. Hydrogen production can be realised temporarily by natural gas steam reforming at a central facility or a remote site. In the long term, hydrogen must be produced from renewables, for example from wind energy using wind turbines, and subsequent electrolysis.

Another option is the use of short-lived hydrogen. Short-lived hydrogen is produced on demand from liquid energy carriers and is directly consumed by a fuel cell for electricity generation. If mobile applications are targeted, on-board reforming of hydrocarbons must be designed prior to the fuel cell. Therefore, it is essential that hydrogen is produced from readily available energy carriers. For stationary applications, natural gas and heating oil are suitable sources, while for mobile applications, gasoline, kerosene and diesel are options. Currently, the energy carriers listed above are mainly produced from crude oil as a fossil primary energy carrier. In the long term, biomass will be used to produce some of the liquid energy carriers needed today.

In the following section, the choice of fuel will be discussed from different perspectives. Table 4.4 shows the energy densities of various energy carriers in $MJ\,kg^{-1}$ and $MJ\,L^{-1}$ as well as the primary energy use based on a well-to-tank analysis and the equivalent CO_2 release during production. The data were taken from a selection from different sources.[52–57] Biofuels in particular could be produced in various process chains. Some are very efficient and release little CO_2 during production. In terms of CO_2 reduction, other fuel chains make less sense; for example, ethanol from sugar beet in combination with process steam from lignite combustion. It is obvious that fossil-based fuels such as gasoline, kerosene and diesel combine the highest energy densities, the lowest primary energy use and very low CO_2 release during production. This is because:

- During fuel production, mainly distillation processes are applied
- Heat exchanger net works for heat recovery are optimised
- Crude oil is split into a broad range of products.

Regardless of the excellent fuel properties, fossil fuel resources are finite and their use leads to higher CO_2 emissions in the application chain, *i.e.* tank-to-wheel, in spite of automotive applications. When considering alcohols, methanol displays half the power density of gasoline and ethanol about two thirds. Efficiencies for bioalcohols range from 34% to 53%. Methanol from natural gas is much more efficient: about 63% at remote sites and more than 72% if it is produced together with dimethyl ether in a chemical plant at central sites.

Fuels like FAME, Fischer–Tropsch diesel (FT diesel) and hydrotreated plant oils (NExBTL) have similar power densities to fossil fuels. The primary energy use during production leads to efficiencies between 44% (NExBTL) and 59% (FAME). Some of these liquid energy carriers presently release more CO_2 during production than fossil fuels. The process chains of wood gasification combined with a Fischer–Tropsch and Bioliq./MT synfuel process are an exception, leading to only $4.8\,g\,CO_2\,MJ^{-1}$ fuel.

The main disadvantage of gaseous fuels like hydrogen and natural gas is their low power density in terms of volume. Unfortunately, the high power densities of hydrogen ($120\,MJ\,kg^{-1}$) and natural gas ($50\,MJ\,kg^{-1}$) decrease if the storage systems are taken into consideration. In comparison, the power density of a full gasoline tank amounts to 27 MJ kg^{-1} which is only slightly below that of the pure fuel (32 MJ kg^{-1}). Liquid hydrogen storage systems (0.5 MPa, 26 K) have a power density of only 7 MJ kg^{-1} and compressed hydrogen systems (70 MPa, 293 K) of 5 MJ kg^{-1}. Natural gas systems have slightly better power densities at 11 MJ kg^{-1} (20 MPa, 293 K) for compressed and 16 MJ kg^{-1} (0.5 MPa, 135 K) for liquid storage.[53] The primary energy use and CO_2 release during production mean that hydrogen must be produced from renewable sources. Natural gas as a primary fuel for hydrogen is only a bridging technology. Efficiency for the LH2 chain is quite low and CO_2 release is extremely high, *i.e.* ~44% and 132.8 g $CO_{2,eq}$ MJ^{-1}$_{Fuel}$. For compressed hydrogen produced by electrolysis from offshore wind power, the values are excellent at 0.79 MJ MJ^{-1}$_{Fuel}$ (~56%) and 9.1 g $CO_{2,eq.}$ MJ^{-1}$_{Fuel}$. These figures are in the same range as liquid biofuels from second-generation gasification processes. Unfortunately, ethanol from second-generation fermentation processes uses twice as much primary energy compared to Fischer–Tropsch diesel production. The combination of wind energy and liquid hydrogen is not included in Table 4.4. The reason for this is that wind mills produce electricity for electrolysis at decentralised sites. Unfortunately, liquefaction requires large-scale facilities which are best suited for hydrogen production *via* gasification of cultivated wood.

Due to the high investment costs and competition with BtL fuels from wood the realisation of such facilities remains an issue to be resolved.

One of the most important issues in fuel processing is the homogeneous mixing of liquid fuels with steam and air. Fuel is injected into a mixing chamber and evaporates simultaneously, extracting evaporation heat from the superheated steam. Therefore, the boiling behaviour of the fuel is very important. Figure 4.4 shows the boiling curves for different fuels. Crude oil quality for two crude oil brands is also depicted for a light and a heavy type. It is important to note that heavy fuel oils are often composed of diesel-like process streams blended with residual crude oils which correspond partially to the original feedstock in terms of boiling behaviour. Biofuels such as hydrotreated vegetable oil (HVO) or fatty acid methyl esters (FAME) have different boiling behaviours.[58] HVO and FAME have to be adapted or blended with fossil fuels to reach the characteristics specified for automotive fuels. Both HVO and FAME have a flat boiling profile which seems to be favourable for reforming. The boiling temperatures of HVO are most attractive with values below 573 K. A further aspect of choosing a fuel is the effort for the storage system. It is always the best choice to apply the fuel which is already on-board for propulsion purposes. The hydrogen production from this fuel must be reliable. Additional fuels require additional tanks. The specific data of liquid fuels are better due to the fuel properties and the simpler vessel design. Liquefied gas storage systems require more volume for insulation and for a suitable evaporation system. Compressed storage systems are targeted today which operate at a pressure of 70 MPa using carbon fibre-reinforced polymers.

Considering the fuel properties listed in Table 4.4, future fuels may be hydrogen from renewables or methanol or middle distillates (diesel and

Table 4.4 Energy densities, primary energy use and CO_2 release for various energy carriers from well-tank analyses.[52-57]

Fuel	Energy densities		Source/chain	Primary energy use $MJ\,MJ^{-1}_{Fuel}$	CO_2 release $g_{CO_2\text{-}eq.}\,MJ^{-2}_{Fuel}$
	$MJ\,L^{-1}$	$MJ\,kg^{-1}$			
LPG[a]	25.3	46.0	Crude oil	0.12	8.0
Gasoline	32.2	43.2	Crude oil	0.14	12.5
Kerosene[a]	34.9	43.2	Crude oil	0.13	10.7
Diesel	35.9	43.1	Crude oil	0.16	14.2
Marine gas oil[b]	36.3	42.7	Crude oil	0.16	14.2–19.8
Marine diesel oil[c]	38.0	42.2	Crude oil	0.16	14.2–19.8
Methanol	15.8	19.9	Natural gas/remote site	0.61	24.2
Methanol	15.8	19.9	Biofermentation from wood	n.s.	n.s.
Methanol	15.8	19.9	Bioliq./MT synfuels from wood or straw	1.07	4.8[b]
Ethanol	21.3	26.8	Fermentation from sugar cane	1.81	13.1[c]
Ethanol	21.3	26.8	Fermentation from sugar beet	0.88	13.9[d]
Ethanol	21.3	26.8	Fermentation from wheat	1.10	28.6[e]
Ethanol	21.3	26.8	Fermentation from complete plant (2nd generation)	1.95	19.0[f]
FAME	33.1	37.2	Rape plant/esterification	0.7	28.2[g]
Hydrogenated plant oil, NExBTL[d]	34.4	44.0	Palm oil/hydrotreatment	1.26	49.6[h]
Fischer–Tropsch middle distillate	34.3	44.0	Wood/FT syngas production (2nd gen.)	1.19	4.8[i]
Hydrocracked middle distillate	n.s.	n.s.	Heavy feed stocks/hydrocracking	n.s.	n.s.
Hydrogen (LH2)[j]	7	120	Natural gas/on-site electrolysis	1.28	132.8[k]
Hydrogen (LH2)	7	120	Farmed wood/large-scale gasification	1.50	7.5 (6.6–21.2)

Hydrogen (CH2)[j]	5	120	Natural gas/centralised electrolysis	0.72	98.2
Hydrogen (CH2)[m]	5	120	Offshore wind	0.79	9.1[n]
Natural gas (CNG)[o]	8	45.1	Natural gas	0.19	14.5[p]
Natural gas (LNG)[q]	19	45.1	Natural gas	0.26	20.8[r]
Biogas (CBG)	n.s.	45.1	Municipal waste	0.87	−39.5

Sources: primary energy use and CO_2 release from reference [52] except (1) from reference [56]; energy densities from reference [52] except (1) from reference [55], (2, 3) from reference [57]; (4) from reference [54].

n.s., not specified.

[a]Liquefied petroleum gas (propane/butane) for automotive application. Tank: 0.8 MPa; 293 K.

[b]Waste wood.

[c]Credit for saving heavy oil by using excess bagasse.

[d]Biomash used for gasification and heat production.

[e]Biogas as by-product biogas.

[f]Straw for BtL production.

[g]Colza cake and glycerine for gasification to biogas.

[h]Methane emissions from waste.

[i]Waste wood.

[j]Liquified hydrogen, stored at 0.5 MPa; 26 K.

[k]Transport by natural gas pipeline (4000 km).

[l]Compressed hydrogen, stored at 70 MPa; 293 K.

[m]Compressed hydrogen, stored at 70 MPa; 293 K.

[n]Central electrolysis.

[o]Compressed natural gas, stored at 20/25 MPa; 293 K.

[p]Transport by natural gas pipeline (4000 km).

[q]Liquified natural gas, stored at 0.5 MPa; 135 K.

[r]Distributed by truck.

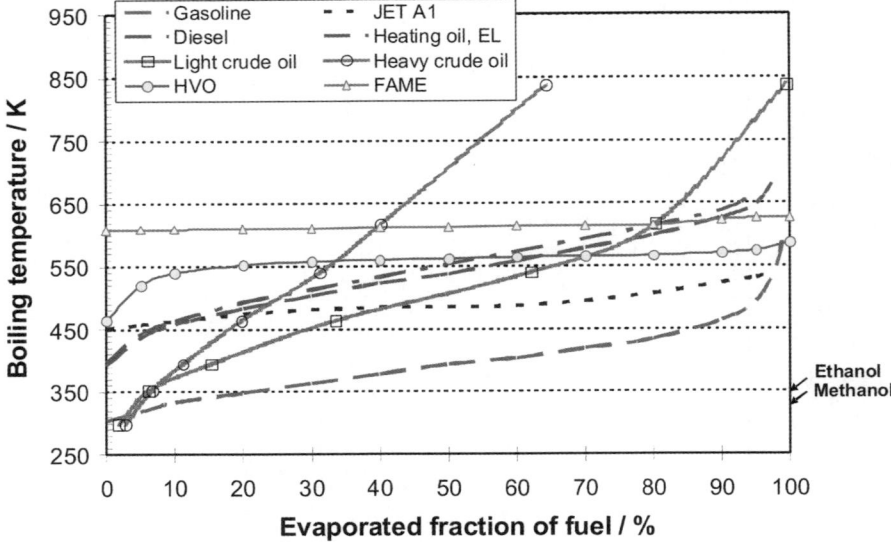

Figure 4.4   Boiling behavior of different fuels.[58,67]

kerosene) from waste wood, although waste wood resources are limited. Therefore, cultivated wood must also be used, which leads to a higher CO_2 release during the process chain. A final decision cannot be made from the present perspective. A diversified fuel strategy enables the use of fossil-based middle distillates today while offering the option of switching to biofuels in the mid-term.

4.4.2.1   Focus on Aeronautical Fuels

Commercial fuels used for different applications lead to various challenges for fuel processing. Jet fuel JET A-1 contains up to 3000 ppmw of S, while a typical reforming catalyst can only tolerate fuels with 10 ppmw S. Commercially available JET A-1 contains less than 1000 ppmw S in Europe and the USA. Predominant species are benzothiophenes and their alkylated forms. Aromatic compounds are found primarily as monoaromatics (15–25%). Generally, desulfurisation should be performed in a refinery rather than on board. Presently, kerosene as the precursor of the later product JET A-1 is not desulfurised to 10 ppmw S as required by fuel cell systems. The dominant sulfur types in JET A-1, *i.e.* benzothiophenes and their derivates, are easier to convert than those in diesel fuels, *i.e.* dibenzothiophenes and their alkylated forms. Desulfurisation in centralised stationary systems at the airport, in smaller decentralised systems at the gate or finally on board must use different technologies than conventional processes in refineries. Desulfurisation should be carried out upstream of the fuel processor in the liquid phase. A possible process for liquid-phase desulfurisation is a modified hydrofining process with pre-saturated hydrogen.[59] The modified hydrofining process requires a hydrogen source which could be hydrogen-rich reformate (about 30% H_2) from the fuel processor. JET A-1 containing 700 ppmw S was pressurised up to 2.4 MPa and heated up to 390 °C. The average sulfur content in the product during a 200 h experiment was 8.3 ppmw S.[59] Hydrogen sulfide must subsequently be separated from the desulfurised fuel by a hydrocyclone. It should be burned in the jet engine and must leave the system as SO_2.

As the variety of available technologies and processes leads to different results, a study of the overall situation on board an aircraft needs to be performed and compared. The requirements of aircraft operation and market demands also have to be taken into consideration. This means that some criteria remain directly conflicting. This is very obvious for the selection of the on-board H_2 supply system. Taking into consideration on-board H_2 storage and supply instead of an on-board reforming system and a suitable fuel cell technology would simplify the entire process chain. The consequence would, however, be a second fuel for the aircraft, which requires adequate airport infrastructure.

4.4.2.2 Focus on Maritime Fuels

In the maritime market, different fuels are in use: marine gas oil (MGO), marine diesel oil (MDO), intermediate fuel oil (IFO), medium fuel oil (MFO) and heavy fuel oil (HFO). They correspond roughly to a series of blends mixed from fuel oil no. 2, *i.e.* diesel, and an increasing part of fuel oil no. 6, which is a residual fuel oil. Approximately 75% of all marine residual oils are heavy and 25% are light; the ratio of marine distillates is 63% for MGO to 37% for MDO.[39] Unfortunately, about 90% of all diesel engines in sea transport are slow-running machines which are operated with HFO.[60] In 2008, the International Maritime Organization passed a resolution on the reduction in sulfur emissions for the shipping industry.[61] From the year 2020, shipping companies either have to use distillate fuels with a limited sulfur content of 0.5% instead of heavy fuel oil or they have to use scrubbing technology to clean their exhaust gases. This should lead to higher market shares for MGO and MDO.

MGO contains up to 1000 ppmw S and 100 ppmw ashes. Typically, marine gas oil evaporates at higher temperatures of 470–660 K. At 633 K, only 85% of the original fuel must be evaporated. Fuel oil no. 4 (MDO) – a blend of nos. 2 and 6 – may contain up to 15% residual process streams and up to 5% polycyclic aromatic hydrocarbons.[62] Desulfurisation is difficult for sulfur-containing di-aromatic components, polyaromatics, long-chain alkanes such as waxes, and sulfur species like di-, tri- and tetramethyl-dibenzothiophenes. In particular, 4,6-dibenzothiophene and its derivatives are very challenging for desulfurisation because two methyl groups on both sides of the sulfur atom sterically hinder chemical cracking of the thiophene ring structure. Fuel cell-based APUs for ships should be envisaged only with marine gas oil as fuel. Other maritime fuel qualities produce a high risk of nozzle plugging, carbon deposition and catalyst degradation by fuel residues. As to fuel quality, applications based on diesel seem to be most attractive.

4.4.3 Fuel Processors

4.4.3.1 Basics of Fuel Processing

Hydrogen can be produced from carbonaceous substances such as hydrocarbons or alcohols using a fuel processor. A fuel processor consists of a reformer – the actual hydrogen formation step – and a downstream gas cleaning unit. If process heat is required, a catalytic burner is frequently used to recover combustion heat from residual gases. Regarding the basic chemical reactions, hydrocarbon reforming is based either on steam reforming (Equation (4.1)) or on partial oxidation (Equation (4.2)). Autothermal reforming constitutes a compromise between heated steam reforming and partial oxidation:

$$C_nH_mO_l + (n - l)H_2O \rightarrow nCO + \left(\frac{m}{2} + n - 1\right)H_2 \qquad (\Delta H_R > 0) \qquad (4.1)$$

$$C_nH_mO_l + \frac{(n - 1)}{2}O_2 \rightarrow nCO + \frac{m}{2}H_2 \qquad (\Delta H_R < 0) \qquad (4.2)$$

In addition, two side reactions can be found, *i.e.* the water-gas shift reaction (WGS in Equation (4.3)) and methane synthesis (Equation (4.4)), which is the opposite of methane steam reforming:

$$CO + H_2O \rightleftharpoons CO_2 + H_2 \qquad (\Delta H_R < 0) \qquad (4.3)$$

$$CO + 3H_2 \rightleftharpoons CH_4 + H_2O \qquad (\Delta H_R < 0) \qquad (4.4)$$

The most important side reactions of the reforming process are those leading to undesired carbon formation, for example the Boudouard reaction (Equation (4.5)) and the pyrolysis reaction (Equation (4.6)):

$$2CO \leftrightarrow CO_2 + C \qquad (4.5)$$

$$C_nH_m \rightarrow \text{'carbon deposit'} + \frac{m}{2}H_2 \qquad (4.6)$$

A more detailed description of the mechanisms of carbon deposition can be found in Rostrup-Nielsen.[63]

Thermodynamic calculations are often used to determine preliminary reaction conditions and to evaluate product quality. If the contact time of the gas molecules at the active sites of the catalyst is long enough, the gas composition in the product gas corresponds to chemical equilibrium. Typical residence times in the catalyst zone of an autothermal reformer are 100–200 ms, corresponding to gas hourly space velocities of 20 000–35 000 $m^3 h^{-1} m^{-3}_{Cat}$.

Table 4.5 lists data from chemical equilibrium calculations for various energy carriers. The equilibrium gas composition of an ATR operated with dodecene ($C_{12}H_{24}$) at 1053 K and 2 bar using an educt mixture with $O_2/C = 0.47$, $H_2O/C = 1.9$ amounts to 8.87% CO, 8.75% CO_2, 0.03% CH_4, 31.21% N_2, 23.76% H_2O and 27.37% H_2. Further information on these calculations can be found in.[64]

As can be seen from Table 4.5, hydrogen concentrations are highest for methanol steam reforming (65%) and lowest for autothermal reforming of ethanol (23%). Methane formation was excluded for methanol steam reforming because currently available Cu/ZnO catalysts are tailored to suppress undesired side reactions. Most of the calculations show low methane concentrations which is disadvantageous for internal reforming in SOFCs and MCFCs. Therefore, pre-reforming is preferable for high-temperature fuel cells instead of complete steam reforming. The product gas is determined by kinetics and not by chemical equilibrium. Pre-reforming on nickel catalysts is well established only for methane and natural gas. For diesel, Krummrich reported a good performance of nickel catalyst compared to precious metal catalysts, but sulfur poisoning is a serious problem.[32] With increasing chain length of hydrocarbons, hydrogen concentration slightly decreases. Partial oxidation of higher hydrocarbons in particular leads to a product with nearly equal concentrations of hydrogen and carbon monoxide. Without treatment, such a product gas is only applicable in SOFCs and MCFCs.

It is obvious from the properties shown in Tables 4.4 and 4.5 that ethanol is less attractive. Due to the relatively high steam-to-carbon ratio of 2.5 required to suppress carbon deposition and the inclusion of the OH group in the molecule, hydrogen concentrations from steam reforming are lower than for other energy carriers. Autothermal reforming of ethanol also results in a very low hydrogen content. Progressive reaction conditions are necessary because ethanol must be prevented from forming ethene, a presumed intermediate step in soot formation. In addition, ethanol consumes a large amount of energy during fuel production and correspondingly releases high amounts of CO_2. Bioethanol is in strong competition with food production because at present bioethanol is obtained exclusively from edible parts of the plant.

4.4.4   Fuel Cell Systems

A broad variety of fuel cell systems is possible if all fuels, fuel cell types and fuel processor configurations are taken into consideration. Figure 4.5 shows a flow sheet of systems based on hydrocarbonaceous energy carriers. Upstream of the fuel cell, a fuel processor converts the fuel into a hydrogen-rich product gas and cleans it from impurities and undesired components. In the fuel cell, short-lived hydrogen is converted into electricity, heat and the chemical product water. Another important item is the air supply of the fuel cell which must include a

Table 4.5 Gas compositions in chemical equilibrium for different energy carriers.

Species	H_2O/C	O_2/C	T_{Start} (K)	T (K)	ζ $(\%)$	x_{H_2} $(\%)$	x_{CO} $(\%)$	x_{CO_2} $(\%)$	x_{H_2O} $(\%)$	x_{CH_2} $(\%)$	x_{N_2} $(\%)$
CH_3OH	1.5	–	–	553	99.99	64.81	1.85	20.37	12.96	0.00	0.00
CH_3OH	1.0	0.3	333	825	100.00	23.82	3.00	12.42	28.78	6.36	25.62
C_2H_5OH	2.5	–	–	923	100.00	44.88	7.38	10.96	34.01	2.76	0.00
C_2H_5OH	1.5	0.5	333	864	100.00	23.06	3.90	11.57	25.29	1.98	34.20
CH_4	3	–	–	1073	99.05	55.46	10.78	5.78	27.82	0.16	0.00
CH_4	1.7	0.45	653	909	84.37	34.20	6.02	7.71	20.85	2.54	28.69
C_3H_8	2	–	–	1073	100.00	58.56	17.28	5.56	18.16	0.44	0.00
C_3H_8	1.5	0.40	673	954	100.00	33.85	9.50	8.29	17.49	1.15	29.71
C_8H_{18}	1.8	0.45	673	1066	100.00	29.48	9.67	7.95	22.08	0.03	30.80
$C_{14}H_{30}$	1.8	0.45	697	1083	100.00	28.75	9.85	7.94	22.36	0.02	31.08
$C_{20}H_{42}$	1.8	0.45	697	1086	100.00	28.26	9.97	7.84	22.47	0.02	31.45
$C_{12}H_{24}$	3.5	–	–	1123	100.00	44.11	10.39	7.78	37.71	0.02	0.00
$C_{12}H_{24}$	1.9	0.47	639	1053	100.00	27.37	8.87	8.75	23.76	0.03	31.21
$C_{12}H_{24}$	–	0.5	639	1372	100.00	25.09	25.20	0.03	0.06	0.05	49.57

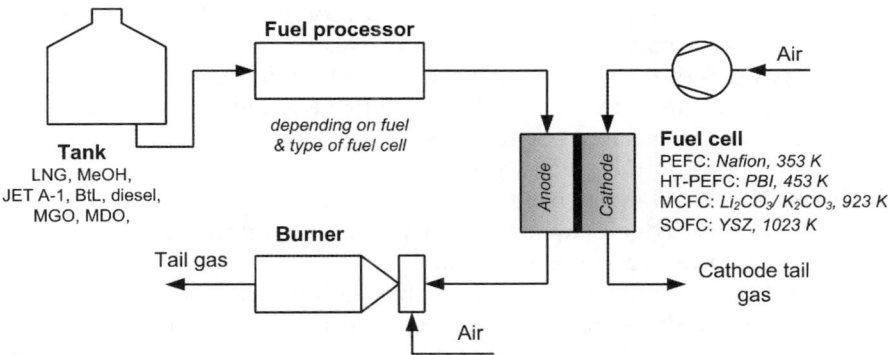

Figure 4.5 Flow sheet for a fuel cell system based on hydrocarbonaceous energy carriers.

moistening system in the case of PEFCs. Hydrogen is not converted completely by the fuel cell. In order to maintain a certain chemical potential between anode and electrolyte at the exit area of the fuel cell stack and to guarantee the proper removal of other gases, a certain amount of hydrogen-rich gas will be purged. Therefore, the anode tail gas still contains hydrogen which must be burnt in a catalytic burner at very low emissions. The released heat should be used in the system.

Not all possible system configurations can be discussed in this chapter. The next section describes some of them with regard to large APU systems for aircraft and vessels.

4.4.4.1 Focus on Aeronautical Systems

Analyses of fuel cell technology showed the following potential benefits:

1. The electric generator is more efficient – about 45–55% – than electrical power generation by turbo engines, especially at partial load.
2. Thermal energy, which is the energy of the supplied H_2 minus the energy converted to electrical energy and minus the parasitic energy consumed by the subsystem elements, could be used for heating components like a wing ice protection system or other consumers of thermal energy, generally replacing heating systems. This depends on the temperature level and the design of the thermal system components.
3. The process water in the exhaust gas is a product of H_2 from the storage system and O_2 supplied by the ambient air. Hydrogen (H_2) is a very light-weight element with an atomic mass of about $2\,g\,mol^{-1}$, while the atomic weight of oxygen is about $16\,g\,mol^{-1}$. That means that $2\,g\,mol^{-1}$ originates from H_2 storage and $16\,g\,mol^{-1}$ is added by the ambient air to form H_2O, the process water. In other words, $1\,kg\,H_2$ reacts with $9\,kg$ water. Assuming that the losses during the

condensation process lead to a recovery of 6–7 kg of weight, an equivalent balance of weight can be assumed.

4. Analyses of the exhaust gas have shown that during normal operation and at an air ratio of $\lambda = 2$, the O_2 content is about 12%. The following concentration should be regarded:
   - 12% O_2 in the exhaust gas
   - 21% O_2 in the ambient air
   - 16% O_2 as a minimum for human beings to survive and to extinguish fire in the cargo compartment
   - 14% O_2 content to extinguish fire and to inert kerosene tanks for extinguishing fire in the cargo compartment (see Figure 4.6).

5. At a certain installed power, it would be possible to completely replace the traditional gas turbine-based APU. Engine start would be carried out by an electrical system, which is state of the art on board the B787 aircraft.

6. As mentioned earlier, a power source independent of the kerosene is required by the certification regulations. A fuel cell generator system based on an independent fuel system could fulfil this requirement.

Assuming that these effects can be used, a first qualitative evaluation may show a preliminary overview of such an unconventional use of fuel cell technology as a permanent power station integrated in a 'closed' area like a commercial passenger aircraft compared with the conventional system.

With the intention of gaining the greatest possible benefit from a fuel cell-based system concept and taking into account that the efficiency of electrical power generation with fuel cell technology is much higher than with conventional technology, this system should be the dominant supplier of electrical power on board. This implies that most of the consumers should be consumers of electricity. The consequence would be to convert traditional power systems like pneumatic and hydraulic power systems into electric systems or at least into electro-hydraulic systems.

Such a concept was partially realised in the Boeing 787 aircraft, which may be used as a reference aircraft for this evaluation. The pneumatic power system was completely replaced by electric systems. The engine start is electrical and the wing ice protection is based on electro-thermal elements. Hydraulic systems are still on board for landing gear retraction and nose gear steering. The starter–generator component has a power generation capacity of 225 kVA AC power.[65] Consumption is presumably at a similar level. Electrical generators with about 1000 kVA are installed at the main engines. It can be assumed that about 500 kVA DC power can be provided by a fuel cell generator. The remaining 500 kVA will be provided as AC power by the starter generators at the main engines. A comparable concept was outlined in the framework of an FP5 EU project called 'Power-Optimized Aircraft POA'.[66]

Table 4.6 shows the preliminary evaluation of an on-board HT-PEM fuel-cell generator used as a multifunctional component. At this stage, the

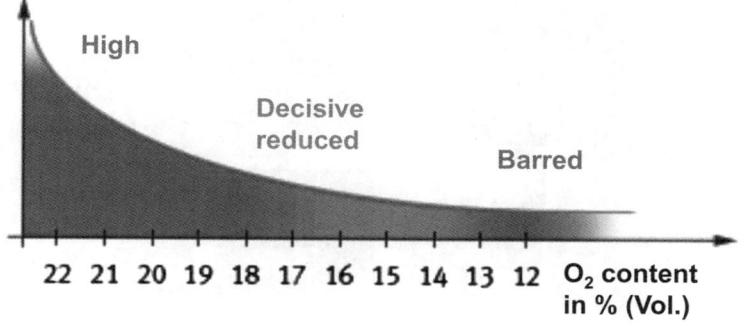

Figure 4.6 Fire risk for solid materials.

Table 4.6 General comparison between conventional electric generation on aircraft and innovative fuel cell-based technologies as multi-functional systems.

Aspect/concept	Unconventional	Conventional
Efficiency of electrical power generation	+	−
Component weight	−	+
Use of thermal energy	+	+
On-board water	+	−
Tank inerting	+	−
Fire protection	+	−
Independence of kerosene	No RAT	−
APU	No	Yes

concept is kept as simple as possible in order to show its full potential at ideal conditions. Requirements which lead to more complexity will be discussed later.

For a more precise validation exercise, various subsystems, such as heat exchangers, condensers, drying components for the exhaust gas before it can be used as inert gas, *etc.*, have to be defined and sized according to the demands. These additional elements may increase the overall weight and volume, while decreasing the efficiency.

One further aspect which should not be underestimated for a system operated on board an aircraft is the aerodynamic drag caused by air intake for the process air and the cooling air. These aspects clearly show that aircraft components are not comparable with stationary ground or mobile ground systems. Ground systems are generally oriented to cost targets while aircraft components must be lightweight, reliable and should not cause additional drag.

In sections 4.3.1 and 4.4.2.1 it has already been mentioned that new systems have to be accepted by the customers, the airlines and the airports. Additional complexity, as in additional service for LH2, may not be possible in the first

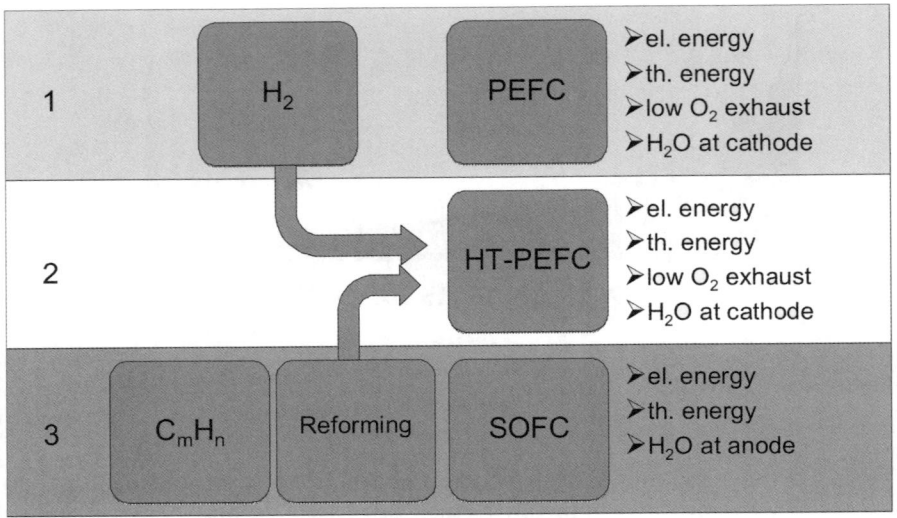

Figure 4.7   Options for an aeronautical fuel cell system design.

step when fuel cell technology enters the aviation market. An on-board reforming system and suitable fuel cell technology could be the answers to these operational constraints. Benefits may be shifted from the downstream end to the upstream end.

Figure 4.7 shows the options for aeronautic system design. In chain one, H_2 is used as fuel and PEFC technology as a converter, whereas the second chain shows the on-board kerosene reforming option in combination with an SOFC power converter. In the center of the diagram, there are two options for a HT-PEFC system. From a technical point of view, more options are possible, but it must be taken into consideration that the lower the operational temperature of the fuel cell, the lower its acceptance of hydrogen gas impurities. However, the by-products of this technology are easy to process for direct use, especially the water and the exhaust gas.

In the case of an on-board process chain starting with standard kerosene as an energy carrier, a converter system should be chosen which accepts impurities of the hydrogen gas from the on-board reforming system. In the case of using SOFC technology the process water cannot be used for drinking as O_2 and H_2 react at the anode side, where the reformed gas is led through the fuel cell. Consequently, all the remaining by-products of the reforming process will contaminate the process water. The only way to use the process water would be either to clean the gas before it enters the fuel cell or to clean the water after it leaves the fuel cell to a degree where it can be used as sanitary water.

Figure 4.8 shows a flow sheet of an HT-PEFC system for kerosene based on information given Peters *et al.*[44,67] The system will be explained in more detail in section 4.5.1.2. A preliminary design study considered heat exchangers and

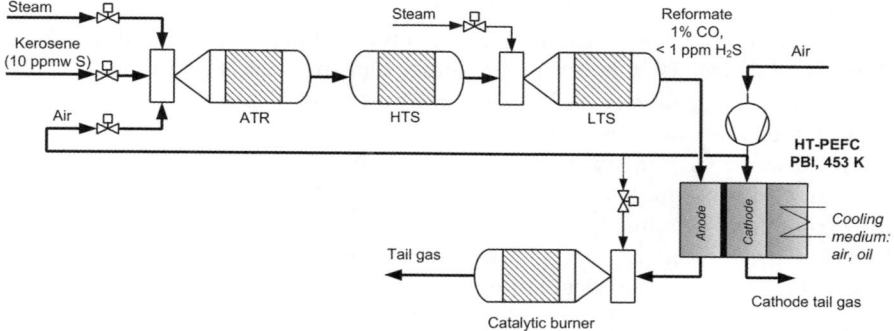

Figure 4.8 Flow sheet for an HT-PEFC system driven by kerosene.[44,67] ATR, autothermal reformer, HTS, high-temperature water gas shift reactor; LTS, low-temperature water gas shift reactor; PBI, polybenzimidiazole.

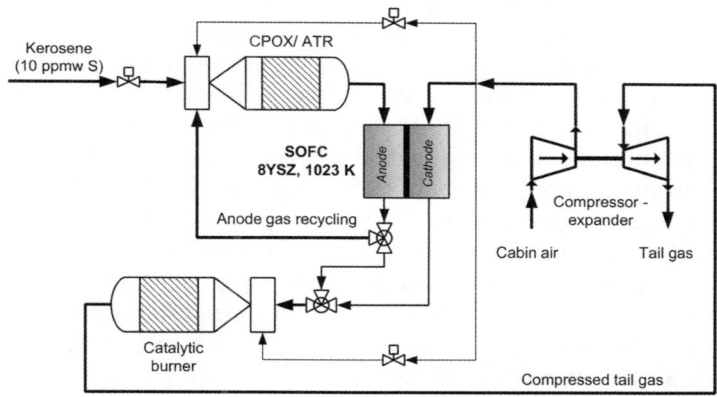

Figure 4.9 Flow sheet for a SOFC system with anode gas recycling and a catalytic partial oxidation (CPOX) of kerosene for an educt mixture according to $H_2O:C = 0.4$, $O_2/C = 0.5$ published after Mak and Meier.[22]

condensers for different systems and evaluated the potential of these systems for multifunctional usage in aircraft. A hybrid system consisting of an SOFC and a gas turbine is sketched in Figure 4.9. Studies[13–15a] based on such a scheme are discussed in section 4.5.1.2. Such systems are no longer at the focus of current development work.

4.4.4.2 Focus on Maritime Systems

Fuel cell systems for marine APUs have been investigated by several research groups.[30–36] Small APU systems were planned with DMFCs[36] or PEFCs combined with LPG reforming.[29] Large APU systems were proposed for nearly all fuel cell types, namely PEFCs, HT-PEFCs, MCFCs and SOFCs.

A PEFC system with NATO fuel F-76 for naval ships was investigated by Krummrich *et al.*[32] Figure 4.10 shows a flow sheet based on their data. NATO fuel F-76 is a military diesel fuel for vessels and contains high amounts of sulfur (about 0.13%, *i.e.* 1300 ppmw S). Therefore, diesel must be evaporated and desulfurised before it is reformed. Desulfurisation was intended to be carried out on a commercial NiMo catalyst in the gas phase at 4 bar with subsequent H_2S adsorption on ZnO. The implementation of a pre-reformer unit simplified the design of the steam reformer because no higher hydrocarbons had to be converted in the steam reformer, which mini-mised the risk of carbon deposition. The gas treatment consisted of two shift stages and two PROX stages with corresponding intercooling steps which are not shown in Figure 4.10. Additionally, the PEFC must be moistened on the cathode side. The flow sheet clearly shows that such a system set-up is quite complex.

Figures 4.11 and 4.12 and show possible flow sheets for methanol as an energy carrier used in HT-PEFCs and SOFCs. Such systems were proposed in a case study from Aachen University[38] as APUs for megayachts. Steam reforming requires a heat source, which is the catalytic burner for the fuel cell tail gases (see Figure 4.11). A shift stage can be omitted if the CO concentration in the product gas is limited to about 1%. According to design data of a methanol steam reformer[64] at 553 K and 2 bar, methanol can be converted into a reformate at rates above 95% with CO concentrations ranging from 1% to equilibrium concentrations of about 1.85% (see Table 4.5). For this reason, the working range is extremely narrow without an additional shift reactor. Therefore, Emonts *et al.*[68] designed a shift reactor included in an HT-PEFC system. They proposed a water injection unit upstream of a shift reactor to reach a molar water-to-carbon ratio of 0.6 at the inlet. Due to the cooling effect of evaporating water an inlet temperature of 493–513 K was reached. By kinetic modelling, a catalyst capacity of $18\,000\,m^3_{gas}\,h^{-1}\,m^{-3}_{cat}$ gas hourly space velo-city (GHSV) can be calculated for a water–gas shift reactor of a methanol steam reforming system, which results in a reactor which is less compact than a shift reactor for autothermal diesel reforming ($30\,000\,m^3_{gas}\,h^{-1}\,m^{-3}_{cat}$). Without additional water supply, the shift reactor in such a methanol-reforming system requires more than 20 times its volume size, *i.e.* $GHSV < 1000\,m^3_{gas}\,h^{-1}\,m^{-3}_{cat}$. These boundary conditions for system design are a result of the higher hydrogen concentration in the reformate and the corresponding state of che-mical equilibrium.

Figure 4.13 shows a possible flow sheet for combining methanol steam reforming with SOFC technology. Hansen[69] proposed introducing a metha-nator between the steam reformer and the SOFC to enable the internal steam reforming of methane. This measure leads to a cooling effect and thus to a lower required air flow rate on the cathode side. Additionally, anode gas recycling means a steam generator can be dispensed with.

A 320 kW$_e$ MCFC stack with LNG as an energy carrier was installed on a supply vessel by CFC Solutions.[48,49] Figure 4.13 shows a flow sheet for the

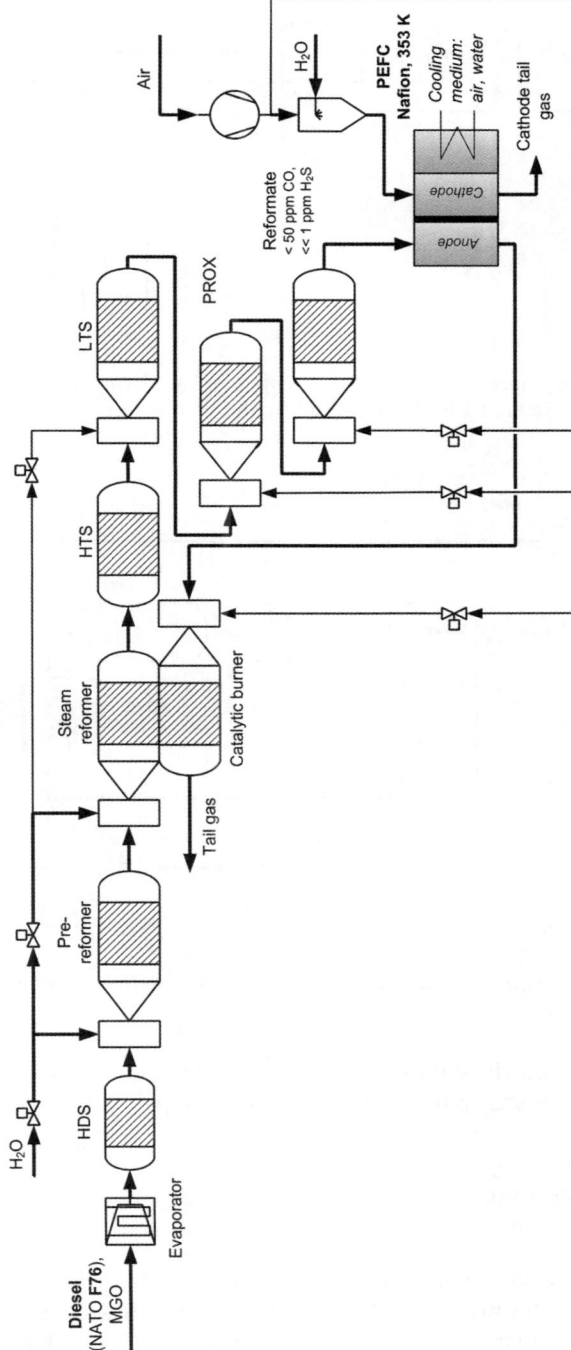

Figure 4.10 Flow sheet of a PEFC system for NATO F76 diesel fuel based on data published by Krummrich *et al.*[32] for the DESIRE project.

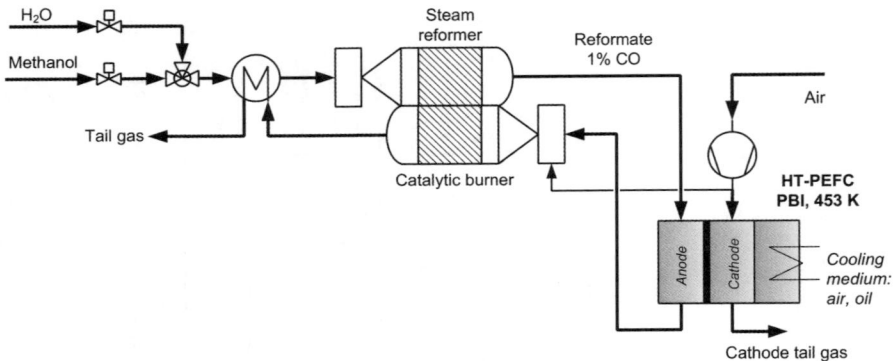

Figure 4.11   Flow sheet for an HT-PEFC system based on methanol as fuel for megayacht APUs as proposed by the EC project Roads2HyCom.[38]

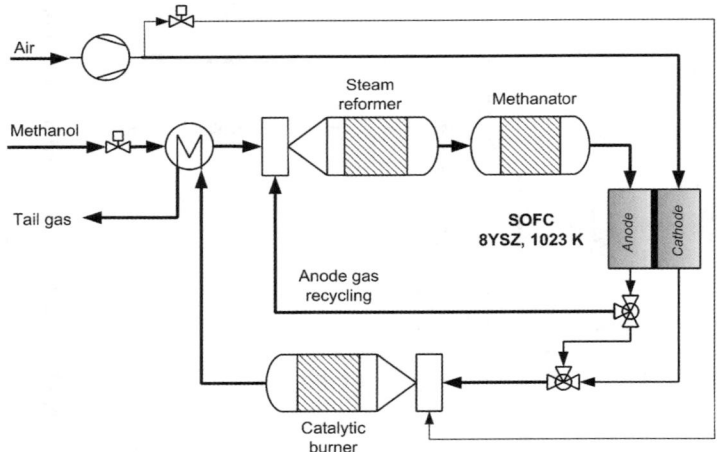

Figure 4.12   Flow sheet for an SOFC system with methanol as fuel based on Hansen.[69]

HotModule based on the data published by Huppmann.[70] In the HotModule, the reforming process is split into three different steps:

- External reforming
- Integrated reforming
- Internal reforming.

Each of these three steps can also be used as an isolated process for methane steam reforming. The principle of the HotModule is based on very effective conversion and transport processes. In the external adiabatic reformer, short-chained hydrocarbons are reformed to methane using heat from the fuel gas, which is heated by the off-gas from the fuel cell. The reformer's operating

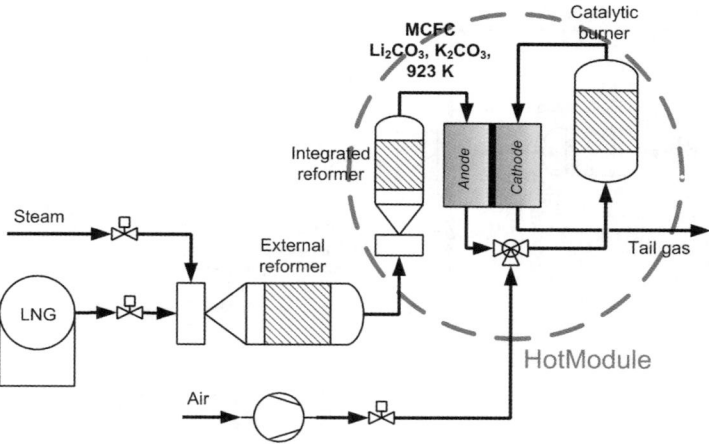

Figure 4.13 Flow sheet of the HotModule MCFC system according to Huppmann.[70]

temperature is lower than the fuel cell temperature. The indirect internal reformer is located between the cells in the cell stack. Due to the thermal coupling between the electrochemical processes in the cell and the integrated reformer, waste heat from the cells is recovered. This reforming step takes place at cell temperature. No mass exchange takes place between the integrated reforming reactions and the electrochemical cell reaction. The hydrogen concentration at the entrance of the cell is increased significantly by integrated reforming. As a last step, direct internal reforming continuously produces new hydrogen from the remaining methane as required by the electrochemical cell reaction. Thus conversion in the reforming process is almost complete. In an MCFC, carbonate ions migrate through the electrolyte from the cathode towards the anode electrode driven by the concentration gradient and the electric potential. The operating principle requires carbon dioxide to be present on the cathode side. This is realised by using the off-gas from the catalytic burner. The anode exhaust gas consists of unreformed feed gas, reforming products and oxidation products. It is mixed with air and then fed into a catalytic burner where all combustible species are completely oxidised. The chosen air ratio should be sufficiently high ($\lambda > 1$) to leave some oxygen in the burner exhaust gas. This mixture is then fed into the cathode channel.

In order to achieve higher efficiency, several studies evaluated hybrid systems consisting of micro gas turbines with PEFCs,[71] SOFCs[17,18] and MCFCs.[31,33] Figure 4.14 shows such a pressurised system for an MCFC with an additional fuel burner to increase the inlet temperature to 1073 K for the gas turbine operated at 3.55/1.1 bar.[33] The system was designed for desulfurised diesel. Liquid desulfurisation was performed *via* reactive adsorption of the sulfuric compounds at 473 K on Ni-based sorbents. Desulfurised diesel was converted into a hydrogen-rich gas by autothermal reforming. A certain amount of diesel was burnt in an additional fuel burner to increase the inlet temperature of the expander unit.

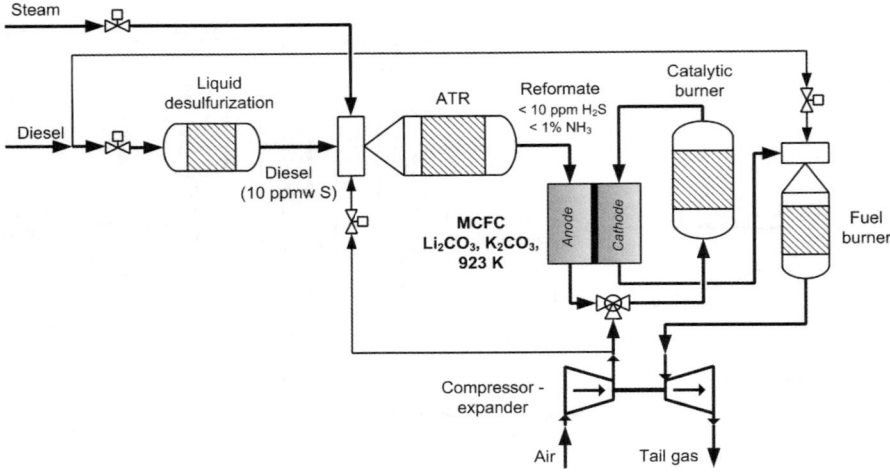

Figure 4.14 Flow sheet for a hybrid system consisting of an MCFC for diesel fuel and a gas turbine based on Bensaid *et al.*[31] and Specchia *et al.*[33]

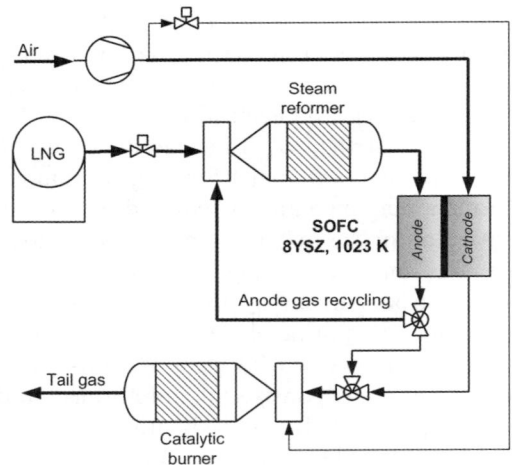

Figure 4.15 Flow sheet for an SOFC system with LNG as fuel.

Similar to MCFC systems, SOFCs can use LNG as an energy carrier. Such a design is sketched in Figure 4.15. All systems will be evaluated further in the next section.

4.5 System Evaluation

The starting point for the evaluation of a new system, technology, concept or a combination of these three is the reference technology, which usually represents the conventional technologies at the time of the design freeze of a product. This takes the development of both conventional and new or unconventional technology into account. Usually technologies which have already been established for a considerable time and thus represent traditional technologies need greater efforts to achieve a certain level of improvement than young technologies do. This may lead to a situation where young technologies achieve higher targets than traditional technologies. On the other hand, physical or chemical characteristics may represent a performance limit. Therefore a solution under ideal conditions, for any case, may be built up and evaluated. A solution may therefore be established under ideal conditions for a particular case and then evaluated. This first evaluation shows the potential under optimum conditions, which will never be found in the first step.

4.5.1 Focus on Aeronautical Systems

Fuel cell technology has already flown in several research and test aircraft. This has demonstrated its safe operation and thus makes it possible to certify H_2 storage systems and fuel cell technology. Even if none of the demonstrators represent the operational environment of a commercial aircraft they significantly lower the risks, which are clearly part of each evaluation for these items. At the same time, these demonstrators have increased the acceptance of H_2 on the part of the public by positive reporting in the press.

4.5.1.1 Systems with Hydrogen as the Fuel

Fuel cell system concepts based on an on-board hydrogen storage system may display quite a simple architecture, but they may have some difficulty in entering the market due to non-acceptance by aircraft and airport operators. Nevertheless, they show their advantages under ideal conditions.

Option 1, whose main components were oxygen and hydrogen storage facilities and a fuel cell, was meant to be a potential replacement for the ram air turbine. For the evaluation it must be ensured that both systems meet the same operational requirements, which is to provide emergency power in the case of all engines being inoperative due to no fuel or contaminated fuel and to be independent of kerosene. In this case, the following parameters have to be compared: maintenance, weight, volume, drag and cost.

According to the different categories of aircraft – short, long, ultra-long range – the advantages of the compared systems may vary. For example, while a long-range aircraft is more sensitive to the aerodynamic quality of the overall aircraft, a short-range aircraft gains on lightweight systems.

Option 2 represents the most complex system architecture. In order to achieve the highest level of benefits, the most suitable concept is chosen. For the first evaluation step, the simplest and most efficient layout was selected for further investigation. The electrical efficiency of a fuel cell system is assumed to be about 50%. The remaining energy is partly transformed into thermal energy and also consumed by periphery systems, like compressors, cooling systems, *etc.*

Table 4.7 Gas composition in chemical equilibrium for different energy carriers.

Case	$2 \times 100 \, kW_e$	$4 \times 100 \, kW_e$
Operational time	1.5 h	8.5 h
Energy (100%)	600 kWh	6.800 kWh
Mass hydrogen	18.2 kg	226.0 kg
Volume (1 atm.; 21 K, 90%)	0.3 m^3	3.58 m^3
Water production @ 60% load of the fuel cell ($\rightarrow \eta = 50\%$)	7.3 kg H$_2$ for 52.4 kg H$_2$O (1 h)	116.4 kg H$_2$ for 837.8 kg H$_2$O (8 h)
Electricity production @ 60% load ($\eta = 47.5\%$)	7.7 kg H$_2$ (1 h)	122.5 kg H$_2$ (8 h)
60% load at $\eta = 40.0\%$	25.2 kg JET A-1 (1 h)	401.9 kg JET A-1 (8 h)

The basic aircraft has an MEA system architecture on board (MEA: more electrical aircraft). It is assumed that 50% of the installed generator capacity is provided by the fuel cell generators. The following assumptions were selected for the preliminary sizing of the components. Two concepts are compared in Table 4.7. One system consists of two 100 kW$_e$ fuel cell units operated more or less continuously on a short-range mission lasting 1 h. The other system is designed with four 100 kW$_e$ systems for a long to ultra-long-range mission. The two conditions represent the limits of a set of possible combinations. The number of units, *i.e.* two to four fuel cell power units of 100 kW$_e$, depends on the size of the aircraft. The mission range determines the operational times of 1–8 h at 100% loading. For safety reasons, full loading during the entire flight must be considered plus 0.5 h for the case of any deviation. Considering H$_2$ storage, the following calculation can be made: (2×100 kW$_e$ at 100% load for 1.0 + 0.5 h)/0.5 = 600 kWh $\rightarrow$ 18.2 kg H$_2$. The most significant aspect of volume and weight storage of H$_2$ for this amount is at the liquid stage.[72] At a filling level of 90%, this leads to volumes of (18.2 kg H$_2$/0.9)/70 kg m^{-3} LH2 $\rightarrow$ 0.30 m^3 LH2 (atmospheric pressure; 21 K).

Assuming that the fuel cell generators are 60% loaded by electrical consumers during the entire mission, water will be produced by the chemical reaction of H$_2$ and O$_2$. According to the chemical equation, the mass of the water produced will be nine times the mass of the consumed hydrogen. On the assumption that 80% of the process water can be obtained by condensation in cruise condition, the following water mass can be used as fresh water: 2×100 kW$_e$ at 60% load for 1 h $\rightarrow$ 7.3 kg H$_2$ $\rightarrow$ 52.4 kg H$_2$O after condensation and 4×100 kW$_e$ at 60% load for 8 h $\rightarrow$ 116.4 kg H$_2$ $\rightarrow$ 837.8 kg H$_2$O after condensation. These values lead to a potential specific value of 55 cL kW$_e^{-1}$ h^{-1} and an expected value of 43.7 cL kW$_e^{-1}$ h^{-1}. The values from this first approximation are already in good agreement with the results from a more detailed analysis given in section 4.5.1.2.

It can be expected that the power train consisting of H$_2$, fuel cell, power electronics and DC consumer is more effective than the power train of gas turbine, gearbox, generator, rectifier, power electronics and DC consumer. This

Table 4.8 Efficiency chain for fuel cells in relation to a turbo engine.

Fuel cell	η (%)	Turbo engine	η (%)
Fuel cell system	50	Turbo machine	45
		Gearbox	98
		Generator	97
		Rectifier	98
Power electronics	95	Power electronics	95
Total	47.5	Total	40

may vary in the case of electrical AC consumers. Table 4.8 shows the efficiency chain of both systems.

Differences between the two concepts can be estimated roughly for the 'real' average power demand for on-board systems per mission, which could be 30% of the entire installed generator power. Taking into account the assumption of section 4.3.1.1 results in the installed generator power at the main engines amounting to 50% of the demand, while the remaining 50% comes from the fuel cell generators. Assuming that the fuel cell generators will be the primary power providers, because of their higher efficiency, would result in an average loading of 60%. For this case, H_2 consumption can be calculated against kerosene consumption in the environment of the traditional system architecture (Figure 4.16).

This results in weight differences due to the different fuels and their consumption rate. For example, the fuel cell system case can be calculated as $2 \times 100\,kW_e$ at 60% load for 1 h at an efficiency of $47.5\% \rightarrow 7.7\,kg\,H_2$ and as $4 \times 100\,kW_e$ at 60% load for 8 h at an efficiency of $47.5\% \rightarrow 122.5\,kg\,H_2$. The corresponding results of the turbo engine case lead to $200\,kW_e$ at 60% load for 1 h at an efficiency of $40.0\% \rightarrow 25.2\,kg$ JET A-1 and $400\,kW_e$ at 60% load for 8 h at an efficiency of $40.0\% \rightarrow 401.86\,kg$ JET A-1. These numbers promise a fairly high advantage of fuel cells and hydrogen as fuel. Following the figures given by Meinert,[53] present storage systems for hydrogen suffer from the poorest ratio between mass of stored fuel and tank weight, *i.e.* $17\,kg_{Tank}\,kg^{-1}{}_{Fuel}$ for LH2 and $24\,kg_{Tank}\,kg^{-1}{}_{Fuel}$ for GH2 relative to $1.4\,kg_{Tank}\,kg^{-1}{}_{Fuel}$ for fuels like gasoline and diesel. Taking these values into account, the advantage in stored fuel weight becomes a drawback of 2100 kg LH2 storage relative to 562.6 kg JET A-1 storage. Research and development for hydrogen storage must be focused on lightweight systems. Latest research results lead to a projected value for the mass of an LH2 storage system including fuel of between 700 and 950 kg. Volumetric and mass specific values will be discussed in more detail in section 4.5.1.2. Unfortunately, the improvement in volume-specific values, *i.e.* $1.75\,L_{Tank}\,L^{-1}{}_{Fuel}$ for LH2 and $1.67\,L_{Tank}\,L^{-1}{}_{Fuel}$ for GH2,[53] is limited and a comparison with liquid fuels shows much lower energy densities (see Table 4.4). Nevertheless, weight aspects are one of the most important items for more efficient short-range missions.

Ambient air consists of about 21% O_2 by weight and about 79% N_2. The chemical reaction process inside the fuel cell usually runs at an O_2/H_2 ratio of

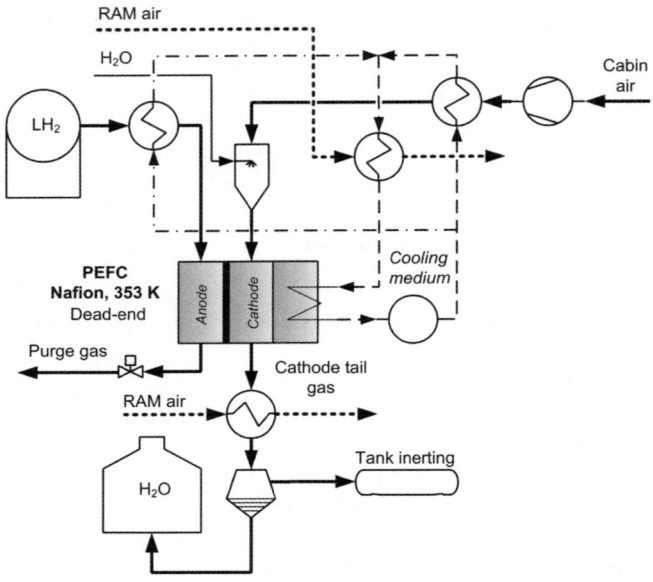

Figure 4.16 Flow sheet for a PEFC system for liquid hydrogen (LH2) with water generation and tank inerting option.

about 2. This leads to oxygen content in the exhaust of about 10.5%. As shown in section 4.4.4.1, 10.5% O_2 in the exhaust would be sufficient to prevent a wide range of materials from burning.

The ratio of thermal power produced by the reaction inside the fuel cell can be considered as 40%. This means that a full cell system with 2 stacks at 100 kW$_e$ at 60% loading will produce about 48 kW thermal power. Depending on the technology selected and the system concept, the thermal power can be used to lower the consumption of electrical or pneumatic power for the wing anti-ice system.

The results of the rough calculations show the feasibility of a multifunctional fuel cell generator on board an aircraft consisting of an LH2 storage system and a fuel cell system. Deficiencies arising from the higher specific weight of such a technology can be compensated by the use of the by-products as shown above. This result shows the benefit of the new technology in a future environment where LH2 is available at airports.

An alternative on-board H_2 supply has to be selected in order to bridge the time gap until an LH2 infrastructure is available. This leads to a technology which extracts H_2 from either kerosene or water or both. Various existing reforming processes used for this purpose are evaluated in the next chapter.

4.5.1.2   Systems with Kerosene as the Fuel

Kerosene fuels such as JET A-1 are commonly used fuels for aircraft applications. If JET A-1 is used in fuel cells, a fuel processing unit must be implemented upstream of the fuel cell. Such a fuel processor must be able to desulfurise JET A1 from 3000 ppmw S as the specified maximum value down to 10 ppmw S. A subsequent reforming process converts the fuel into a hydrogen-rich gas with air and/or steam. Figure 4.17 shows a flow sheet of an HT-PEFC system for kerosene.[44,67] It resembles Figure 4.8, but additionally indicates all required heat exchangers and the devices needed for water production and tank inerting. Water is obtained in this process from the anode and cathode tail gases after a cooling and condensation process.

The most important issue of a fuel-processing system is a closed water balance. It must be checked whether this condition can be fulfilled or if it is possible to produce additional water. In order to evaluate the limits of a system, determining parameters have to be found and reliable but challenging parameters must be defined. It needs to be considered that dry air must be used for fuel cell systems and that water production is restricted by thermodynamic conditions, *i.e.* system pressure and condensation temperature. The heat exchanger area determines the temperature of the cooling unit. Water condenses at a given pressure and temperature until a humidity of 100% is reached. The use of air with humidities below 100% always leads to a certain loss of water. As an additional challenge, PEFCs must be fed with completely wet – *i.e.* nearly 100% humidified – streams. As a consequence, water must be recycled either with a membrane or by condensing water from the off-gas and subsequent water injection and evaporation in the feed gas.

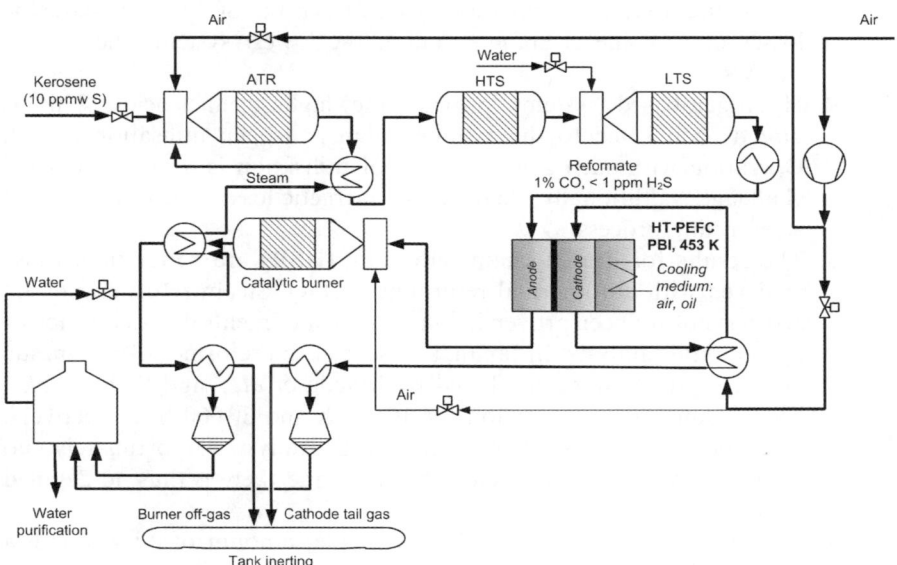

Figure 4.17 Flow sheet of an HT-PEFC system for kerosene[44,67] including all the required heat exchangers and the devices needed for water production and tank inerting.

The most challenging data are defined for ambient temperature (55 °C) and humidity (65%). The critical temperature at which a closed water balance is possible for these conditions has been determined (see Table 4.9). In addition, the amount of excess water can be calculated after cooling and condensing at 40 °C. The amount of tail gas and its oxygen concentration was calculated for the mixture of cathode and the burner off-gases (see Figure 4.17). Finally, the electric efficiency is given for all cases assuming more or less challenging operation conditions for the fuel cell, *i.e.* cell voltage, $U_{cell} = 750\,mV$, cathode air ratio, $\lambda_{Air} = 1.5$ and system pressure, $p_{sys} = 1.1$ bar. These values are the best results from a statistical evaluation[41] varying five decisive parameters such as cell voltage, air ratios for cathode and burner air, system pressure and hydrogen utilisation. Hydrogen utilisation depends strongly on the overall heat balance and amounts to as much as 75% for steam reforming-based systems. The statistical variation leads to a broad range of systems. For example, an HT-PEFC system in combination with autothermal kerosene reforming has a maximum efficiency of 38.5%, which must, however, be interpreted as a potential. However, under disadvantageous conditions, a low efficiency of only 20% would be achieved. At the beginning of system development, efficiencies might start in the range between 25% and 30%.

Comparing the results from Table 4.9, the following general findings were made:

1. Of ATR-based systems, SOFC systems deliver the highest system efficiency followed by HT-PEFC systems. This can be explained by the fact that CO is also used as a fuel in the SOFC. The parasitic losses of CO fine cleaning result in the lowest system efficiency in PEFCs.

2. Hydrogen-based systems provide the highest efficiencies. This is directly connected to the assumed high hydrogen utilisation (99%). Operation with such a high degree of utilisation is a great technical challenge. Additionally, there are no energetic losses due to the lack of a reforming process.

3. The results for heated-steam reforming (HSR) are better than those achieved with autothermal reforming (ATR). Steam reforming of Jet A-1 has not yet been proven in long-term experiments due to the lack of appropriate catalysts. In Japan, kerosene steam reformers for domestic heating systems were developed by Maeda *et al.*[73] and Saito *et al.*[74] They claim 100% conversion for 10 000 h and 30 000 h, respectively. They both used a Ru-based reforming catalyst. Unfortunately, no attention was paid to residual substances and there is thus no detailed analysis.

4. Hydrogen-based systems produce a smaller amount of off-gas with a higher oxygen concentration.

5. SOFC off-gases are not suitable for tank inerting despite having the highest amount of gas available due to the high air ratio which is required for adequate stack cooling.

Table 4.9 Optimum results from a statistical approach for different system configurations with regard to high efficiencies[41,44]

Fuel	Process	Fuel cell	Hydrogen utilisation (%)	Efficiency (%)	Water ($cLkW_e^{-1}h^{-1}$)	Critical temperature (°C)	Tail gas ($m^{-3}h^{-1}kW_e^{-1}$)	O_2 (%)
JET A-1	ATR	PEFC	84	36.1	16.3	51.1	3.4	4.7
H_2	–	PEFC	99	55.5	38.2	67.2	1.9	7.3
JET A-1	ATR	HT-PEFC	90	38.5	15.1	51.0	3.2	4.9
JET A-1	HSR	HT-PEFC	75	42.3	56.6	65.8	3.2	6.0
H_2	–	HT-PEFC	99	55.2	38.4	–	1.9	7.4
JET A-1	ATR	SOFC	90	40.3	20.3	66.4	20.2	17.7
JET A-1	CPOX	SOFC	90	39.1	20.8	–	21.8	17.8

Cell voltages: 750 mV; system pressure: 1.1 bar; cathode air ratio: 1.5 for PEFC; 12.3/ATR and 13.2/CPOX for HT-PEFC at a temperature difference between anode gas inlet and SOFC cell temperature of about 100 K.

6. The system efficiency is maximised by using lower system pressure (1.1 bar), higher fuel utilisation and higher cell voltage.

The results of the process analyses are discussed in more detail by Pasel et al.[41] and Peters et al.[44] Further results which are not reflected in this chapter are:

1. The water recovery in HT-PEFC systems can be realised with smaller heat exchanger areas. Even using ATR, the required area is much smaller than for PEFC technology.
2. To maximise the water recovery rate, a higher system pressure must be selected. Therefore, the systems cannot be optimised for maximum efficiency and water recovery rate at the same time.
3. In some cases, fuel utilisation is limited because of the system configuration.
4. The major disadvantage for the combination of an SOFC stack with a diesel reformer is caused by a very low methane concentration in the reformate eliminating the cooling effect of internal reforming. Consequently, the heat of reaction must be withdrawn in SOFC by warming the inlet flows, i.e. mainly the cathode inlet air. The latter option is limited by material stresses to a temperature difference of about 100–150 K. Finally, the air flow must be increased to cool the stack adequately. High air ratios of 12–13 for $\Delta T = 100$ K combined with high system pressures result in very unsuitable energy balances in SOFC systems. Therefore, these systems are equipped with expanders to achieve reasonable efficiencies.[17,18,31,33,71]

Although SOFC systems should be excluded as an option for a multifunctional fuel cell design a brief comparison with other studies[13–16] should be made. Due to the different system architectures and load profiles assumed, the results cannot be compared directly. According to the calculation by Srinivasan et al.,[17] SOFC system efficiencies amount to between 53% and 68% at cruise and between 39% and 48% on ground. The lowest efficiency corresponds to an architecture pressurised only to 1 atm; other system architectures work at higher pressure delivering higher efficiencies. Gummalla et al.[18] achieved efficiencies of 64.3% at cruise and 44.8% on the ground for their hybrid system including a 300 kW SOFC. Mak and Meier[22] reported efficiencies between 32% and 36% for ground operation and MES. The poorest results were calculated for an SOFC/CPOX design with reduced size, whereas better efficiencies were obtained for an SOFC/ATR design at full power operation, i.e. 164 kW$_e$. At cruise efficiencies are even higher, i.e. between 46% and 53%. This effect is caused by the different pressure levels during ground and flight operation. Considering ground operation, the best results were achieved with compression to 2 bar at the gate and 3 bar for MES and expansion to 1.1 bar. During a mission, cabin air was compressed from 0.9 bar to 2.2 bar and expanded after the fuel cell system down to 0.46 bar resulting in an efficiency of 53%. Slightly lower efficiencies were calculated for compression of ambient air from 0.3 bar to 0.9 bar at partial load and from 0.3 bar to 1.6 bar at full load resulting in efficiencies of 46% and 48%, respectively. Expansion was always considered down to 0.26 bar. The necessity of a turbine was shown by Dollmeyer *et al.*[23] They calculated efficiencies of 24–26% for architecture without a turbine and a water recovery system and 34–59% for two system designs incorporating a turbine. The minimum efficiency of 34% occurs in the special case where the compression works with ambient air and expansion is limited to 1 bar. Using cabin air for this design an efficiency of 47% was achieved.

Decisive parameters for system calculations are cell voltage, current density, hydrogen utilisation, cathode air ratio and the efficiencies for compression and expansion. Cell voltages were chosen by Srinivasan *et al.*[17] for a defined architecture between 700 mV on ground and 800 mV during mission leading to stack efficiencies of 52% and 65%, respectively. The cell voltages assumed by Mak and Meier[22] are even higher, *i.e.* between 780 mV and 890 mV, while Dollmayer *et al.*[23] used 800 mV in all calculations. Hydrogen utilisation differs between the cited studies, namely 85% in Srinivasan *et al.*[17] and Gummalla *et al.*[18] 80% in[23] and 70% (besides one positive case with 85%) in.[22] A hydrogen utilisation of 90%, see Table 4.9, offers the maximum potential.

Air ratios on the cathode side were given by Mak and Meier[22] and by Dollmeyer *et al.*[23] as 4–5 and 2, respectively. Values of about to 4–8 are reasonable for an SOFC system with internal reforming and somewhat too low for a system design with an external reformer. Therefore, Pasel *et al.*[41] and Peters *et al.*,[44] used air ratios of 12–13 ($\Delta T = 100$ K) and 6–8 ($\Delta T = 150$ K) for their analysis. Higher air ratios led to increased peripheral energy consumption. Important input values for the peripheral energy demands are the efficiencies of the compressor and the turbine. Mak and Meier[22] varied them between 73% and 80% for compression and assumed a constant value of 85% for the turbine. Dollmeyer *et al.*[23] used higher efficiencies for the turbine with 90% and 85% for the compressor. The calculations in the studies by Pasel *et al.*[41] and Peters *et al.*[44] were based on isotropic efficiencies of 70% for compressor and expander units and took into account a mechanical efficiency of 95% for their coupling.

It must be noted that the ATR system design by Mak and Meier[22] included anode gas recycling leading to an educt mixture characterised by an oxygen to carbon ratio of 0.5 and a steam to carbon ratio of 0.4. The process flow sheet corresponds more or less to a modified CPOX system. However, the analysis by Dollmayer *et al.*[23] is based on an ATR with an air ratio of 0.3 (O_2:C $\sim$ 0.5) and a steam to carbon ratio of 1.5.

Finally, studies[13–16] considered an optimistic view of the development of SOFC technology. For example, they stated performance data for 2015/2020[22] leading to an efficiency of 48–53%. The analysis clearly shows that efficiencies of 75% are rather too high even for hybrid systems consisting of fuel cells and gas turbines. The values in Table 4.9 cannot be optimised for maximum efficiency and water recovery rate at the same time.

The evaluation methodology for different designs of multi-purpose fuel cell systems in aircraft application was extended[41,44] to account for ambient conditions in an even more accurate process. A first criterion can be obtained by defining the most challenging data set and subsequently calculating a critical temperature for closing the water balance of the system. These results are somewhat misleading if ambient conditions are given only for a small portion of operation time. An improved methodology takes into account the meteorological data at the departure and the arrival airports depending on the time of day and the month, the time schedule of the missions and APU use on the ground and during the mission. The improved methodology shown in Figure 4.18 included the following elements:

1. The monthly average values of the maximum, minimum and average temperatures of the selected locations between 1961 and 1990 were obtained from the German meteorological service (DWD). Using this data, a mathematical function was derived in order to describe the daily temperature profiles.
2. Six flight routes with mission durations between 1.25 and 5 h were selected. The airports were chosen to cover different ambient conditions. In order to evaluate water production during a mission, a seventh route was selected to represent a long-range mission from Frankfurt to New York lasting around 9 h. Missions 1–6 were simulated with 6932 data sets, whereas mission 7 was simulated with 10 573 data sets. The large amount of data results from the combination of missions, for example assuming an initial departure from San Francisco or from Phoenix, and from monthly repetitions.
3. For on-ground APU operation, the data of Srinivasan *et al.*[17] were used. The fuel cell-based APU was also operated during mission 7 at a maximum power of 260 kW$_e$.
4. The water requirements for fuel processing amount to 1.255 mol H$_2$O mol^{-1} H$_2$.

The required heat exchanger areas were defined considering the following conditions:

- 65% Humidity and 55 °C ambient temperature
- 2 bar system pressure
- Minimum temperature difference of 15 K and closed water balance.

Finally, a 260 kW$_e$ PEFC system with hydrogen requires a heat exchange area of 28 m^2 for the anode side and 96 m^2 for the cathode side for cooling and condensation. The HT-PEFC system with ATR using Jet A-1 requires 34 m^2 and 33 m^2, respectively. The heat exchange area for the PEFC system is twice as large as the area required for the HT-PEFC system. According to the balances, a maximum of 800 L h^{-1} can be recovered in the PEFC system, whereas up to 650 L h^{-1} is used for humidification of the membranes. As an alternative to the

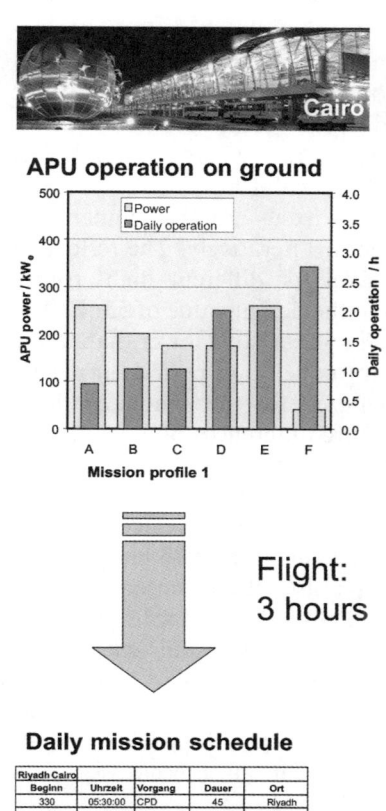

APU operation on ground

Mission profile 1

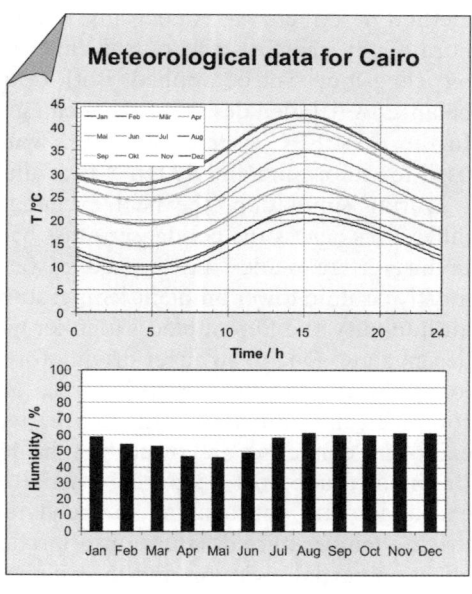

Meteorological data for Cairo

Flight:
3 hours

Daily mission schedule

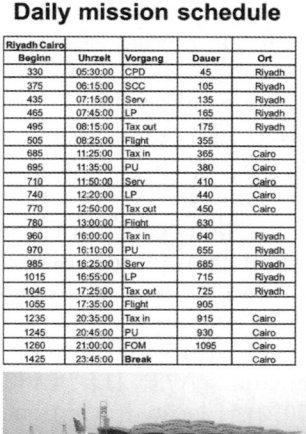

Riyadh Cairo Beginn	Uhrzeit	Vorgang	Dauer	Ort
330	05:30:00	CPD	45	Riyadh
375	06:15:00	SCC	105	Riyadh
435	07:15:00	Serv	135	Riyadh
465	07:45:00	LP	165	Riyadh
495	08:15:00	Tax out	175	Riyadh
505	08:25:00	Flight	355	
685	11:25:00	Tax in	365	Cairo
695	11:35:00	PU	380	Cairo
710	11:50:00	Serv	410	Cairo
740	12:20:00	LP	440	Cairo
770	12:50:00	Tax out	450	Cairo
780	13:00:00	Flight	630	
960	16:00:00	Tax in	640	Riyadh
970	16:10:00	PU	655	Riyadh
985	16:25:00	Serv	685	Riyadh
1015	16:55:00	LP	715	Riyadh
1045	17:25:00	Tax out	725	Riyadh
1055	17:35:00	Flight	905	
1235	20:35:00	Tax in	915	Cairo
1245	20:45:00	PU	930	Cairo
1260	21:00:00	FOM	1095	Cairo
1425	23:45:00	Break		Cairo

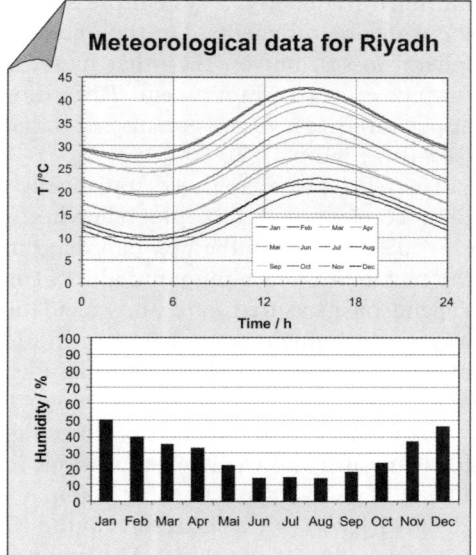

Meteorological data for Riyadh

Figure 4.18 Flow chart of the simulation method for a multi-purpose fuel cell system with electricity, heat and water production and utilization of tail gases for tank inerting.

method described, *i.e.* condensing water from the cathode off-gas and re-evaporating it again in the outside air, a humidification membrane and/or a recycling loop can be applied. Both options do not change the overall water balance. A detailed design analysis can give indications of an optimum process. In the HT-PEFC system, $300 \, \text{L} \, \text{h}^{-1}$ water is recovered, whereas $225 \, \text{L} \, \text{h}^{-1}$ is used for reforming and $75 \, \text{L} \, \text{h}^{-1}$ is available for the cabin.

During simulation, the heat exchanger size is fixed and the temperature differences across the condenser must be determined iteratively. The following findings were made: the respective weather data for different flight routes mostly contain lower ambient temperatures than the design value of $55 \, °\text{C}$. The air humidity at different places is either higher or considerably lower than in the design state. Due to stronger driving forces and to operation periods at partial load, the defined heat transfer areas are oversized for the simulated flight routes. In the simulations, the driving temperature differences of 3–$8 \, \text{K}$ at the exit of the condenser are lower than the design-point value of $15 \, \text{K}$. Considering the maintenance period during partial load operation (15% load), the heat exchangers are approaching the behaviour of infinitely large devices and the driving temperature difference is approaching 0. System pressure has a strong effect on partial pressures and therefore on the quantity of condensed water. In several cases, a system pressure of 2 bar is necessary to close the water balance, but most frequently, a system pressure is of 1 bar is already sufficient to fulfil the water requirements. On the one hand, pressurisation leads to higher peripheral losses, but on the other hand, a higher pressure improves stack performance to a certain extent. Regarding the ambient conditions for aircraft applications, a system pressure of 1 bar was tested in an alternative design (see also Figure 4.19).

The results were analysed statistically for clearer representation. In contrast to a conventional box plot showing only the lower and upper quantiles (25–50%, 50–75%), the box plot diagrams in Figures 4.19 and 4.20 also show the maximum and minimum values. This is based on the fact that the results depend on measured data and calculations instead of experimental measurements which might contain errors. In order to be able to interpret the amount of water produced, additional quantiles were added to the box plot, representing the results in the ranges (min – 10%) and (90% – max).

Figure 4.19 shows the box plot for flight routes 1–6 together with the results for Hamburg and worldwide locations for the combination (ATR; HT-PEFC; 1 bar). The average value for water production (50%) is located at around 28–$30 \, \text{cL} \, \text{kW}_\text{e}^{-1} \, \text{h}^{-1}$. The results in the 10–50% quantile show a large deviation between different data sets. The amounts of water produced are remarkably low for the flight routes San Francisco–Phoenix and Riyadh–Cairo. The values are even negative in a limited number of cases. The negative values are observed in Phoenix and Riyadh in summer with extremely low humidity ($<20\%$) and high temperatures ($>40 \, °\text{C}$) at noon time. As a result, it can be concluded that a fuel cell APU with the ATR/HT-PEFC combination delivers a positive water balance even at 1.1 bar system pressure instead of the design pressure of 2 bar in all locations up to some extreme cases. The average value for the Riyadh–Cairo

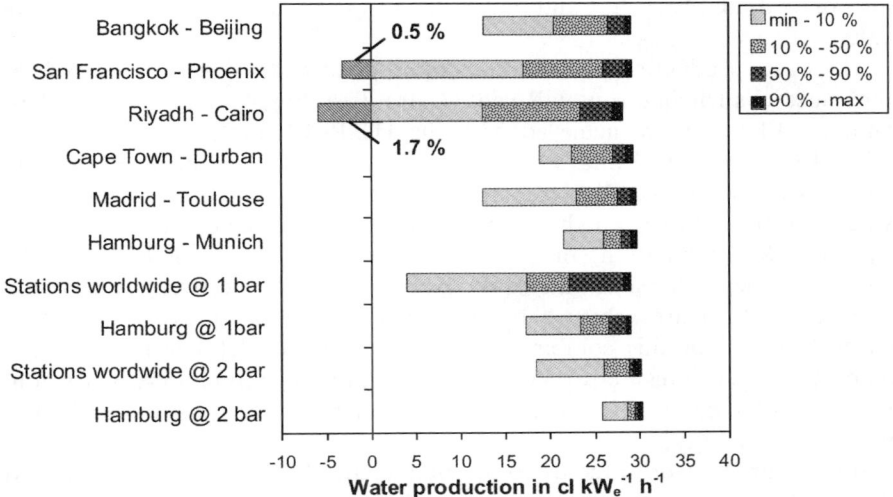

Figure 4.19 Box plot with results from a water balance for APU ground operation with HT-PEFC based on kerosene at different locations worldwide and on airports for six different missions at 1 bar.[41,44]

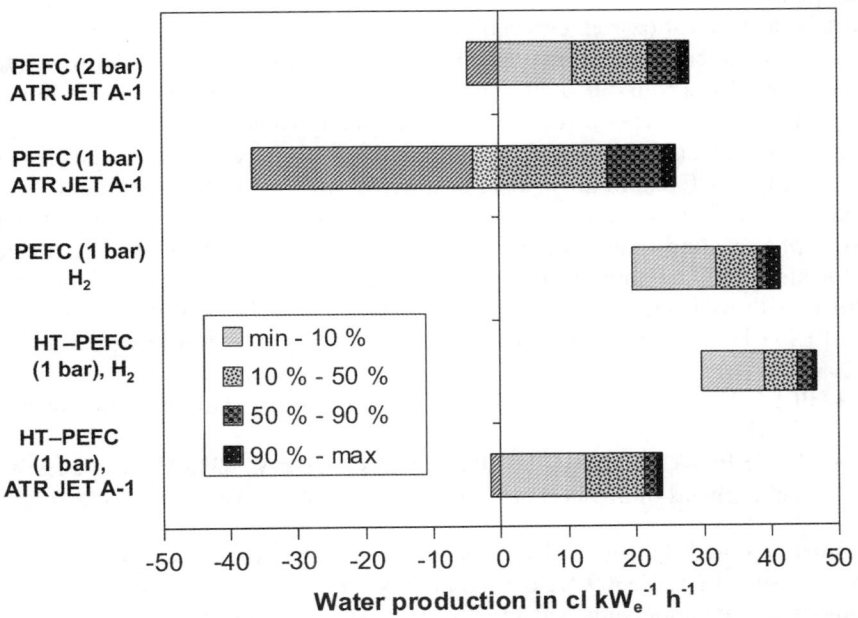

Figure 4.20 Box plot with results from a water balance for APU ground operation on the Riyadh–Cairo mission for different system designs.

mission is $27\,cL\,kW_e^{-1}\,h^{-1}$ and only 1.7% of the conditions result in negative values.

The water production performances of different systems were compared for the most challenging route, Riyadh–Cairo (see Figure 4.20). The highest amounts of water are achieved with the HT-PEFC system with hydrogen ($57\,cL\,kW_e^{-1}\,h^{-1}$ max. and $50\,cL\,kW_e^{-1}\,h^{-1}$ average) and the PEFC system with hydrogen ($52\,cL\,kW_e^{-1}\,h^{-1}$ max. and $50\,cL\,kW_e^{-1}\,h^{-1}$ average). These values are about twice as high as the values which can be achieved with the HT-PEFC/ATR combination. In the case of the PEFC/ATR combination, the amount of negative values increases. Acceptable water recovery values (15–$30\,cL\,kW_e^{-1}\,h^{-1}$) are achieved by increasing the system pressure to 2 bar. Critical values for this combination are observed for 1.1 bar pressure and ambient temperatures above 40 °C with humidity of around 15% in Riyadh. For similar temperatures in Cairo, the water balance is even positive for 1.1 bar with 50% humidity.

If large amounts of water are to be obtained the APU must also be operated during the mission. The most important question is where the system is mounted, *i.e.* the pressurised area of the cabin or the unpressurised area outside the cabin. For a first approximation, an ambient temperature of 223 K and a system pressure of 0.2 bar was assumed. The incoming air offers hardly any humidity and it must take up a huge amount of water before entering a PEFC. The operating temperature must be reduced to 60 °C to prevent the membrane material from drying out because at 0.2 bar and 80 °C water exists only in the gas phase. Additionally, the cell characteristics are strongly affected by the partial pressure of oxygen especially at low pressure. Kallo *et al.*[75] reported on low-pressure performance investigations with a H_2 PEFC system. Stack voltage is reduced by about 30% at 0.62 bar relative to atmospheric conditions. Separate air compression for the fuel cell system leads to yet another component which demands electric power and additional space. It is more efficient to use a supply with cabin air instead of ambient air.[22,75] The process analysis by Peters *et al.*[44] shows that for the pressure range 0.075–0.5 bar (relative to atmospheric conditions) water production for an HT-PEFC with kerosene reforming could be possible in a range between 0 and $30\,cL\,kW_e^{-1}\,h^{-1}$ and for a PEFC with hydrogen between 25 and $57\,cL\,kW_e^{-1}\,h^{-1}$. Two fuel cell systems, *i.e.* PEFC/H_2 and HT-PEFC/ATR Jet A-1, were analysed at low pressures:

- PEFC/H_2 at 0.6 bar abs., 700 mV, 90% H_2 usage leading to an efficiency of 45%
- HT-PEFC/ATR JET A-1 at 0.8 bar abs., 650 mV, 83% H_2 usage leading to efficiencies of 36% (H_2 to electricity) and 29% (kerosene to electricity).

It must be noted that these deviations of about 10% (abs.) in efficiency relative to the results in Table 4.9 are caused mainly by the high electrical demand of a compressor expander unit working at pressure ratios of 1:3 or 1:4. The compressor expander unit is evaluated with an isotropic efficiency of 70%. Improving these values to 85% leads to efficiencies of 39% (H_2 to electricity) and 31% (kerosene to electricity) for an HT-PEFC with kerosene reforming. Additionally, at sub-atmospheric pressures a lower fuel cell performance was taken into account. The assumption that higher air ratios can improve the oxygen supply on the cathode side was disproved in the experimental investigation reported by Kallo *et al.*[75] An air ratio of $\lambda = 3$ leads to a decrease in current at a constant voltage of 25% relative to an air ratio of 1.6. Efficiency and electric power would each be about 19% lower at constant current density.

In order to close the water balance at low pressures, *i.e.* 0.2–0.8 bar for the PEFC, a large amount of water must be recycled. The resulting heat exchanger size is incompatible with the design made for ground operation. In principle, two options are possible:

1. With improper design, data balances leading to higher driving forces for heat transfer, for example a temperature difference of 44 K for the condenser on the cathode side instead of 15 K and to a lower water production rate of $30 \, cL \, kW_e^{-1} \, h^{-1}$ instead of $45 \, cL \, kW_e^{-1} \, h^{-1}$.
2. Implementing a larger heat exchanger/condenser, for example increasing the heat exchanger from $1.8 \, m^2$ to $7 \, m^2$ to reach the full potential or implementing a corresponding membrane size for water recovery between cathode air entry flow and exit flow.

This effect is less pronounced for HT-PEFC because PBI membranes should be operated with dry or unhumidified cathode air. The calculated average specific water production decreased from $27 \, cL \, kW_e^{-1} \, h^{-1}$ on ground to $19 \, cL \, kW_e^{-1}$ h^{-1} during a mission at a system pressure of 0.8 bar. This is caused by the lower pressure and the extremely low humidity of the incoming air.

Finally, the question must be posed of how to evaluate the impacts of introducing fuel cell technology in aircraft systems. The fuel cell system was designed for a maximum electric power of $260 \, kW_e$ on ground. The load profile for ground operation was based on the data published by Srinivisan *et al.*[17] Tables 4.10 and 4.11 show basic calculations for efficiency, water production and CO_2 reduction. Due to the different efficiencies of the fuel cell systems considered during on-ground operation and during flight the maximum electric power varies between 229 and $251 \, kW_e$ during the short-range mission in Table 4.10 and between 245 and $257 \, kW_e$ for the long-range mission, in Table 4.11. The specific water production rates were taken from Peters *et al.*[44] and amount to between 27 and $50 \, cL \, kW_e^{-1} \, h^{-1}$ for ground operation and between 20 and $50 \, cL \, kW_e^{-1} \, h^{-1}$ for flight operation. The reasons for the smaller production rates during flight have been discussed in detail in the previous section. For the short-range mission, the system design focused on minimised space and mass requirements for the fuel cell system components and on high efficiency, and for the long-range mission on high water production rates. As described in section 4.5.1.1, cabin air should be used for the fuel cell system. About 40% of the cabin air was recycled and mixed with bleed air from the turbines. During the flight, the residual 60% is released by the cabin outflow valves. In a modern aircraft the complete cabin air is replaced every 3 min[76] leading, for a cabin volume between 100 and 600 m^3, to an air flow between 2000 m^3 h^{-1} and nearly 10 000 m^3 h^{-1}. Relative to these flows, a 260 kW$_e$ fuel cell system requires about 1200 m^3 h^{-1} air. Considering the water balance of the fuel cell system, the cabin air exhibits a humidity of 2% for an empty cabin and 15% during a long-range mission.[76] The results of the calculations presented by Pasel *et al.*[41] and Peters *et al.*[44] are based on a humidity of 1% while Tables 4.10 and 4.11 take the data from Reference 76 into consideration.

The specific tank storage values were taken from Meinert.[53] The power density of PEFC stacks was assumed to be 1 kW$_e$ L^{-1}. For present HT-PEFCs, a value of 0.5 kW$_e$ L^{-1} was assumed. Future stacks should gain 1 kW$_e$ L^{-1}. The power densities of present fuel processors in the 250 kW$_e$ power class were approximated by 1 kW$_e$ L^{-1} and by 1.33 kW$_e$ L^{-1} for future components.

Six different process routes were compared: PEFC with hydrogen from natural gas stored as LH2 (a) and GH2 (b), HT-PEFC with hydrogen from wind energy *via* electrolysis stored as GH2 (c), HT-PEFC combined with autothermal reforming of JET A-1 (d) and BtL from wood (e) and finally an improved system design (f) of option (e).

For a typical short-range mission from Hamburg to Munich with a flight time of 75 min and an assumed number of five flights per day[17,18] the fixed APU operation periods, namely cabin pull down, sustaining cool cabin and flight maintenance, were divided equally between the five flights leading to an APU energy for on-ground operation of 168 kWh per flight. During the flight the fuel cell-based APU was operated at full power, *i.e.* mass flows corresponding to 260 kW$_e$ on ground, resulting in an APU energy of 287–314 kWh on mission. The conventional architecture includes electricity production on ground by the turbine-based APU with an efficiency of 20% and during flight by the main engines (see section 4.5.1.1) with an efficiency of 40%. Fuel cell-based APUs always offer an advantage for on-ground operation. During the mission, hydrogen-operated fuel cell systems show higher efficiencies relative to systems with kerosene reforming. Therefore, hydrogen-based fuel cell APUs display lower fuel consumption in terms of energy content than the conventional system while reformer systems show a slight increase in fuel consumption. Additionally, PEFC and HT-PEFC systems with hydrogen as fuel achieved a water production rate twice as high as kerosene-operated fuel cells. HT-PEFC delivers higher water quality while condensed water from PEFC can contain microorganisms due to the lower operating temperatures of between 333 and 353 K. Unfortunately, PEFCs demand a high effort for humidification and thereby an appreciable part of the recovered water. Finally, the consumables saved must be evaluated together with the space demands of all required components, the conventional APU that is no longer necessary and the hydrogen storage system. The specific effort as L$_{Tank}$ L^{-1}$_{Fuel}$ was taken from Meinert[53] and is listed in Tables 4.10 and 4.11. By taking these numbers into account it can be seen that none of the fuel cell systems offers an advantage with respect to the space required.

Additionally, the potential amount of water produced by a hydrogen-based fuel cell APU need not be recovered completely. An A 320 aircraft is equipped with a 200 L water tank.[42] A higher amount would enable additional consumers. If the system mass of different systems was considered, most components were evaluated by a specific value of about 1–1.5 kg kW$_e^{-1}$ corresponding to a lightweight design. The fuel tank must be evaluated separately. In accordance with the data in section 4.5.1.1, conventional APUs need a fuel mass (JET A-1) which is 16 times heavier than the hydrogen mass required for the fuel cell system. This advantage for fuel cell systems is cancelled out if the mass of storage systems presently in use is considered. As a conclusion, more compact hydrogen storage systems must be developed in future.

Further evaluation criteria are CO_2 reduction and energy consumption of the complete energy chain labelled as a 'well-to-wheel' analysis for automotive applications. Therefore, the specific fuel data given in Table 4.4 were taken into account. A conventional APU system will release about 800 kg CO_2 on one short mission (Table 4.10) and requires 3000 kWh primary energy such as crude oil. An HT-PEFC system with kerosene reforming has the lowest demand for primary energy, *i.e.* 2173 kWh. PEFC systems with LH2 as fuel (from natural gas as primary energy carrier) cannot offer large improvements for this short-mission profile due to the low efficiency for hydrogen liquefaction. BtL-kerosene is analysed as a product from wood gasification and a subsequent Fischer–Tropsch process.

As shown in Table 4.4, the fuel production efficiency for BtL amounts to 45.7% and is thus much less than for kerosene from crude oil at 87.7%. Therefore, the amount of primary energy is highest for BTL kerosene chains, *i.e.* 3832 kWh and 4212 kWh, respectively. The reason for introducing such fuels is CO_2 reduction. With BTL kerosene or hydrogen from wind energy *via* electrolysis, a remarkable CO_2 reduction of 748 kg CO_2 to 767 kg CO_2 is possible. Due to the better efficiencies, all fuel cell-based APUs offer an advantage for CO_2 release.

Table 4.11 shows the result of a long-range mission – 9 h from Frankfurt to New York – and for an ultra-long-range mission from New York to Singapore, which needs a flight time of 18 h 45 min. The main items of system design are high efficiency and a large water production rate. Especially for the long- and ultra-long-range mission the heat exchanger area was chosen according to requirements during flight operation. Each area was multiplied by an approximated factor of 200 m^2 m^{-3} resulting in a volume for a heat exchanger block and by a factor for the housing, *i.e.* 1.25 m^3$_{HEX}$ m^{-3}$_{HEX,block}$. The space requirements in Table 4.10 for fuel cell, reformer and heat exchanger minus APU amount to 783 L. In Table 4.11, values of 816 L and 872 L can be found for the HT-PEFC/reformer system. The amount of water produced is in the range of 628–2883 L. These values correspond well to the water tank volume of an Airbus A 380 with 1700 L and an extra 500 L on demand.[43]

Concerning the total space requirements, HT-PEFCs with JET A-1 or kerosene offer an advantage in volume, *i.e.* 670 L on an ultra-long-range mission for HT-PEFC/JET A-1 and 1018 L for an improved HT-PEFC with BtL kerosene. On a long-range flight, the change in space demand is fairly balanced.

Table 4.10 Analysis of volume requirements for different fuel cell systems and different fuels based on fossil and renewable primary sources for a short-range mission.

Fuel cell type	PEFC	PEFC	HT-PEFC	HT-PEFC	HT-PEFC	HT-PEFC
Fuel	LH2	GH2	GH2	JET A-1	BtL	BtL
Primary source	*Natural gas*	*Natural gas*	*Wind energy*	*Crude oil*	*Wood*	*Wood*
Electric power (mission), in kWe	239	239	229	231	231	251
Efficiency (ground)	51.6%	51.6%	46.8%	36.0%	36.0%	39.6%
Efficiency (flight)	47.4%	47.4%	41.3%	32.1%	32.1%	38.2%
Water production (ground) $V_{H_2O,ground}$, in cL kWe^{-1} h^{-1}	50.0	50.0	50	27	27	27
Water production (mission) $V_{H_2O,mission}$, in cL kWe^{-1} h^{-1}	48.8	48.8	55	24	24	23
Fixed space, ΔV_{fix}						
Fuel cell stacks, in L	260	260	520	520	520	260
Fuel processing, in L	0	0	0	260	260	195
Heat exchanger and condenser, in L	249	249	95	133	133	123
Commercial APU (0.5 L kWe^{-1}), in L	-130	-130	-130	-130	-130	-130
Short-range mission such as:		Hamburg to Munich (5 times/day, 75 min flight time, 613 km)				
Energy APU ground, in kWh	168	168	168	168	168	168
Energy APU mission, in kWh	299	299	287	289	289	314
Water produced/mission, in L	327	327	345	156	156	152
Δ Kerosene/mission, in L	-267	-267	-264	-66	-63	-87
Hydrogen/mission, in L	695	973	945	—	—	—
Variable space, ΔV_{var}						
Water tank, $\Delta V_{var,H_2O}$, in L	-376	-376	-397	-179	-179	-175
JET A-1 tank, $\Delta V_{var,JET\ A-1}$, in L	-316	-316	-313	-78	-74	-103
Hydrogen tank, $\Delta V_{var,H_2}$, in L	1208	1611	1778	0	0	0
$(\Delta V_{fix} + \Delta V_{var})$; in L	957	1360	1577	558	563	200
Energy well to el., in kWh	3061	2309	2651	2173	4212	3832
CO_2 reduction, in kg	155	322	748	219	764	767

$P_{el,max} = 260$ kW. Specific values; energy densities for different fuels: 9.69 kWh $L_{JET\ A-1}^{-1}$; 1.94 kWh L_{LH2}^{-1}; 1.39 kWh L_{GH2}^{-1}; 9.53 kWh L_{BtL}^{-1}; tank storage systems: 1.185 L_{tank} $L_{JET\ A-1}^{-1}$; 1.750 L_{tank} L_{LH2}^{-1}; 1.667 L_{tank} L_{GH2}^{-1}; 1.185 L_{tank} L_{BtL}^{-1}; 1.250 L_{tank} L_{water}^{-1}; heat exchanger: 1.250 L_{HEX} $L_{HEX\ block}^{-1}$.

Systems based on hydrogen as fuel suffer from the low tank storage performance. For the ultra-long distance about 323 kg H_2 were required corresponding to a fuel volume of 5533 L LH2 or 7746 L GH2. Considering the storage performance data of Magna Steyr presented by Meinart in 2008[53] the storage volumes inclusive hydrogen shown in Table 4.11 can be calculated, *i.e.* 9683 L LH2 storage or 12 911 L GH2 storage. The corresponding masses of LH2 and GH2 storage amount to 5533 kg and 7746 kg, respectively.[53] The disadvantage in total volume of 5213 L for the complete assembly on the ultra-long distance and for liquid storage (LH2) and of 9623 L for gaseous storage (GH2) can be converted to additional mass of 1160 kg and 4000 kg, respectively. These figures lead again to an urgent need for more compact and lightweight hydrogen storage systems especially for aeronautic applications. The 2010 targets for automotive applications were published in the final report of the integrated StorHy project funded by the European Commission[77] with specific values of 7.2 MJ kg^{-1} and 5.4 MJ L^{-1}. These figures correspond to the specific values given by Meinert.[53] Obviously, hydrogen storage systems applied in automotive applications cannot be used for avionics systems. Recent results from StorHy clearly showed an improvement beyond the automotive goals in mass-specific energy density for LH2, but not in volume-specific energy density. Unfortunately, increased energy densities for GH2 are not expected in future, neither per volume nor per mass. A mass-specific analysis will be discussed at the end of this section.

The best chain efficiency and the lowest primary energy demand can be realised by HT-PEFC and autothermal reforming of JET A-1. For the ultra-long-range mission, a PEFC with GH2 storage is the next best option, apart from the fact that this design needs more space. CO_2 reduction is best achieved by BtL fuels or hydrogen from wood. Due to the lower efficiency of 35.4% for an HT-PEFC system combined with JET A-1 reforming relative to 40% for the conventional system the advantage of more efficient ground operation disappears with increasing flight time. Finally, improved components together with a liquid fuel from renewables offer the best package for a multifunctional system. Unfortunately, in contrast to Tables 4.10 and 4.11, the efficiency and power density of stacks cannot be chosen without any interdependencies. With regard to their voltametric performance, *i.e.* their U, i characteristics, a trade-off analysis between power density and efficiency must be investigated. The system analysis clearly shows that further improvements in power density of the components hydrogen storage and/or fuel processing and stack are necessary. CO_2 reduction can be realised by a fuel switch to bioliquids or to renewable hydrogen. Switching from JET A-1 to BtL kerosene will be easier to realise than building up a hydrogen infrastructure at each airport. Additionally, it must been noted that in future a minimum aromatic content in jet fuels of about 8% is mandatory.[78] Therefore, BtL fuels must be blended with aromatic compounds or with fossil jet fuels. Higher efficiencies for electricity production are easy to realise for on-ground operation and possible for hydrogen-based fuel cell APUs during a mission.

Table 4.11 Analysis of volume requirements for different fuel cell systems and different fuels based on fossil and renewable primary sources for a long-range and an ultra-long-range mission.

Fuel cell type	PEFC	PEFC	HT-PEFC	HT-PEFC	HT-PEFC	HT-PEFC
Fuel	LH2	GH2	GH2	JET A-1	BtL	BtL
Primary source	Natural gas	Natural gas	Wind energy	Crude oil	Wood	Wood
Electric power (mission), in kW_e	257	257	245	255	255	250
Efficiency (ground)	51.6%	51.6%	46.8%	36.0%	36.0%	39.6%
Efficiency (flight)	50.9%	50.9%	44.0%	35.4%	35.4%	38.1%
$V_{H_2O, \text{ground}}$, in $cL\,kW_e^{-1}\,h^{-1}$	50.0	50.0	50.0	27.0	27.0	27.0
$V_{H_2O, \text{mission}}$, in $cL\,kW_e^{-1}\,h^{-1}$	48.5	48.5	55.0	21.0	21.0	20.0
Long-range mission such as	Frankfurt to New York (9 h flight time, 6209 km, 3353 n.m.)					
Energy APU, in kWh	2987	2987	2880	2975	2975	2931
Water produced/mission, in L	1460	1460	1561	669	669	628
Δ Kerosene/mission, in L	-945	-945	-918	-78	-63	-131
Hydrogen/mission, in L	3007	4210	3318	0	0	0
Fixed space, ΔV_{fix}, in L	440	440	512	816	816	516
$\Delta V_{\text{var,H2O}}$, in L	-1679	-1679	-1795	-769	-769	-722
$\Delta V_{\text{var,JET A-1}}$, in L	-1120	-1120	-1087	-93	-75	-155

$\Delta V_{var,H_2}$, in L	5263	7017	7741	0	0	0
$(\Delta V_{fix} + \Delta V_{var})$; in L	2904	4658	5371	-46	-28	-361
Energy well to el., in kWh	13332	10058	11547	9465	18344	16686
CO_2 reduction, in kg	-37	696	2589	292	2633	2640
Ultra-long-range mission such as New York to Singapore (18 h 45 min flight time, 15 350 km)						
Energy APU, in kWh	5488	5488	5266	5464	5464	5373
Water produced/mission, in L	2689	2689	2883	1347	1347	1242
Δ Kerosene/mission, in L	-1590	-1590	-1533	6	34	-89
Hydrogen/mission, in L	5533	7746	8546	0	0	0
Fixed space, ΔV_{fix}, in L	507	507	512	872	872	516
$\Delta V_{var,H2O}$, in L	-3093	-3093	-3316	-1549	-1549	-1428
$\Delta V_{var,JET\ A-1}$, in L	-1884	-1884	-1817	7	40	-105
$\Delta V_{var,H2}$, in L	9683	12911	14243	0	0	0
$(\Delta V_{fix} + \Delta V_{var})$, in L	5213	8441	9623	-670	-637	-1018
Energy well to el., in kWh	24530	18505	21246	17415	33752	30702
CO_2 reduction, in kg	-464	876	4291	50	4414	4438

For specific values see Table 4.10, $P_{el,max} = 260\,kW$.

In future, a multifunctional APU system based on fuel cell technology must offer an efficiency of at least 40% for on-ground and flight operation. Additional benefits will be gained by water production during the complete mission. This would enable enhanced passenger comfort and a further technical utilisation of the water produced. On condition that the technical targets are reached, a full performance APU with $850\,kW_e$ for a very large aircraft could be envisaged. Calculations such as given in Table 4.11 lead to an advantage of 3200 L for an HT-PEFC/BtL configuration.

The power density of the fuel cell-based APU system with stacks, fuel processor and water recovery is estimated here to be $3\,L\,kW_e^{-1}$ and $3\,kg\,kW_e^{-1}$ for an HT-PEFC/ATR system design and $2\,L\,kW_e^{-1}$ and $2\,kg\,kW_e^{-1}$ for a PEFC/H_2 and for a further improved HT-PEFC/ATR system. The specific power density of SOFC systems from other studies ranges between $2\,kg\,kW_e^{-1}$ and $3\,kg\,kW_e^{-1}$.[13–15a] Concerning the power density of the entire system, the break-even point amounts to $6\,L\,kW_e^{-1}$ for an ultra-long-range mission (15 000 km) and $3.5\,L\,kW_e^{-1}$ for a long-range mission (6200 km). These numbers are in good agreement with the $4\,kg\,kW_e^{-1}$ published by Dollmeyer *et al.*[23] and $5.5\,kg\,kW_e^{-1}$ for a long-range aircraft.[17] The benefits of the fuel cell system studied by Pasel *et al.*[41] and Peters *et al.*[44] result from water production, while the advantages of the SOFC system analysis[13–16] were based on better efficiencies and lower fuel consumption. Therefore, Gummalla *et al.*[18] found a break-even point at a power density of $14\,kg\,kW_e^{-1}$ for a short-range aircraft. As stated in this chapter, and also in the respective studies,[13–14] these calculations must be analysed critically because input parameters were chosen from an optimistic point of view resulting in high system efficiencies.

According to the figures given in Table 4.11 a PEFC/hydrogen configuration with $1.5\,L\,kW_e^{-1}$ and $1.5\,kg\,kW_e^{-1}$ for $260\,kW_e$ again including a water recovery system must reach $2.4\,kWh\,L_{Tank}^{-1}$ and $2.5\,kWh\,kg_{Tank}^{-1}$ to equal the break-even point for the ultra-long-range mission. The index 'tank' summarises the volume and mass for the storage compartment and the fuel. The values for the specific volume and the power density for liquid hydrogen storage of state of the art technology are $1.75\,L_{Tank}\,L_{Fuel}^{-1}$ and $1.94\,kWh\,L_{Fuel}^{-1}$ and $1.667\,L_{Tank}\,L_{Fuel}^{-1}$ and $1.39\,kWh\,L_{Fuel}^{-1}$ for gaseous storage at 70 MPa, respectively.[53] Due to the thermodynamic properties a realisation of $2.4\,kWh\,L_{Tank}^{-1}$ is never possible.

With regard to the mass specific quantities, the specific energy density of hydrogen is excellent, *i.e.* $33.3\,kWh\,kg_{Fuel}^{-1}$ but the high mass of the hydrogen compartment determines finally the mass difference between conventional and fuel cell architecture. The break-even point from a mass specific analysis can only be reached if the tank fulfills a ratio of $13\,kg_{Tank}\,kg_{Fuel}^{-1}$ for a long-range mission corresponding to 7.7% hydrogen mass/total mass. Present gaseous storage systems require $24\,kg_{Tank}\,kg_{Fuel}^{-1}$, *i.e.* 4% hydrogen mass, and cryogenic storage systems $17\,kg_{Tank}\,kg_{Fuel}^{-1}$, *i.e.* 6% hydrogen mass. Air Liquide has developed a reduced-weight cylindrical tank made of aluminium, which weighs 50% less than the standard steel design and achieves a gravimetric storage density of 13.3% hydrogen corresponding to $7.5\,kg_{Tank}\,kg_{Fuel}^{-1}$.[77] The volume-specific energy density amounts to $5.15\,MJ\,L_{Fuel}^{-1}$ and is only slightly increased in comparison to the values given by Meinert.[53] As part of the StorHy project, a lightweight free-form cryogenic hydrogen storage system using fibre-reinforced materials was developed.[77] The performance data are 18% hydrogen mass, *i.e.* 5.6 $kg_{Tank} kg_{Fuel}^{-1}$ and 21.4 MJ kg_{Tank}^{-1}, and 4.3 MJ L_{Tank}^{-1} indicating a slightly higher volume-specific energy density relative to Table 4.4. The size of an additional hydrogen storage system in an aircraft could be envisaged as approximately 3.5 m^3. Volumes of up to 6 m^3 could be made possible with some changes in aircraft construction. Larger space demands such as described above require more extensive changes in the aircraft architecture. This would be the case for long-range to ultra-long-range missions with a projected LH2 tank size between 6 and 10.5 m^3 for an HT-PEFC system. Additionally, the power demand of very large aircraft is much higher than 260 kW_e. Assuming a maximum electric power of 850 kW_e an LH2 storage system of 35 m^3 is required for an ultra-long-range mission. Analogously, the LH2 storage system for 400 kW_e HT-PEFC during a long-range mission can be approximated by 10 m^3. The corresponding hydrogen masses amount to 1164 kg and 308 kg, respectively. It should be noted that recent results from StorHy[77] have been obtained for liquid storage of 10 kg. In regard to the examples cited above, it is necessary to scale-up the lightweight free-form cryogenic hydrogen storage system demonstrator by a factor between 30 and 120.

Besides the great effort required for hydrogen storage for long-range missions, hydrogen-based fuel cell systems will release higher amounts of CO_2 in comparison to conventional technologies, *i.e.* 20% for an HT-PEFC during an ultra-long-range mission and 10% during a long-range mission. This effect is caused by the less efficient process route with natural gas reforming and liquefaction. This drawback will be eliminated in the long term if renewable hydrogen becomes available. The application of kerosene-based fuel cell APUs for long-range missions could be preferable for several reasons: a lower volume for the complete multifunctional system, usage of the existing infrastructure, a short- to medium-term CO_2 reduction and a gradual switch to biofuels. This could be more in line with a more general use of biofuels for jet engines.

Another chain of considerations could relate to short-range missions. As mentioned above, short-range missions require low system mass. Table 4.12 shows, basic calculations for three different routes with flight times between 1 h and 3 h. Seven different system designs were compared for HT-PEFC in the 260 kW_e power class according to the data given in Tables 4.10 and 4.11.

The first four rows show the results for hydrogen as fuel and a switch to renewable hydrogen. In addition, the hydrogen storage system is improved from row 1 to rows 2 and 3 by choosing the mass-specific data from StorHy[77] instead of those of conventional systems,[53] *i.e.* 5.6 $kg_{Tank} kg_{Fuel}^{-1}$ and 17 kg_{Tank} kg_{Fuel}^{-1}. Row 4 considers the gaseous storage of hydrogen. The calculations in rows 5 to 7 are based on autothermal reforming of Jet A-1 and BtL. Additionally, rows 3, 4 and 7 estimate an improved power density of HT-PEFC systems of about 1 $kW_e L^{-1}$ and 1 $kW_e kg^{-1}$ instead of 0.5 $kW_e L^{-1}$ and 0.5 $kW_e kg^{-1}$ for rows 1, 2, 5 and 6. Focussing on the mass analysis, only an improved stack design for future lightweight LH2 storage systems would lead to advantages for very short mission duration. The mass benefits can be extended to 200–450 kg for lightweight LH2 storage systems on a 3 h mission. Besides higher volume requirements, conventional LH2 and GH2 storage systems also demand higher mass in comparison to reformer systems showing an increasing discrepancy with extended mission duration. As mentioned above, reformer systems with system efficiencies of about 40% can in future only offer improved system performance on long-range missions by means of water production. Considering CO_2 release, reformer systems show an improvement of 18.5–27.5% while hydrogen-based systems with natural gas as the primary energy source offer an advantage of 11% for very short missions and a CO_2-neutral balance for a flight time of 3 h. In general, efficiencies for the complete chain from well to APUs based on renewable fuels are rather low, *i.e.* 15–18%. In comparison, efficiencies of conventional electricity generation for mixed ground and flight operation amount to 23–26%. The best values from these predictions of 27.5–30% can be obtained for JET A-1 reforming. The calculated figures are not corrected for increased mass of the fuel cell system and subsequently higher fuel consumption, mainly during take-off.

4.5.2   Focus on Maritime Systems

The system evaluation for maritime systems considers two cases:

- A megayacht APU with 500 kW$_e$ operated for 3000 h per year
- A container vessel APU with 3×500 kW$_e$ operated for 54 days per year in port.

The annual electrical energy for both systems amounts to 1500–2000 MW$_e$ h. Different system configurations have already been discussed in section 4.4.4.2. The most important reasons for introducing fuel cell technology in maritime applications cannot be properly evaluated by a technically oriented system analysis. The advantages of the fuel cell are based on comfort-related items and economics (see section 4.3.2). Passengers on a megayacht expect a system with low noise, no vibrations, no odour nuisance and no pollution. Energy must be produced by simple and reliable devices which offer low emissions mainly to guarantee the desired comfort.

Different fuel cell systems in combination with various energy carriers were considered for a possible application as megayacht APU. Table 4.13 shows the results regarding efficiency of the fuel cell system and the energy chain from well, water production rate and CO_2 release from well. An analogous analysis was presented in the case study by Aachen Universtiy,[38] but CO_2 release was only calculated from tank to electricity production. Especially efficiencies for HT-PEFC systems with methanol as energy carrier were set rather low, *i.e.* 25–35%.[38] Emonts *et al.*[68] presented values of 45% for an HT-PEFC coupled to a methanol steam reformer. Wichmann *et al.*[79] published efficiencies of about 38.5% for a small system of 30 W$_e$ for a golf caddy.

Table 4.12 Analysis of mass requirements for HT-PEFC systems based on hydrogen (LH2, GH2), JET A-1 and BtL analysing four short-range missions.

Fuel Primary source Storage technology	LH2 Natural gas Ref. 53	LH2 Natural gas Ref. 77	LH2 Wood Ref. 77	GH2 Wind energy Ref. 53	JET A-1 Crude oil Ref. 53	BtL Wood Ref. 53	BtL Wood Ref. 53
Hamburg to Munich	5 times per day, 75 min flight time, 613 km, 700 kWh max.						
$(\Delta V_{\text{fix}} + \Delta V_{\text{var}})$, in L	1161	1068	1040	1577	558	563	200
$(\Delta m_{\text{fix}} + \Delta m_{\text{var}})$, in kg	586	73	−188	630	563	567	206
Energy well to el., in %	19.4	19.4	17.7	24.7	30.2	15.6	17.8
CO_2 reduction, in %	11.1	11.1	95.0	93.9	27.5	95.8	96.2
San Francisco to Phoenix	4 times per day, 120 min flight time, 1052 km, 900 kWh max.						
$(\Delta V_{\text{fix}} + \Delta V_{\text{var}})$, in L	1365	1243	981	1685	517	523	152
$(\Delta m_{\text{fix}} + \Delta m_{\text{var}})$, in kg	637	−35	−297	774	521	526	157
Energy well to el., in %	19.1	19.1	17.5	24.4	29.9	15.4	17.7
CO_2 reduction, in %	4.7	4.7	94.6	93.5	22.2	95.5	95.9
Riyadh to Cairo	3 times per day, 180 min flight time, 1640 km, 1200 kWh max.						
$(\Delta V_{\text{fix}} + \Delta V_{\text{var}})$, in L	1752	1587	1520	2469	453	460	77
$(\Delta m_{\text{fix}} + \Delta m_{\text{var}})$, in kg	717	−189	−452	992	457	464	83
Energy well to el., in %	19.0	19.0	17.3	24.2	29.7	15.3	17.7
CO_2 reduction, in %	0.2	0.2	94.4	93.2	18.5	95.3	95.7

For specific values see Tables 4.10 and 4.11, $P_{\text{el,max}} = 260\,\text{kW}$.

Table 4.13 Evaluation criteria for different fuel cell systems based on various energy carriers for maritime APUs.

Fuel cell type	Reference	Fuel	Efficiency from tank (%)	Δ fuel (L h^{-1})	Water production (m^3 day^{-1} @ cL kW$_e$$^{-1}$ h^{-1})	CO$_2$ release (kg h^{-1})	Efficiency from well (%)
HT-PEFC	41,44	Diesel	36.0	+ 3.3 diesel	1.1 @ 27	453	31.1
HT-PEFC	–	MGO	35.0	+ 1.8 MGO	1.1 @ 27	400	30.7
HT-PEFC	–	BtL	39.6	−140 diesel; + 130 Btl	1.1 @ 27	22	18.3
HT-PEFC	68	MeOH	45.0	−140 diesel; + 253 MeOH	1.9 @ 46	372	28.0
SOFC	44	Diesel	38.4	−5.5 diesel	0.28 @ 6.7	425	33.1
SOFC	38	MeOH	55.0	−140 diesel; + 207 MeOH	–	304	34.2
SOFC	–	LNG	50.0	−140 diesel; + 189 LNG	–	294	39.7

Input data: $P_{el,max} = 500$ kW; operating time: 3000 h year^{-1}; $\xi_{ICE} = 36.9\%$.
ICE, internal combustion engine; here a diesel generator.

Hydrogen usage is much lower for such a system, *i.e.* 73% relative to 83% for a 10 kW$_e$ system.[68]

The highest efficiencies were achieved by SOFCs with fuel cell efficiencies between 50% and 55% for methanol and LNG as energy carriers. The lowest efficiency from well results for the use of BtL diesel, which is caused by the poorest energy efficiency of fuel production, *i.e.* lower than 50%. As an advantage, CO_2 release is extremely low because CO_2 from combustion of biofuels is not included, so these figures are only determined by CO_2 release during fuel production. Also a fuel switch to methanol and LNG decreases CO_2 release slightly. The main drawback of methanol and LNG is the lower power density of the fuel relative to diesel and BtL. Due to improved system efficiencies, the storage effort is only increased by about 48% for the case of methanol and 35% for LNG. The fuel storage of the megayacht analysed amounts to 234 455 L diesel.[38] Refilling time is of no significance if only pure APU operation with a 500 kW$_e$ generator, *i.e.* 1150 L diesel day^{-1}, is assumed.

Considering water production rates, the HT-PEFC system based on methanol offers the most promising data. A 500 kW$_e$ system could produce as a maximum potential of nearly 2 m^3 water day^{-1}. German citizens require 112 L day^{-1}, which is a low figure in relation to other European countries.[80] For the 36 passengers of a megayacht a water demand of about 4.4 m^3 day^{-1} can be calculated.[1] A megayacht offers a water capacity of 110 m^3.[38] Based on these figures, the tanks would have to be refilled every 25 days without additional water production from the fuel cell system and every 46 days with water from a fuel cell system. A cruise ferry produces 3300 kg h^{-1} demi water by osmotic membrane devices for 2800 crew members and passengers.[31] Maximum APU power amounts to 4.3 MW$_e$. Ro-Ro and Ro-Pax ships produce 1250 kg h^{-1} demi water and have an APU power of 2.9–5 MW$_e$. Those figures show a much lower water availability of 28 L day^{-1} per passenger for 2800 people. A 500 kW$_e$ system produces 3.2–5.5 m^3 water day^{-1} resulting in 1–2 L day^{-1} per passenger. Even a full-sized system with 4300 kW$_e$ reached 28–47 m^3 water day^{-1}, *i.e.* 10–17 L day^{-1} per passenger. As can be seen from Table 4.13 water production by fuel cell systems cannot fulfill the demands of maritime applications completely, *i.e.* 25–77 cL kW$_e^{-1}$ h^{-1} water related to the APU power of a conventional system architecture with separated systems for water and electricity production. The allocation of stored amounts of water for aircraft passengers results in specific values between 1 and 3 L per passenger and flight. HT-PEFC systems with diesel as fuel could produce only one third of the maximum amount of water. An SOFC system based on diesel would only achieve 6.7 cL kW$_e^{-1}$ h^{-1} leading to only one quarter of the amount relative to the HT-PEFC system. Calculations with methanol were not performed. MCFC system design calculations offer a plus of 28 kg h^{-1} for a 500 kW$_e$ system,[33] which is a comparable figure to 33.5 kg h^{-1} for SOFC. The latest results of Bensaid *et al.*[31] considered a CHP configuration of MCFC systems. Unfortunately, the water balance was not closed and a demand for 77 kg water h^{-1} appeared, which is 2% of the maximum capacity of the osmosis installation.

Finally, efficiency, water production, storage effort and CO_2 release for the complete chain, *i.e.* well to end consumer, give no clear recommendation for one process route. A cost analysis must be performed with respect to economic aspects. The choice of fuel is difficult to decide. Regarding the quality of maritime fuels, MGO would be most suitable, but the issue is still challenging. An additional tank must be installed for large systems needed by ocean-going vessels. Methanol must be evaluated under specific safety aspects for the maritime sector. LNG could be an additional option because as the main constituent methane is easier to convert to hydrogen and suitable process technology is already available.

4.6 Conclusions

The demand for electricity in mobile applications increases in nearly all future prospects. The reasons for such a development are electric devices for more comfort and a guaranteed energy supply during idling mode. Especially for the avionics sector the 'more electric aircraft' (MEA) has been envisaged. Today combustion engines and turbo jet engines are applied as auxiliary power units (APUs) on board trucks and airplanes. In this context, fuel cells are considered as an environmental friendly and highly efficient energy conversion system for future systems. For logistical reasons, APUs usually have to operate on the same fuel as the main engine. This will be kerosene or JET A-1 for airplanes and diesel for trucks and ships. As an alternative option, a fuel switch to hydrogen or biofuels has been analysed. The present contribution has shown the requirements of different applications and an overview of current developments.

Fuel cell technology offers the opportunity to merge the individual tasks of electricity generation, water production, tank inerting and supply of heated media in one system. The dimensions of conventional systems like generators and batteries can be reduced. Others like the conventional turbine-based APU of an aircraft can be omitted if the development of such a system has been finished. Today the amount of water which must be transported is limited due to weight restrictions. More water would increase passenger comfort and could also improve the technical performance of the main engines. New systems like tank inerting must be designed for future aircraft whereas for the time being there is no unique solution. These arguments lead to the idea of a multi-functional fuel cell system.

If different fuel cell technologies are compared the advantages and drawbacks must be balanced. The starting point for such an evaluation of a new system technology is the reference technology, which usually represents the conventional technologies at the present time. Developments of both technologies must be taken into account. Established technologies require a higher effort for a certain level of improvement than a new technology. New technologies must be evaluated in terms of present status and future potential.

As described in this paper, SOFC-based APUs are not applicable for multifunctional usage in aircraft. Due to the high air ratios on the cathode side and the residual oxygen content of nearly 18%, tank inerting is not possible. Additionally, material stresses by temperature difference, shock and vibrations are challenging issues for SOFC. Several studies[13-16] offered a high potential for reducing fuel consumption in hybrid systems consisting of SOFC and gas turbines. These calculations implied a cell voltage of >800 mV – also for reformate operation, pressurised and tight stacks up to 3 bar – and thereby reduced air ratios from 4 to 5 on the cathode side. First figures show the potential for efficiency under optimum conditions between 53% and 65%, which will never occur in the first step. Results under more realistic conditions for an intermediate development stage offer an efficiency of about 40% (see Table 4.9).

HT-PEFCs at operating temperatures of 160 °C offer a simplified system design compared to PEFCs if fuel processing is required. The advantage of HT-PEFC systems is the option for multifunctional usage in aircraft. An efficiency of about 38.5% (see also Table 4.9) shows no improvement in relation to commercial electricity generation during an aircraft mission. Water production and tank inerting are important tasks for an optional market introduction of fuel cell technology by the avionics industry. Also for a future development stage, increased efficiency must be facilitated by the introduction of compressor/expander units. Peters *et al.*[67] reported on Monte Carlo simulations varying decisive parameters of both SOFCs and HT-PEFCs between minimum and maximum values. Efficiencies must be improved by an optimisation of the different system designs. Considering, for example, the degree of heat recuperation for SOFC and HT-PEFC systems, about 95% must be recovered for the former cell type and 25% for the latter, respectively. The chance to realise system efficiencies between 33% and 37% was assessed to be higher for HT-PEFCs than SOFCs.[67] Contestabile[3] used a simulation model for market penetration and concluded that PEFC will prevail over SOFC if PEFC reformers reach the cost target of 150 US\$ kW_e^{-1} for long-haul heavy-duty trucks. Such considerations have led to an increased interest in HT-PEFC systems.

Fuel cell systems based on PEFC at operating temperatures of 80 °C are well established as an option for future automotive propulsion systems. Such systems use hydrogen as the energy carrier. Hydrocarbon fuel processing and especially the demands of PEFCs on system design are regarded as too complex. Fuel cell system concepts based on an on-board hydrogen storage system may display quite a simple architecture, but may have some difficulty entering the market due to non-acceptance by aircraft and airport operators. A system analysis, see section 4.5.1.2, has shown that a clear benefit can only be achieved with greatly improved hydrogen storage systems. An increase in system efficiency of about $\Delta\eta = 30\%$ for on-ground and $\Delta\eta = 7\text{-}10\%$ for flight operation of the APU leads to lower fuel consumption. Water production rates are nearly doubled in relation to systems with fuel processing although a high amount of condensed water must be used for moistening the NAFION® membranes used. The required humidification for PEFC is the major drawback in terms of mass and volume and especially for operation under subatmospheric conditions related to HT-PEFC. Taking advantages and drawbacks into account, a break-even point in terms of volume and mass can be calculated for different flight missions. The specific tank data for long-range to ultra-long-range missions differ only slightly and amount to $2.4\,kWh\,L_{Tank}^{-1}$ and $2.5\,kWh\,kg_{Tank}^{-1}$. For short-range missions, the values vary between 3.5 and $5.7\,kWh\,kg_{Tank}^{-1}$, *i.e.* 12.5–$20.5\,MJ\,kg_{Tank}^{-1}$, corresponding to a flight time of between 1 h and 3 h. Volumetric targets cannot be realised due to the fact that the energy density of the liquid hydrogen alone amounts to $1.94\,kWh\,L_{Fuel}^{-1}$ (see Table 4.4, $7\,MJ\,L_{Fuel}^{-1}$) and a compartment for the liquid fuel requires additional space. The mass-related targets can be achieved for a specific weight of the complete storage system including fuel of about 12–$13.5\,kg_{Tank}\,kg_{Fuel}^{-1}$ for long-range missions and 6–$11\,kg_{Tank}\,kg_{Fuel}^{-1}$ for a 2 h short-range mission depending on stack power density, *i.e.* in the range of 1 to $0.5\,kW_e\,kg^{-1}$. Obviously, hydrogen storage systems applied in automotive applications cannot be used for avionics systems. These conventional systems achieved only $17\,kg_{Tank}\,kg_{Fuel}^{-1}$ for LH2 and $24\,kg_{Tank}\,kg_{Fuel}^{-1}$ for GH2. Within the integrated project StorHy, funded by the European Commission,[77] a lightweight free-form cryogenic hydrogen storage system using fibre-reinforced materials was developed offering specific values of $5.6\,kg_{Tank}\,kg_{Fuel}^{-1}$, $21.4\,MJ\,kg_{Tank}^{-1}$ and $4.3\,MJ\,L_{Tank}^{-1}$. In regard to the latter application, it is necessary to scale up this technology by a factor between 30 and 120. Considering present stack sizes of $5\,kW_e$ a scale-up by a factor of 100 is required. The advantage of fuel cell stacks is their modular design and their simpler scalability. Nevertheless, intermediate development steps supported by present computational fluid dynamic (CFD) calculations must be investigated.

The application of kerosene-based fuel cell APUs for long-range missions could be preferable due to the lower volume required for the complete multifunctional system, the usage of the existing infrastructure, a short- to medium-term CO_2 reduction and a gradual switch to biofuels. Short-range missions require a low system mass which lead to the preference of hydrogen-based HT-PEFCs. Therefore, fuel cell systems with liquid hydrogen on board an aircraft are important demonstration platforms for this technology.

A CO_2 reduction for avionics applications cannot be realised if natural gas is used for hydrogen production. Due to mass restrictions hydrogen must be stored in liquid form. In the long term, hydrogen must be produced by renewable primary energy sources. In particular, renewable LH2 will be produced in large facilities preferably from synthesis gas from wood gasification.[52] Finally, BtL reforming and LH2 from wood gasification both lead to rather low system efficiencies from well, *i.e.* 17–18%. It must be considered that low efficiencies are coupled to a high technical effort and ultimately lead to high fuel costs.

Maritime applications of fuel cell-based APUs are focused on MCFC technology[48,50] due to the lack of an alternative stack technology in the $100\,kW_e$ power class. Sail and leisure boats can be operated with DMFCs[36] or PEFCs and LPG as the energy carrier.[29] HT-PEFC- and SOFC-based APUs could possibly play a role in the long term. As previously described concerning the development of SOFCs and HT-PEFCs for maritime APUs, a certain activity must be set in motion to generate such progress. Considering the space requirements of MCFC technology, it is reasonable to place a 40 foot container with contents of about $77\,m^3$ on a container ship transporting more than 10 000 other containers from one harbour to the next. CFC Solutions[81] already has a stationary system design in the power range 250–320 kW_e with $67\,m^3$. A successful demonstration could accelerate market introduction for ocean-going vessels.[48] Small distributed power systems on board cruise vessels are subject to space requirements, which may offer an opportunity of applying other fuel cell types.

Acknowledgement

Part of the work was funded by the Ministry of Economy and Technology within the National Aerospace Research Programme (ELBASYS project). The authors thank Th. Grube, R.C. Samsun, S. Göll and C. Döll for their assistance in process analytic simulations. J. Pasel and B. Emonts are thanked for valuable discusssions concerning fuel cells and hydrogen technologies.

References

1. N. Lutsey, C. J. Brodrick and T. Lipman, *Energy*, 2007, **32**(12), 2428–2438.
2. S. Jain, H. Y. Chen and J. Schwank, *J. Power Sources*, 2006, **160**(1), 474–484.
3. M. Contestabile, Analysis of the market for diesel PEM fuel cell auxiliary power units onboard long-haul trucks and of its implications for the large-scale adoption of PEM FC. *Energy Policy*, 2009, doi: 10.1016/J.enpol.2009.03.044.
4. F. Baratto and U. W. Diwekar, *J. Power Sources*, 2005, **139**(1), 188–196.
5. A. Docter, G. Konrad and A. Lamm, *VDI-Berichte*, 2000, No. 1565, 399–411.
6. T. Grube, B. Höhlein and R. Menzer, *Fuel Cells*, 2005, **7**, 128–134.
7. C.B. Diegelmann, PhD thesis, University of Munich, 2008, p. 159,.
8. A. S. Patil, T. G. Dubois, N. Sifer, E. Bostic, K. Gardner, M. Quah and C. Bolton, *J. Power Sources*, 2004, **136**(2), 220–225.
9. G. Sattler, *J. Power Sources*, 1998, **71**(1-2), 144–149.
10. N. Lutsey, C. J. Brodrick, D. Sperling and H. A. Dwyer, *J. Transport. Res. Board*, 2003, **1842**, 118–126.
11. http://www1.eere.energy.gov/hydrogenandfuelcells/fuelcells/systems.html (22 November 2009).
12. S. Sriramulu, K. Isherwood, S. Lasher, C. J. Broderick and N. Lutsey, in *Proceedings of the Fuel Cell Seminar, San Antonio, Texas, 1–5 November 2004*, Courtesy Associates, Washington, 2004, on CD.
13. J. A. Seidel, A. K. Sehra and R. O. Colantonio, *NASA Aeropropulsion Research: Looking Forward, Proceedings of the 15th ISABE, Bangalore, India, 2–7 June 2001*, NASA/TM-2001-211087, National Technical Information Service, Springfield, IL (2001).

14. H. J. Heinrich, *presented at Symposium der Wasserstoffgesellschaft Hamburg*, 18 October 2007, http://www.h2hamburg.de/index.php?page = download (18 November 2009).

15. (a) L. Faleiro, Beyond the More Electric Aircraft, *Aerospace America* 2005, **9**, 35–40; (b) B. Glover, Presented at Aircraft Noise and Emissions Reduction Symposium, 24 May 2005, Monterey, CA, 2005.

16. D. Daggett, J. Freeh, C. Balan and D. Birmingham, in *Proceedings of the Fuel Cell Seminar, 3–7 September 2003, Miami, FL*, Courtesy Associates, Washington, 2003, p. XX (on CD).

17. H. Srinivasan, J. Yamanis, R. Welch, S. Tilyani and L. Hardin, *Solid Oxide Fuel Cell APU Feasibility Study for a Long Range Commercial Aircraft Using UTC ITAPS Approach, vol. 1: Aircraft Propulsion and Subsystem Integration Evaluation*, NASA/CR-2006-214458/VOL1, National Technical Information Service, Springfield, IL, 2006.

18. M. Gummalla, A. Pandy, R. Braun, T. Carriere, J. Yamanis, T. Vanderspurt, L. Hardin and R. Welch, *Fuel Cell Airframe Integration Study for Short-Range Aircraft, vol. 1: Aircraft Propulsion and Subsystems Integration Evaluation*, NASA/CR-2006-214457/VOL1, National Technical Information Service, Springfield, IL, 2006.

19. D. Stolten, P. Biedermannn, L. G. J. de Haart, B. Höhlein and R. Peters, in *Brennstoffzellen in Energiehandbuch*, ed. E. Rebhan, Springer-Verlag, Berlin, 2002, pp. 391–510.

20. M. C. Williams, J. Strakey and W. Sudoval, *J. Power Sources*, 2006, **159**(2), 1241–1247.

21. Department of Energy, *Multi-year Research, Development and Demonstration Plan, Hydrogen, Fuel Cells and Infrastructure Technologies Program*, Revision 2007, http://www.eere.energy.gov/hydrogenandfuelcells/mypp (23 November 2009).

22. A. Mak and J. Meier, *Fuel Cell Auxiliary Power Study, vol. 1: Raser Task Order 5*, NASA/CR-2007-214461/VOL1, National Technical Information Service, Springfield, IL, 2007.

23. J. Dollmeyer, N. Bundschuh and U. B. Carl, *Aerospace Sci. Technol.*, 2006, **10**, 686–694.

24. F. de Bruijn, *Green Chem.*, 2005, **7**, 132–150.

25. C. J. Brodrick, T. E. Lipman, M. Farshchi, N. P. Lutsey, H. A. Dwyer, D. Sperling, S. W. Gouse, D. B. Harris and F. G. King, *J. Transport. Res. Part D*, 2002, **7**, 303–315.

26. Research Bureau of Transportation Statistics, US Department of Transport, Washington, http://www.bts.gov/publications/national_transportation_statistics/2008/ (18 November 2009).

27. H. Lim, *Study of Exhaust Emissions from Idling Heavy-Duty Diesel Trucks and Commercially Available Idle-Reducing Devices*, US Enviromental Protection Agency, EPA420-R-02-025, Office of Air and Radiation, Washington, 2002, p. 12.

28. http://www.eia.doe.gov/cneaf/electricity/epa/epaxlfilees1.pdf (20 November 2009).

29. P. Beckhaus, M. Dokupil, A. Heinzel, S. Souzani and C. Spitta, *J. Power Sources*, 2005, **145**(2), 639–643.
30. G. Sattler, *J. Power Sources*, 2000, **86**(1), 61–67.
31. S. Bensaid, S. Specchia, F. Federici, G. Saracco and V. Specchia, *Int. J. Hydrogen Energy*, 2009, **34**(4), 2026–2042.
32. S. Krummrich, B. Tunistra, G. Kraaij, J. Roes and H. Olgun, *J. Power Sources*, 2006, **160**(1), 500–504.
33. S. Specchia, G. Saracco and V. Specchia, *Int. J. Hydrogen Energy*, 2008, **33**(13), 3393–3401.
34. P. Lebutsch, G. Kraaij and M. Weeda, *Analysis of Opportunities and Synergies in Fuel Cell and Hydrogen Technologies*, publishable report, R2H 4007PU.3, http://195.166.119.215/roads2hycom/ (16 November 2009).
35. T. Aicher, B. Lenz, F. Gschnell, U. Groos, F. Federici, L. Caprile and L. Parodi, *J. Power Sources*, 2006, **154**(2), 503–508.
36. SFC bringt Brennstoffzelle für Segelyachten auf dem Markt, http://www.innovations-report.de/html/berichte/energie_elektrotechnik/bericht-38 537.html (16 January 2010).
37. Z. Karin, N. Leavitt, T. Costa and R. Grijalva, *Marine Fuel Cell Market Analysis*, CG-D-01-00, US Coast Guard Research and Development Center, Washington, 1999.
38. Case study MEGA Yacht APU, RWTH Aachen University, Institut für Kraftfahrzeuge http://www.ika.rwth-aachen.de/r2h/Case_Study:_Mega_Yacht_APU (11 January 2010).
39. J. J. Corbett and J. J. Winebrake, *J. Air Waste Management Assoc.*, 2008, **58**, 538–542.
40. V. Hiebel, in *Proceedings of the European Strategic Research Agenda (SRA), Brussels, 17–18 March 2005*, https//:www.hfeurope.org/uploads/700/812/CELINA.pdf (18 November 2009).
41. J. Pasel, R. C. Samsun, R. Menzer, R. Peters and D. Stolten, in *Proceedings of the Lucerne PEFC Forum, 29 June to 2 July 2009*, ed. F. De Bruijn, Lucerne, European Fuel Cell Forum, Oberrohrdorf, Switzerland.
42. www.euromat-online.de/download_files/Trinkwassertank%20A320.pdf (24 January 2010), p. 150.
43. http://www.flightglobal.com/articles/2008/08/01/226345/emirates-plan-for-recycled-a380-shower-water-scuppered-by.html (24 January 2010), p. 525.
44. R. Peters, J. Latz, J. Pasel, R. C. Samsun and D. Stolten, Abschlußbericht Verbundvorhaben, ELBASYS, Elektrische Basissysteme in einem CFK-Rumpf, Teilprojekt: Brennstoffzellenabgase zur Tankinertisierung, Schriften des Forschungszentrums Jülich, Reihe Energie & Umwelt, Band 46, Forschungszentrum Jülich, Jülich, 2010.
45. Department of Energy, United States of America on-board fuel processing go/no-go decision, 2004, pdf file downloaded @ www.eere.energy.gov/hydrogenandfuelcells/ news_fuel_processor.html (16 November 2009).
46. B. J. Bowers, J. L. Zhao, M. Ruffo, R. Khan, D. Dattatraya, N. Dushman, J. C. Beziat and F. Boudjemaa, *J. Hydrogen Energy*, 2007, **32**(10–11), 1437–1442.

47. F. N. Büchi, G. Paganelli, P. Dietrich, D. Laurent, A. Tsukada, P. Varenne, A. Delfino, R. Kötz, S. A. Freunberger, P.-A. Magne, D. Walser and D. Olsommer, *Fuel Cells*, 2007, **7**(4), 329–335.
48. Hotstrøm, MTU-Report 03/09, pp. 23–27, http://www.mtu-online.com/fileadmin/fm-dam/mtu-global/pdf/mtureport/0903/0903_MTU-Report_Hotstrom.pdf (18 January 2010).
49. Bordstromversorgung für Schiffe – das HotModule lernt schwimmen, BWK 2009, **61**(11), 20–21.
50. K. Klinder, Das Leuchtturmprojekt e4ships, Rostock-Warnemünde (2009), www.now-gmbh.de/uploads/.../e4ships_090701_PK-Praesentation.pdf (18 January 2010).
51. C. J. J. Reijerkerk, Hydrogen filling stations commercialisation, Master's thesis, University of Hertfordshire, 2001, p. 158.
52. Concawe, Well-to-Tank report Version 3.0 (2008).
53. M. Meinert, Wasserstoff Speichertechnologien für Fahrzeuge, presented at Seminar Wasserstoff und Brennstoffzellen im Automobil, Haus der Technik Essen, 11.03.2008, Essen.
54. H. Aatola, M. Larmi, T. Sarjovaara and S. Mikkonen, *Hydrotreated Vegetable Oil (HVO) as a renewable Diesel fuel: Trade-off between NOx, Particulate Emission and Fuel Consumption of a Heavy Duty Engine*, SAE International, 2008.
55. Chevron Corporation, *Alternative Jet Fuels*, Addendum 1 to Aviation Fuels Technical Review (FTR-3/A1) (2006).
56. C. Wilson, Environmental assessments/initiatives (life cycle assessment, air quality measurement; international initiatives), Presentation at the Workshop on Aviation Alternative Fuels 2009, International Civil Aviation Organization, http://www.icao.int/WAAF2009/Documentation.htm (16 June 2010).
57. European Commission, *Quantification of Emissions from Ships Associated with Ship Movements Between Ports in the European Community*, Final report, July 2002, ec.europa.eu/environment/air/pdf/chapter1_ship_emissions.pdf and ec.europa.eu/environment/air/pdf/chapter2_ship_emissions.pdf (16 June 2010).
58. R. Linnaila, Status of Neste oil's biobased NExBTL diesel production for 2007, presented at SYNBIOS, 19.05.2005, Stockholm, Sweden, Ecotraffic ERD AB, TPS - Termiska Processor AB, http://www.ecotraffic.se/synbios/konferans/presentationer/19_maj/automotive/synbios_linnaila_raimo.pdf (16 June 2010).
59. J. Latz, R. Peters, J. Pasel, L. Datsevich and A. Jess, *Chem. Eng. Sci.*, 2008, **64**, 288–293.
60. Cargo shipping and enviromental, http://www.skysails.info/index.php?id = 579&L = 2 (18 January 2010).
61. Prevention of Air Pollution from ships, http://www.imo.org (18 January 2010).
62. http://www.globalsecurity.org/military/systems/ship/systems/diesel-fuel.htm (16 November 2009).

63. J. R. Rostrup-Nielsen, Catalytic steam reforming, in *Catalysis, Science and Technology*, ed. J. R. Anderson and M. Boudart, Springer, Berlin, 1984, p. 129.
64. R. Peters, Fuel processors, in *Handbook of Heterogeneous Catalysis*, ed. G. Ertl, H. Knözinger, F. Schüth and J. Weitkamp, Wiley-VCH, Weinheim, 2008, pp. 3045–3080.
65. M. Sinnet, 787 No-bleed systems: saving fuel and enhancing operational efficiency, *Aero Quarterly*, 2007, **4**, 6–11.
66. The Power Optimised Aircraft Project, http://www.dlr.de/rm/en/desktop-default.aspx/tabid-3837/5985_read-8790/ (1 March 2010).
67. R. Peters, T. Grube, J. Pasel and R. C. Samsun, Einsatzgebiete und technische Vorraussetzungen für Brennstoffzellen in APU-Anwendungen, in *Proceedings 4. Deutscher Wasserstoff Congress 2008*, ed. D. Stolten, B. Emonts and T. Grube, Schriften des Forschungszentrums Jülich, Reihe Energie & Umwelt, vol. 12, pp. 133–148.
68. B. Emonts, J. Pasel, R. Menzer, A. Tschauder and R. Peters, Hydrogen production from methanol and diesel for small fuel cell APUs, presented at the 17th World Hydrogen Energy Conference, 15–19 June 2008, Brisbane, 2008, ICMS, SouthBank, Queensland, Australia.
69. J. B. Hansen, Oxygenates as fuels for SOFC auxiliary power units, presented at the 15th International Symposium on Alcohol Fuels, 9 May 2005, San Diego, CA, 2005, http://www.eri.ucr.edu/JSAFXVCD/JSAFXVAF/OxFSOFC.pdf (16 June 2010).
70. G. Huppmann, MTU's carbonate fuel cell hot module, in *Molten Carbonate Fuel Cells*, ed. K. Sundmacher, A. Kienle, H. J. Pesch, J. F. Berndt and G. Huppmann, Wiley-VCH, Weinheim, 2007, pp. 3–26.
71. S. Campanari, G. Manzolini, A. Beretti and U. Wollrab, *J. Engineering Gas Turbines Power*, 2008, **130**, 021701-1-8.
72. S. Rau, Wasserstoffspeicherung – Technologien, Sicherheit, Kosten, Anwendung, 1. Deutscher Wasserstoff-Energietag am 12–14 November 2002, Essen, Germany, EE Energy Engineers, Essen, 2002.
73. S. Maeda, S. Kikunaga, T. Akoi, S. Nishikawa, S. Yamamoto, J. Akimoto, I. Anzai and T. Ikeda, Development and operational study of kW-class PEFC cogeneration system using kerosene, in *Proceedings of a Fuel Cell Seminar*, San Antonio, Texas, 2004, Courtesy Associates, Washington, 2004, on CD.
74. K. Saito, H. Matsumoto, T. Kisen, O. Takahashi and H. Katsuno, Development of fuel processing technologies for kerosene fuel cell cogeneration system, in *Proceedings of a Fuel Cell Seminar*, San Antonio, Texas, 2004, Courtesy Associates, Washington, 2004, on CD.
75. J. Kallo, P. Schumann, C. Graf and K. A. Friedrich, presented at *Fuel Cell Seminar, Phoenix, Arizona, 2008*, http//www.fuelcellseminar.com/assets/pdf/2008/presentations.aspx (20 November 2009).
76. Klima an Bord von Verkehrsflugzeugen, http://www.lh-regional.com/print/view.php?nprint = yes&nsub = LSY&nart = 1714&nlang = de (16 February 2010).

77. European Commission, *Hydrogen Storage Systems for Automotive Applications*, Publishable final activity report, Project No. 502667, 2008, http://www.storhy.net/pdf/StorHy_FinalPublActivityReport_FV.pdf (16 June 2010).

78. Ministry of Defence, United Kingdom, *Turbine Fuel, Aviation Kerosene Type, Jet A-1, 2008*, http://www.dstan.mod.uk/data/91/091/00000600.pdf (22 February 2010).

79. D. Wichmann, K. Lucka, H. Köhne, A. Klausmann, K. Martin and S. Köhne, Dampfreformierung von Methanol in einem kompakten Wärmeübertrager für eine HT-PEM-Brennstoffzelle mit einer Leistung von 30 W$_{el}$, in *Proceedings 4. Deutscher Wasserstoff Congress 2008*, eds. D. Stolten, B. Emonts, T. Grube, Schriften des Forschungszentrums Jülich, Reihe Energie & Umwelt, vol. 12, pp. 71–80.

80. http://www.destatis.de/jetspeed/portal/cms/Sites/destatis/Internet/DE/Presse/pm/2009/10/ PD09__377__322,templateId = renderPrint.psml (20 January 2010).

81. M. Bischoff, *J. Power Sources*, 2006, **154**, 461–466.

Part 3
Novel Fuels

Introduction

Fuel supply is critical for fuel cell long-term perspectives in view of low- and zero-carbon policies. Although fuel cells for stationary applications are a technology that offers higher efficiencies of electricity production than conventional power producing devices, they generally run on the same fuels (mostly natural gas), in contrast to those for transport applications where hydrogen is the fuel of choice. Since the limitations of fossil resources are well known and the volatility of market prices poses considerable threats for the economies, the options of alternative fuels need to be thoroughly explored. This again is a challenge in technology development. In particular, biomass-derived gases (fermenter or gasification) as well as coal gasification syn-gas carry a large number of compounds with them which might have negative effects on the fuel electrode (anode). Within this context the influence of fuel impurities on the anode is of extreme interest in order to reconcile the necessity for materials development for improved impurity tolerance with the effort needed for gas clean-up, whichever is easier to achieve. On the other hand, the composition of the fuel can lead to carbon deposition on the anode. Finally, carbon itself may be a fuel, although this option is still seen sceptically due to the difficulties of handling solid fuels. Nevertheless, thermodynamically, this fuel may deliver some extremely interesting results.

CHAPTER 5

Going Beyond Hydrogen: Non-hydrogen Fuels, Re-oxidation and Impurity Effects on Solid Oxide Fuel Cell Anodes

MARK CASSIDY,[a] JAN PIETER OUWELTJES[b] AND NICO DEKKER[b]

[a] University of St Andrews, Fife KY16 9ST, UK; [b] Energy Research Centre of the Netherlands (ECN), P.O. Box 1, 1755 ZG Petten, Netherlands

5.1 Introduction

Fuels cells and hydrogen have become almost inseparably linked; indeed the term 'hydrogen fuel cell' has become probably the most popular descriptor of the technology in popular science communications. In part this has been due to the dominance of polymer exchange membrane (PEM) fuel cells in the popular literature due to high levels of interest in this type of fuel cell for transportation applications and secondly that PEMs can only run effectively on hydrogen.[1] However, this link can cause issues both with the perception of fuel cells and where they will fit in future energy scenarios. The link implies that fuel cells will only find widespread application in conjunction with the development of a hydrogen delivery infrastructure. The development of a hydrogen infrastructure is not without issue, as hydrogen itself is not a first choice energy vector. The low density of the gas leads to problems with compression and therefore storage and distribution.[2]

RSC Energy and Environment Series No. 2
Innovations in Fuel Cell Technologies
Edited by Robert Steinberger-Wilckens and Werner Lehnert
© Royal Society of Chemistry 2010
Published by the Royal Society of Chemistry, www.rsc.org

Another issue with hydrogen is that although being the most abundant element in the universe, the molecular form does not naturally occur on Earth and must be manufactured. This is often accomplished by the processing of a higher energy density fuel to release the hydrogen such as the steam reforming of methane (Equation (5.1)).

$$CH_4 + H_2O \leftrightarrow 3H_2 + CO \tag{5.1}$$

Some of the logistical issues of hydrogen can be overcome by combining the reformer into the fuel cell system balance of plant allowing the higher energy density carrier to be utilised as close as possible to the point of use. However, in the case of a PEM system this must be decoupled from the fuel cell itself as the optimal temperature for the steam reforming reaction is 600–700 °C while the PEM operates at around 80 °C. The reformation reaction is also endothermic and requires considerable heat input. Finally, the CO must then be removed from the product stream as this is a significant poison for the Pt-based catalysts in the PEM. All of this reduces the system efficiency while adding to the complexity and cost.[3]

One of the advantages of solid oxide fuel cells (SOFCs) is the higher temperature of operation, currently in the range 500–950 °C, which permits the use of a nickel-based catalyst in the fuel electrode (anode). This allows for better integration of fuel processing and also mitigates the CO poisoning issue (indeed this can be used as a fuel in SOFCs) although this does not solve all of the issues of using more complex fuels. These issues and possible solutions will form the bulk of the discussion in this chapter. However, this better compatibility of SOFCs with existing hydrocarbon fuels, such as natural gas, opens the possibility that this technology could offer the bridge between the existing fuel supply infrastructure and new energy frameworks based on alternative sources of fuel. Furthermore, a fuel flexible SOFC would be able to accept more than one fuel vector so long as the energy carried was in a gaseous form when entering the fuel cell.

The most straightforward of the hydrocarbon fuels is natural gas (CH_4). When used in an SOFC the CH_4 is basically a hydrogen carrier with the gas being reformed to hydrogen and carbon monoxide (as per Equation (5.1)) with both of these capable of undergoing electrochemical oxidation at the triple phase boundary by the reactions detailed in Equations (5.2) and (5.3).

$$H_2 + O^2 \leftrightarrow H_2O + 2e^- \tag{5.2}$$

$$CO + O^{2-} \leftrightarrow CO_2 + 2e^- \tag{5.3}$$

As well as the production of hydrogen *via* the steam reforming reaction further hydrogen can be produced *via* the gas shift reaction of steam and CO (Equation (5.4)).

$$CO + H_2O \leftrightarrow CO_2 + H_2 \tag{5.4}$$

The reformation can either take place in a separate reactor, known as pre-reforming or within the SOFC anode chamber known as internal reforming.[4,5] It is important to note that in the internal reforming much of the CH_4 will most likely undergo the reformation reaction at the outer surfaces of the anode before the products are oxidised at the triple phase boundary. It is not oxidised directly. However, direct utilisation is an important aspect of SOFC research but is often confined to anode systems based on materials other than Ni and the exact reaction mechanisms are still under discussion.[6,7] The conventional Ni-based anode systems are very active for the reforming reaction and it is likely that reformation would take place before any CH_4 could be utilised directly. One important aspect of reformation is the presence of steam in the system. Many metal surfaces, especially those of Ni, are also highly active for the cracking of the hydrocarbon fuel which will result in the formation of solid carbon (coking) (Equation (5.5)). As well as cracking, coking can also occur through the disproportiation of CO (Equation (5.6)) or by the reaction of CO and H_2 (Equation (5.7)). If coking occurs it can have catastrophic effects in terms of blocking gas flow not only in the pores of the anode itself but also in high temperature fuel supply pipe work, as these can often be fabricated from iron and nickel based alloys, leading to rapid performance degradation.

$$CH_4 \leftrightarrow C + 2H_2 \qquad (5.5)$$

$$2CO \leftrightarrow C + CO_2 \qquad (5.6)$$

$$CO + H_2 \leftrightarrow C + H_2O \qquad (5.7)$$

To avoid coking, fuel streams need to contain an excess of steam, usually with a minimum steam to carbon ratio of at least 2.[8] However, this dilutes the fuel concentration and results in reduced open circuit voltages and performance efficiencies of the cell, while at the same time adding system complexity and therefore cost.

As well as the simpler hydrocarbons the use of more complex higher hydrocarbons such as gasoline, alcohols, diesel and other liquid-based logistical fuel is receiving a great deal of attention.[9,10] In all of these cases the fuel must be gasified before it is introduced to the SOFC; often this gasification will break down the hydrocarbon to give a mixture of H_2, CO, CO_2, CH_4 and H_2O, the exact composition will depend on the gasification technique and conditions used.[11] However, these fuels are even less stable than methane towards cracking and can result in coking over a wider range of operational conditions, which in turn puts further constraints on operational flexibility.

So long as they can be gasified, solid fuel sources can also be used in SOFCs. The most important of these is coal, with the application of coal gas in fuel cell plants forming an important part of the clean coal effort in the US.[12] As with the hydrocarbon fuel sources steam must also be present in the coal gas mixture (usually a mixture of CO and H_2) to prevent coking *via* Equations (5.5), (5.6) and (5.7).

As well as the issues surrounding carbon formation, the carbonaceous fuels will also contain various impurities which are detrimental to cell performance. The most significant of these are sulfur-based compounds, either added deliberately (such as odorising agents in natural gas) or occurring naturally (such as various sulfur compounds in coal or diesel). In the higher temperature environments encountered in SOFCs most of these sulfur compounds will have generally reacted to form H_2S by the time they have reached the anode chamber. This has been shown to have a serious deleterious effect on the performance of nickel-based anodes even at very low concentrations (≤ 10 ppm), well below the concentrations found in untreated fuel streams.[13,14] The higher operating temperature SOFCs (900–1000 °C) tend to be more resistant to sulfur poisoning than lower temperatures due to the decomposition of H_2S into H_2 and S at these temperatures. While H_2S can be removed from gas streams through the use of fairly straightforward adsorption beds, this adds another layer fuel processing to a system, as well as further maintenance issues, where H_2S breakthrough can result in significant and long-lasting damage to the cells. This again adds complexity and cost with reduced operational flexibility.

All of the fuels discussed so far are derived from geological sources (oil, coal), which, although they will be more efficiently utilised in a fuel cell so extending their availability, are still a finite source and are not sustainable. However, the SOFC can also accept gaseous fuel streams from biological sources such as fuel crops, output from anaerobic digestion systems, algae, *etc.*[15,16] As with geological fuels these are again based on organic compounds and will have many similar issues relating to coking as already discussed. While many of these fuel sources may not offer a complete panacea to future energy security, having issues such as the displacement of food crops for fuel crops and the energy per hectacre of the latter, the inherent fuel flexibility of the SOFC should allow it to take advantage of multiple fuel vectors. This will also be highly valuable in a situation where the fuel vector may change due to physical availability or price volatility and allowing the most appropriate vector to be utilised at any given time. This may be most significant in smaller remote or distributed energy scenarios and again emphasises the value of the technology as a bridge between evolving fuel supply frameworks.

Any compound that can be oxidised can effectively be used as a fuel; this includes inorganic compounds as well as the more familiar organic materials. Of these inorganic materials probably the most significant is ammonia (NH_3), which can be readily liquefied, easily stored and has a good energy density. Although there are some toxicity issues it is a common industrial chemical so development of safe handling protocols should be feasible. If used in conjunction with an oxide ion conducting SOFC then NO_x formation would result in the anode chamber. Ammonia fuelled systems would most likely be used with a proton conducting oxide such as barium cerate where nitrogen would be the product on the anode and the hydrogen would react with oxygen at the cathode to form steam.[17] However, oxide-based proton conductors are at a far earlier stage of development than oxide ion conductors and so ammonia fuelled systems are still some way from commercial demonstration and deployment.

H_2S was discussed earlier as a serious poison in Ni-based anodes; however, to underline the level of fuel flexibility that can be attained in an SOFC, given the correct anode catalysts this gas has also been used as a fuel, although, to date, the work has been limited to small laboratory demonstrations.[18,19] Due to the high toxicity of this gas it may not find widespread application; however, it may be useful in certain niche applications such as the utilisation of output from sour gas wells where there is a significant proportion of H_2S ($\sim 1\%$) that must normally be removed before SOFC use or in the utilisation of industrial by-products or waste streams in an SOFC. Both of these may provide useful local power for either remote drilling operations or energy intensive chemical plant where the maximum utilisation of all local energy vectors becomes vital to offset the rising costs of importing energy.

In this short introduction the potential advantages and barriers of fuel flexible SOFC anodes have been outlined. The one underlying theme that is central to the realisation of such a device, is a robust anode technology. This must be resistant to coking in carbonaceous fuels, immune to poisoning by fuel impurities yet still provide sufficient electrochemical activity. Such a material would minimise the level of fuel processing required and permit wider operational parameters to be utilised; this in turn would reduce system complexity and therefore cost. In the remainder of this chapter current work into the search for such anodes will be discussed, looking at the limitations of current Ni-based systems, how these can be improved and what other emerging anode materials and technologies are also being explored to improve fuel flexibility.

5.2 Carbonaceous Fuels

5.2.1 Fuel Resources and Processing Options

Carbonaceous fuels may be derived from many different feedstocks of either geological or biological nature. In order to utilise such fuels in a fuel cell system, fuel processing is required. On the one hand, the processing should establish efficient conversion of the fuel in the fuel cell for long periods of time. On the other hand, it may add bulk and cost to fuel cell systems and may alter their response unfavourably. Careful design of the fuel processor with respect to application and economics is therefore required.

Important conversion technologies for SOFC application are steam reforming, partial oxidation, autothermal reforming, plasma reforming, and internal reforming. They all aim for the conversion of organic compounds into a mixture of hydrogen, carbon monoxide, carbon dioxide and steam. Steam reforming involves the catalytic reaction of hydrocarbon fuels with steam in order to liberate fuel-bound and water-bound hydrogen, whereas partial oxidation and autothermal reforming involves the partial combustion of the fuel. Plasma reforming utilises the enhanced reactivity of chemical species in the excited states that are present in plasmas. Internal reforming involves the direct use of organic fuel in the fuel cell stack. Each of these conversion technologies has its specific advantages and disadvantages. A key advantage of catalytic partial oxidation (CPO) is that this reactor can be of very simple design while the relatively low operating temperature, compared to non-catalytic partial oxidation (POX), has the advantage that NO_x emissions are low. However, partial oxidation, by definition, produces less hydrogen per mole of fuel than does steam reforming. The net result is that partial oxidation results in lower fuel processor efficiency than steam reforming. Also, the injected air dilutes the hydrogen product with nitrogen so that CPO typically delivers fuel gas of around 43% to the fuel cell, whereas steam reforming can be around 76%. This means that the fuel cell stacks need to be around 25% larger for the same power output when fuelled by CPO. Similar, but slightly less severe disadvantages exist for autothermal reforming. Set against these disadvantages is the relatively small size and consequently fast start-up and dynamic characteristics of these types of reactor. Internal steam reforming allows for very compact balance-of-plant while the endothermic nature of the conversion reaction provides heat management benefits.[20]

When we look at the availability of organic fuels, it has to be recognised that fossil fuels take the major share. Gasoline and diesel, obtained from crude petroleum, are both complex hydrocarbon mixtures that include polycyclic aromatic and polynuclear naphthenic compounds, and likely require pre-reforming before the fuel stream is fed to the fuel cell system in order to reduce the risk of coking. Logistic fuels such as kerosene have the additional problem of containing relatively high concentrations of sulfur, which brings along large SO_2 emissions to the environment due to the (post-)combustion of the fuel. The most abundant and at the same time the most complex of all fossil fuels is coal. For use in a fuel cell, coal first has to be gasified to produce coal gas. Along with valuable fuel compounds the gasification process may release contaminants such as sulfur, arsene, phosphorus, silicon, volatile metals and halogens. Other compounds that could be utilised as fuels are alcohols, liquid petrol gas (LPG) and plastics. The first two are relatively clean fuels, while the latter could contain halogens.

Besides organic fuels obtained from fossil feedstocks, biogases have received increasing attention lately due to their renewable nature. Biogases are obtained from biomass, a catch-all term for natural organic material associated with living organisms, including terrestrial and marine vegetable matter, together with animal tissue and manure. In view of its high energy content, biomass represents an important source of renewable fuel. Utilisation of biomass in fuel cells can be realised by pyrolysis to produce a gas similar to coal gas, or by anaerobic digestion in digesters to produce a largely carbon dioxide and methane mixture. Gasification of biomass results in a syngas that contains tars (*e.g.* toluene, benzene, naphthalene, up to 2500 ppm) and contaminants (*e.g.* ash, dust, sulfur, alkalis, earth alkalis, halogens, heavy metals, acids, ammonia). Anaerobic digestion is a series of processes in which microorganisms break down biodegradable material in the absence of oxygen. It may also contain contaminants such as hydrogen sulfide and siloxanes.[20]

5.2.2 Conventional Solid Oxide Fuel Cell Anodes

SOFC anode must satisfy five basic requirements:[21]

1. *Catalytic activity.* Oxidation of hydrogen and hydrocarbons begins with a chemisorption and dissociation at the surface of the anode. The anode must facilitate this reaction with whatever fuel is to be used. The dissociative chemisorption needs to be followed by a reaction of the products of the dissociation with oxygen ions from the electrolyte, a step that may involve either transport of products to the electrolyte or of oxygen ions to the products. If a metal catalyst is used, the dissociation products must be transferred to the electrolyte, and the reaction with oxygen ions takes place near a linear triple-phase boundary (TPB) consisting of metal catalyst, oxide electrolyte, and fuel.

2. *Electronic conductivity.* Electrons from the chemical reaction at the anode surface must be transported to the external circuit. Since the electrolyte has a large surface area, a metallic screen current collector is used to reduce the distance that electrons must travel in the anode itself; the electrons are transported long distances to the external circuit by the current collector.

3. *Thermal compatibility.* Since an SOFC is cycled between room temperature and the fuel cell operating temperature, the thermal expansion of the anode must be matched to that of the electrolyte with which it makes chemical contact and, preferably, also to that of the current collector with which is makes physical contact.

4. *Chemical stability.* The anode must be chemically stable at operating temperature not only in the reducing atmosphere at the anode, but also with respect to the electrolyte and the current collector with which it makes contact.

5. *Porosity.* Since the gaseous fuel must reach the TPB of the anode, the anode must be fabricated as a porous structure that retains its physical shape over time under operating conditions and the current collector must not cover the entire anode surface.

Since the anode operates in a reducing atmosphere, metallic catalysts are candidate materials. Comparison of the electrochemical activity of several metals showed that Ni has the highest activity for H_2 reduction[22] and high electronic conductivity while the melting temperature is considerably higher than the SOFC operating temperature. Conventional SOFC anode is, however, made of NiO and yttria-stabilised zirconia (YSZ) or scandia stabilised zirconia (ScSZ) particles, which has four important benefits compared to pure nickel:[21,23,24]

1. The thermal expansion of the cermet may be better matched to that of the YSZ electrolyte.
2. Exposure of the NiO to fuel creates a porous YSZ structure with metallic nickel particles on the surface of the pores. With a proper

fabrication process, a strong YSZ framework contains an internal porous space percolating in three dimensions with contacting Ni particulates partially on the surface of this space also percolating in three dimensions. This configuration introduces a long TPB for the catalytic reaction and provides electronic conduction from the TPB to the current collector.

3. Coarsening of the nickel film at operating temperature that would break the conductive pathways for the electrons is inhibited.[25–27]

4. Ni and YSZ are essentially immiscible in each other and non-reactive over a wide temperature range, which simplifies synthesis of a Ni-YSZ cermet.

5.2.2.1 Impact of Organic Species

Fuel cells operated with hydrocarbon fuel involve the risk of carbon deposition (*cf.* also Equations (5.5) and (5.7)). Two mechanisms for carbon deposition have been reported in literature: (1) carbon deposits formed *via* reactions over the anode catalyst, or (2) pyrolytic carbon formed *via* free-radical gas-phase condensation reactions. Carbon formation on nickel has been studied for many years.[28,29] It involves the formation of adsorbed carbon species due to dehydrogenation reactions, which then give rise to the formation of other carbon species that may deposit on or in the vicinity of the catalyst. Pyrolytic carbon is the result of C–C bond scission at high temperatures.[30–35] This phenomenon originates from free radical gas phase reactions, which are favoured at high temperatures and are supposed to have a high affinity for the formation of higher hydrocarbons. Especially tars, which include a variety of oxygenated aromatics that are formed in the pyrolysis step of a gasification process, are considered problematic for fuel cells.

Although the addition of steam to the fuel may be beneficial for the prevention of catalytic carbon deposits, this is somewhat undesirable in fuel cell operation, as addition of steam to the anode side of the cell reduces the electrical potential according to the Nernst equation. Further, lower steam concentration would reduce thermal gradients across the cell due to gradual internal reforming.[36] In order to know if carbon deposits due to catalytic reactions are preferentially formed, it is very helpful to reveal the carbon deposition in a C–H–O ternary diagram. Once the ratio among these three components is determined, we can know if carbon formation is significant by assuming complete equilibria between the fuel constituents. At 1000 °C, no carbon deposition is expected if the carbon-to-oxygen ratio is less than unity. With decreasing temperature, it can be found that the deposition region expands deeply into the carbon-poor region if we consider thermodynamic equilibrium.[37]

When hydrocarbons adsorb on a Ni catalyst they are dehydrogenated, leading to atomic carbon and hydrogen. Depending on the other surface adatoms, the atomic carbon might react, *e.g.* with oxygen, in order to form a desorbable product or other intermediates. Atomic carbon itself is not stable on the surface and if it does not react with any other ad-species it may either give rise to the formation of carbon clusters (graphitic carbon, encapsulating carbon, carbon whiskers, carbon filaments), or dissolve into the bulk of the nickel particle and cause the nickel particle to swell. Two mechanisms have been proposed for the formation of graphitic carbon. The first is based on a dissolution–precipitation mechanism driven by concentration gradients due to a difference in the activity of carbon dissolved in the catalyst and carbon on the gas–catalyst interface.[29] The second mechanism is based on surface migration of atomic carbon.[37] Graphitic carbon may, in turn, lead to the formation of encapsulating carbon or carbon whiskers. The formation of encapsulating carbon can be ascribed as an advanced state of the graphite formation. At high temperatures, graphitic islands can coalesce to form a super-structure.[38] In this case the term 'encapsulating carbon' defines a state where the catalyst is completely covered by one or more graphite layers.[37] Carbon whiskers are formed when a critical graphite cluster size is reached, which then may grow around the catalyst particle in order to form a step and to cover it partly. This process is continuing on the next step in order to form an additional layer. The collective of these planes is located perpendicular to the step surface on which they are formed and are thus called carbon whisker. The formation of carbon filaments, with their typical hollow nature, is thought to be related to concentration gradients of carbon along the nickel particle due to super saturation on the catalyst–gas interface with respect to graphite saturation.[39] After reaching a sufficiently high carbon concentration in the nickel particle, carbon precipitation at the catalyst–support interface may occur as carbon filaments. This mechanism implies that it may depend on the nickel particle size before a critical carbon concentration is reached and the growth of carbon filaments starts to occur.

Amongst all hydrocarbons that could be fed to an SOFC, methane has been investigated most. The reason lies in the fact that methane is the main fuel constituent of natural gas as well as biogas. The activity of a Ni-YSZ anode for methane steam reforming was found to be high.[40–45] Although for pure methane a steam-to-carbon (S/C) ratio of 1.0 should in theory be enough to prevent carbon deposition, it has been found that in normal practice, the S/C ratio should be as high as 2 to prevent sooting during heating as well. Later it was shown that this anode material is also active for dry methane reforming.[46] The conversion of CO most likely occurs due to the water gas shift reaction,[26] although direct oxidation of CO could not be excluded.[47,48]

The effect of exposing a Ni-YSZ anode to methane with insufficient steam has been reported in several papers. Triantafyllopoulos and Neophytides found that the nature of the carbon species was dependent on the exposure temperature.[49] Below 700 K, carbon species (carbide species, adsorbed carbon species, and adsorbed CH_x species) were observed that were all reactive with H_2 and O_2, while above 700 K graphitic carbon layers were formed that were not reactive with H_2 but only with O_2. He *et al.* found that upon exposure to methane up to 973 K, carbon formulations developed in the form of carbon nanotubes, while at 1073 K all of the carbon appeared to have dissolved into the nickel particles and destroyed the structure of the particles.[50] Nikooyeh *et al.* found that when hydrogen was added to methane, less carbon appeared to have dissolved into the nickel while the amount of carbon filaments increased.[51] A similar effect has been reported when Ni-YSZ was polarised.[52] Alzate-Restrepo and Hill observed that carbon formed under polarised conditions contained hydrogen and was more weakly bound than carbon formed further away from the electrolyte. Visual inspection of the test samples further revealed that the strongly bound carbon is likely dissolved carbon, while the less strongly bound carbon is probably related to carbon filaments.[53] Another way to suppress detrimental carbon formation is the reduction of the SOFC operating temperature. Liu *et al.* found that detrimental amounts of carbon were found in Ni-YSZ exposed to methane of natural gas including 3% steam at 800 °C, while less carbon was formed at 700 °C.[54] This observation is in reasonable agreement with measurements by Finnerty *et al.*, showing that the carbon deposition rate on Ni-YSZ becomes quite low below 650 °C,[55] and those of Lin *et al.*, who found that as the SOFC operating temperature increased, an increasingly large current was required to avoid coking and cell failure.[52]

Exposure of Ni-YSZ to hydrocarbons other than methane under conditions where equilibrium calculations predict carbon may be very different from those observed with methane. Yamaji *et al.* found that stable operation without severe carbon deposition depended on the hydrocarbon type, temperature and fuel utilisation. For methane they reported safe operation of Ni-ScSZ anodes at 700 °C and 50% utilisation, for ethane at 550 °C and 50% utilisation, and for propane at 500 °C and 75% utilisation.[56] Eguchi *et al.* reported that the carbon deposition rate on Ni-YSZ and Ni-ScSZ anodes was significantly larger with propane than with methane at 700 °C. At 600 °C, kinetics were slower and it was found that a Ni-ScSZ anode was less active for carbon formation than Ni-YSZ anodes.[57] Timmermann *et al.* investigated the performance of Ni-YSZ anode between 650 and 850 °C in the presence of acetylene and found that carbon deposition became more problematic at lower operating temperature.[58] Saunders *et al.* observed only small amounts of carbon deposits when a polarised Ni-YSZ anode was injected with methanol or methanoic acid. Interestingly, less carbon was found as the temperature was increased. In contrast to these results, authors observed rapid cell deactivation when the anode was fed with ethanol, ethanoic acid, butanoic acid or *n*-octane. In the case of *n*-octane, little carbon was found at 700 °C while severe coking occurred at 800 °C.[59] Kishimoto *et al.* experienced carbon deposition and considerable damage of the anode microstructure when operating Ni-ScSZ anodes with *n*-dodecane.[60–62] Kim *et al.* operated Ni-YSZ anodes with toluene at 700 °C and experienced severe carbon deposition and cell failure.[63]

In some cases, carbon deposition was observed even when this was not thermodynamically favoured. This may be related to diffusion limitation of the involved hydrocarbon inside the porous anode structure or its high thermal stability. Further, it has been shown by Coll *et al.* that for aromatic compounds the catalytic reactions leading to coke formation are faster than the rates of the reforming and carbon gasification reactions. The order of reactivity was benzene > toluene ≫ anthracene ≫ pyrene > naphthalene.[64]

5.2.2.2 Impact of Inorganic Species

With respect to inorganic fuel impurities, sulfur has always drawn much attention. It is known that H_2S as well as sulfide-based odorants (thiophenes, mercaptans and organic sulfides) are typically contained in commercial gases such as natural gas, liquid petroleum gas, gasoline or diesel, and thus the major impurity in SOFC anode poisoning studies has been sulfur species so far. Little attention has been paid to the poisoning effects by other inorganic impurities, although various kinds of impurities are contained in practical fuels such as coal-derived syngas, biomass-derived syngas or digester gas. Coal syngas is typically a 'dirty' fuel that contains a lot of impurities, and although recent studies have shown the feasibility of fueling SOFC systems with coal syngas,[65–67] the performance and durability of SOFCs may be affected by the presence of the impurity species in the fuel.[68] The impurity species that are present in the solid state are not considered to be problematic since they can be removed *via* filtration to satisfactory levels. Impurities such as Sb, As, Cd, Hg, Pb, Hg, P and Se usually form vapour phases, which poses a potential threat.[69] To take the case of impurities in bio syngas or digester gas, mainly obtained from excrement of livestock or sewage from human life, several ppm of chlorine or siloxanes are often contained as minor constituents.[70] Finally, fuel contaminants may be released by gas tubings, gas valves and glass seals. Here, volatile Si species may pose a threat to the SOFC anode. Below, the available literature on the impact of inorganic species on the SOFC anode is summarised.

5.2.2.2.1 Sulfur. Sulfur is considered to be one of the most detrimental impurities to solid oxide fuel cells.[71] The quantity of sulfur contained in various fuels ranges from a few ppm to over 1000 ppm. Several parts per million of sulfur-containing impurity are added as odorant to natural gas. Commercial LPG contains 10–30 ppm sulfur depending on the supplier. Sulfur contained in typical gasoline varies from 100 ppm to 500 ppm depending on the national regulations, while sulfur levels in biofuel can be over 1000 ppm. Sulfur affects Ni activity in numerous ways. It degrades the catalytic activity of Ni through competitive adsorption with reactant species.[72] It can also react with Ni, forming sulfurous compounds on the surface leading to the blockage of active sites.[73] At critically high sulfur coverage, nickel sulfides are formed as a bulk phase, decreasing the electronic conductivity compared to Ni. For example, Ni_3S_2 has an electronic conductivity of approximately $0.01\,S\,cm^{-1}$ at room temperature.[74]

Although several papers about sulfur poisoning effects on the SOFC electrochemical performance were already published,[75–79] it took until the year 2000 before the impact of sulfur on the electrochemical performance was systematically investigated.[80] In this work of Matsuzaki and Yasuda it was reported that the poisoning effect with 0.05–2 ppm H_2S was dependent on temperature, the equilibrium partial pressure of sulfur and the total sulfur concentration in the fuel. The time needed to achieve a metastable cell voltage after sulfur poisoning was almost independent of the sulfide concentration and was found to increase with decreasing temperature. These authors suggested that the poisoning they observed was most likely related to an adsorption effect. This was later confirmed by Rasmussen and Hagen, who carried out in-plane voltage measurements on an SOFC while sulfur was introduced in the fuel stream and upon recovery, and found that the sulfur poisoning effect was well related to the local gas composition which could only be explained by a fast chemisorption effect.[81] Later, other researchers found that at increased sulfur concentrations irreversible effects were also possible. In this case, the poisoning consisted of at least two stages, *i.e.* a reversible initial voltage drop within a few minutes to a metastable cell voltage, followed by gradual and irreversible voltage degradation, which was sometimes followed by fatal performance loss. Sasaki *et al.* measured irreversible sulfur poisoning at 5 ppm H_2S in H_2 or H_2-CO, depending on temperature and fuel composition.[13] Zha *et al.* carried out tests with fuel including 2 and 50 ppm H_2S and measured a fast drop in cell performance in the first several minutes, followed by a slow but continuous drop in the next 120 h.[14] The sulfur poisoning effect was accumulative and became more severe with increasing H_2S concentration. Another observation, which was also made by Cheng *et al.*,[82] was that the sulfur poisoning effect became smaller with increasing current load, which implied that the poisoning process could be alleviated by oxygen ions passing through the electrolyte that would oxidise adsorbed sulfur by electrochemical oxidation. Haga *et al.* investigated the poisoning effect by sulfur compounds other than H_2S but found that both CH_3SH and COS have the same effect as H_2S.[70] This, however, makes sense, as thermo-chemical calculations indicate that CH_3SH and COS are readily converted to H_2S at SOFC anode conditions.[83]

These experimental results encouraged others to propose the mechanisms responsible for the sulfur poisoning process. The most likely reason behind the dramatic impact of sulfur on the catalytic activity of nickel at low sulfur concentrations appears to be the isosteric heat of adsorption for H_2S on nickel, which value is much higher than the heat of formation for bulk nickel sulfide (Ni_3S_2). In other words, the low sticking effect of H_2S on nickel allows the H_2S to quickly reach the anode/electrolyte interface, where it will adsorb on nickel, hinder the adsorption of other active species and thereby the electrochemical oxidation of hydrogen.[84] Hansen was the first to relate the sulfur coverage of nickel with the impact of sulfur on the electrochemical performance.[85] He did this by using a Temkin isotherm, by which the sulfur coverage was successfully related to the performance drop for Ni-YSZ anodes reported in previous literature. The irreversible performance drop caused by sulfur was less easy to explain. Sasaki *et al.* associated it with the agglomeration of Ni particles.[13] This would be in line with observations from Dong *et al.*[73,86] and Braun *et al.*,[87] who noted the formation of nickel sulfides, which then promote sintering because of their low melting temperature compared to nickel. However, Cheng and Liu suspect, on the basis of *in situ* Raman spectroscopy experiments that these sulfides have been formed during the cooling procedure and hence are not representative for exposure at SOFC operating temperature.[88] There are, however, other possible explanations for irreversible degradation as well. First, it could be that the adsorbed sulfur species could lead to surface reconstruction of the nickel particles.[81,89] Such an effect is expected to be much slower than the surface adsorption process, and may cause a much slower degradation in fuel cell performance by changing the exposed crystal planes to less active ones.[14] Second, it might be that the nature of the oxide that is mixed with nickel to form the anode cermet is important.[13] It is then assumed that oxygen spillover from the oxide to the nickel helps to prevent permanent nickel deactivation. Such an effect would become more pronounced as the oxygen ion conductivity of the oxide increases or when the physical contact between the cermet particles becomes more intimate.

Besides the poisoning effect of sulfur on the electrochemical oxidation of hydrogen, sulfur has an impact on the activity of nickel for direct steam reforming of hydrocarbons. The first indications that the impact on the reforming process was dramatic were reported by Bartholomew *et al.*, who found that sulfur inhibited the ability of the Ni-YSZ anode to reform methane.[90] Similar effects were reported in more recent literature.[91,92] It is important to realise that the impact of sulfur on the electrochemical performance is *only* related to the anode electro-catalytic activity near the anode/electrolyte interface, while *all* Ni surfaces available in the anode microstructure are expected to be involved in the reforming process. In the case of anode supported cell structures, the anode substrate is approximately 50 times thicker than the electrochemically active area, and the substrate layer therefore requires more H_2S for sulfur saturation coverage. It may therefore be observed that the onset of electrochemical degradation is observed before the onset of degradation of the reforming activity.

5.2.2.2.2 Other Non-metals: Phosphorus and Selenium.

Thermo-chemical calculations indicate that phosphorus, a typical coal contaminant, is present in coal syngas in the form of phosphine (PH_3). Under SOFC operating conditions, phosphine is hydrolysed to form HPO_2 or HPO_3 vapour.[93] In contrast to sulfur, the volatile phosphorus species are highly reactive with Ni and YSZ to form phosphides, *e.g.* Ni_3P, Ni_5P_2, Ni_2P, $Ni_3(PO)_4$ and ZrP_2O_7, that may cause degradation of the cell performance.[94-98] Bao *et al.*[99] have combined their experimental observations with those from Zhi *et al.*[98] and concluded that the phosphorus-containing species preferentially segregate into a grain boundary-like phase near the surface of the Ni-YSZ anode, which can break down the interconnectivity of the percolating nickel network which is needed for in-plane electrical conduction due to the formation of nickel phosphates, and/or inhibit the transport of oxygen ions transported through the YSZ network in the anode layer, resulting in performance degradation. Xu *et al.* have made interesting observations on an anode supported cell exposed to 10 ppm PH_3 during 250 h at 800 °C.[100] They found that, as a result of exposure to PH_3, nickel in the anode layer had migrated to the anode surface to become a separate layer and had fused together in pores existing in the anode substrate. Further, they observed that this effect became more pronounced in the presence of an electric field. Altogether, it can be concluded that volatile phosphorus species are detrimental to nickel cermet anodes and that the PH_3 concentration should be cleaned to the sub-ppm level in order to keep the cell degradation rate at an acceptable level.

Information about the poisoning effects of selenium is still limited. Thermochemical calculations indicate that selenium could be present in the forms of H_2Se, AsSe and PbSe in coal syngas.[95] The introduction of 0.5 ppm H_2Se causes the power density to degrade steadily. Increasing the H_2Se concentration to 5 ppm leads to a significant power loss at the initial stage and about 20–25% power loss after 75 h exposure. Further, there are indications that the observed degradation is partly reversible upon re-exposure of the anode to pure H_2.[101] This indicates that part of the degradation is related to surface adsorption at electrochemically active sites in the anode, but more investigation is needed to confirm this.

5.2.2.2.3 Metalloids: Silicon, Arsenic, Antimony. Silicon may be present in the form of $Si(OH)_4$ in coal syngas or be released by silicon containing grease in fuel supply tubing and valves or by glass seals upon exposure to moist gas. Further, it may be present as siloxanes in biomass derived fuels. Haga *et al.* investigated the impact of $[SiO(CH_3)_2]_n$, the major siloxane species in digester gas.[70] It was found that 10 ppm siloxane caused a gradual decrease of the cell performance with time, and at each investigated operational temperature (800, 900 and 1000 °C) the exposure to siloxane resulted in fatal degradation of the cell performance. This degradation is most likely related to formation of silica deposits in the anode, preferentially near the anode surface, that cause mass transport limitation.

Arsenic may be present in the form of AsH_3 (arsine), and to a lesser extent as As_4 or As_3Pb in coal syngas.[102] Trembly *et al.* carried out a long-term test with 0.1 ppm AsH_3 and measured slight deactivation after 800 h testing.[103] Experiments by Krishnan with 10 ppm As_2 caused significant degradation during the first 10 h followed by a period of relatively slower but steady cell performance degradation.[104] Perhaps the most detrimental studies have been carried out by Coyle *et al.*, who found that arsenic strongly interacted with nickel to form Ni_5As_2 above 700 °C and $Ni_{11}As_8$ at 500–600 °C.[105] For anode-supported cells, a loss of electrical connectivity was the principal mode of degradation, whereas electrolyte supported cells, using thinner anodes, failed more quickly due to the lower nickel inventory. Due to the detrimental effect of arsenic, they recommend As concentrations of approximately 10 ppb or less to obtain acceptable rates of fuel cell degradation.

Antimony is yet another compound that may exist in coal syngas in the form of SbO_2H_2.[95] Bao *et al.* have investigated the effect of SbO, which is claimed to be the stable species under fuel cell anode conditions.[99] They found that 8 ppm of SbO in the fuel stream did not cause a significant decline of the cell voltage at 800 and 850 °C after 100 h exposure. This may be related to the low conversion ratio of Sb with Ni to form a NiSb alloy, but more investigation is however needed to get this confirmed.

5.2.2.2.4 Volatile Metals: Zinc, Mercury, Cadmium.

Zinc is expected to form a condensed metal-oxide at increased temperature.[106] Hence, zinc should not exist in significant concentrations in the gas that reached the SOFC anode. Further, it is reported that Zn vapour may be carried over the SOFC anode without deposition. Only if it coexists with HCl, the carry-over effect could be enhanced. Bao *et al.* tested an SOFC in presence of Zn for 270 h, and measured no degradation.[99] This could be explained by the high solubility of Zn in Ni. Therefore, the Zn would only have an impact on the electrical properties of the anode, but not the electrochemical properties.

Mercury could appear in the form of metal vapour in coal syngas due to its high vapour pressure. Krishnan and Krishnan *et al.* reported no significant cell performance degradation of an SOFC exposed to 7 ppm Hg vapour. However, the cell performance declined more rapidly at 800 °C.[104,106] It was thought that HgO could be condensed and deposited on the anode surface, leading to blockage of the fuel supply. However, further work has revealed that Hg shows no tendency to form oxides.[95]

Cadmium could appear in the form of metal vapour in coal syngas as well. Krishnan has tested an SOFC anode in presence of 5 ppm Cd at 800 and 850 °C. At 800 °C no significant degradation was observed. However, at 850 °C 25% cell performance loss was measured over a time period of 120 h.[104] This may be due to the formation of secondary phases since Ni and Cd were found to form compounds at the Ni rich regions at 800 °C,[107] but more experiments will be needed to get this confirmed.

5.2.2.2.5 Halogens: Chlorine.

Chlorine is typically present in the forms of HCl and CH_3Cl in coal syngas. Krishnan *et al.* studied the effect of CH_3Cl and found no significant degradation of the performance after 140 h of exposure to 40 ppm CH_3Cl at 800 °C.[106] However, when the temperature increased to 850 °C, degradation also increased. Trembly *et al.* have investigated the effect of HCl, which has indicated that 20–160 ppm HCl leads to a performance loss of 13–52% at 800 and 900 °C.[103] There seems only weak evidence of performance recovery after HCl has been removed from the gas stream. After testing no chlorine was found in the anode, suggesting that no detrimental reactions took place, and which makes an adsorption effect the most likely process responsible for voltage degradation.

Surprisingly, Haga *et al.* have carried out thermo-chemical calculations which predict that Cl_2 readily reacts with nickel to form nickel chloride, even at concentrations as low as 100 ppb.[70] Sluggish but measurable voltage degradation was observed for H_2 containing 5 ppm Cl_2, whereas fuels containing

100 ppm and 1000 ppm Cl_2 exhibited almost constant degradation, which rate increased with contaminant concentration. Post-test observations suggest that at low Cl_2 concentrations solid $NiCl_2$ is formed, whereas at higher concentrations $NiCl_2$ may also promote sublimation of nickel to gaseous $NiCl_2$ (sublimation temperature 985 °C), and cause deterioration of the metallic network in the anode.

5.2.2.2.6 Synergetic Effects. There are indications that typical coal syngas contaminants may exhibit synergetic effects inside the anode structure that could worsen or alleviate the poisoning effect. Bao *et al.* tested anode supported cells in presence of different coal gas contaminants.[108] During one experiment a cell was first exposed to H_2S, which caused an immediate performance drop and then remained at a lower but constant cell voltage. After superimposition of AsH_3, the power density declined after a certain time period, which became even worse after superimposition of PH_3. During another test, AsH_3 was first introduced, followed by superimposition of CH_3Cl. In contrast to the first experiment, no deactivation was observed after contaminant introduction. From these experiments it can be observed that the presence of H_2S, which poisons the electrochemically active sites at the anode/electrolyte interface, had made the anode more vulnerable to contaminants that readily react with nickel at the anode surface, *i.e.* arsenic and phosphorus. The role of CH_3Cl remains unclear, however.

Besides synergetic effects of different fuel contaminants, there might also be synergetic effects between fuel impurities and impurities that exist in the anode bulk materials. In particular, the formation of glassy phases are considered detrimental as these are highly mobile at SOFC anode conditions and tend to deposit at the anode/electrolyte interface.[54,109–116]

5.2.3 Improved Anodes

5.2.3.1 Suppressed Carbon Deposition

Besides adaptation of the Ni-YSZ or Ni-ScSZ microstructure,[117–119] two strategies for improved carbon tolerance of SOFC anodes have been reported in literature. The first strategy aims for reduced affinity of nickel for carbon adsorption by alloying the nickel or by replacing the support oxide. The second strategy aims for replacing nickel with a catalyst that possesses both activities for hydrocarbon dehydrogenation as well as oxidation of adsorbed carbon due to mixed electronic/ionic behaviour. In order to sustain in-plane electrical conduction within the anode, metal or oxide is added that possesses high electrical conductivity as well as low activity for carbon formation.

5.2.3.1.1 Modified Nickel Cermet Anodes. Alkali and earth alkali metals have been reported to effectively prevent carbon formation on nickel catalyst due to their basicity.[120–125] The reaction kinetics observed suggest that the surface of the nickel catalyst doped with basic metal oxides is abundant in adsorbed H_2O and CO_2, while the surface of undoped nickel is abundant in adsorbed hydrocarbons. Addition of the basic metal oxides thereby makes nickel unfavourable for carbon deposition and, as a result, carbon deposition is strongly mitigated.

The use of alkali metals is considered to be less favourable for SOFC application due to their low melting temperature. Takeguchi *et al.* measured suppressed carbon deposition on Ni-YSZ doped with calcium or strontium.[126] The addition of Mg, however, had an inverse effect, which might be related to improved nickel dispersion due to the presence of Mg. Srinakruang *et al.* reported excellent catalytic activity and anti-coking character for steam gasification of toluene and naphthalene of dolomite supported nickel catalyst.[127]

The addition of transition metals have also been found to suppress carbon deposition. Finnerty *et al.* found that addition of 1 wt% MoO_3 to NiO-YSZ electrode prior to sintering resulted in a fourfold reduction of the quantity of carbon compared to the undoped anode.[55] Temperature programmed reduction (TPR) revealed that the MoO_3 completely reduces to metallic molybdenum upon exposure to reducing atmosphere, indicating that the dopant suppresses carbon formation due to alloying with nickel. Triantafyllopoulos suggests that Mo suppresses the formation of non-reacting adsorbed graphitic layers.[49]

The effect of nickel alloying with copper has been investigated by Kim *et al.*[128] It was found that the amount of carbon that was formed increased with increased copper concentration in the alloy. Sato *et al.* have investigated nickel–cobalt alloys and found that $Ni_{0.5}Co_{0.5}$-YSZ cermet had considerably higher electro-catalytic activity for direct oxidation of methane, which was suggested to be related to a change in the nickel crystallographic orientation.[129]

The use of precious metals in SOFC anodes have been investigated by several authors. Addition of ruthenium to a Ni-YSZ anode resulted in substantial performance improvement in methane, ethane and propane.[130–134] A similar effect has been reported for iridium[135] and rhodium.[136] Takeguchi, however, found that Rh promoted carbon deposition, while Ru and Pt suppressed coking.[126,137] Gavrielatos *et al.* reported positive results for gold at interestingly low concentrations.[138] The positive effect of palladium has been reported by Nabae *et al.*[139] Authors further claim that Ce, Sm, La and Sr may have diffused into the Pd-Ni alloy during their experiments, which may have inhibited carbon dissolution into the bulk, and thereby further suppressing carbon formation at the catalyst surface.

Finally, a theoretical study has shown that tin (Sn) might be a good alloying metal for nickel in order to improve carbon tolerance,[140] but no experimental results have yet been reported.

The use of ceria in SOFC anodes as catalytically active compound has already a long history. Already in 1990, Steele *et al.* have shown that ceria is able to oxidise methane.[141] The suggested mechanism for oxidation of methane and other hydrocarbons is by the Mars–van Krevelen mechanism, with ceria being reduced by the fuel which is then re-oxidised by oxygen ions crossing the electrolyte.[142] This would imply that doping ceria with lower valence cations, which increases the amount of oxygen vacancies in the oxide lattice, would enhance the catalytic activity. Palmqvist *et al.* measured the activity of doped ceria for methane oxidation and found that doping of CeO_2 with Ca, Mn and Nd increased the activity compared to undoped ceria.[143] Ramirez-Cabrera *et al.*, on the other hand, have measured lower catalytic activity for gadolinia-doped ceria than for undoped ceria.[144,145] Similar observations were made by Zhao *et al.*[146]

At high temperatures, ceria can deactivate from sintering or poor redox properties and low resistance to strongly reducing conditions.[147] Besides doping with lower valence cations, it has been reported that partial substitution of Ce^{4+} with Zr^{4+} in the lattice results in solid solution formation which results in better redox properties and sustained electro-catalytic activity.[148-155]

Doped ceria has also been investigated as support oxide for nickel, with a positive effect on carbon deposition. Livermore *et al.* reported that Ni-GDC has the tendency to form less carbon which is less strongly bound and easier to remove than carbon that was previously observed in Ni-YSZ anode.[156] As for Ni-YSZ anodes, the amount of carbon that is deposited seems to depend strongly on the microstructure. Rösch *et al.* measured substantial carbon deposits in Ni-gadolinia-doped ceria (GDC) anodes with 70% nickel in CH_4-3%H_2O at 850 °C.[157] Wang *et al.* studied Ni-samaria-doped ceria (SDC) anodes for direct oxidation of methane and claimed that optimised distribution and connection between Ni-Ni, SDC-SDC and Ni-SDC particles are important factors to obtain high performance.[158] Asamoto *et al.* found that dispersion of 20 wt% Ni in SDC resulted in high performance in dry methane at 700 °C, while 10% and 40% Ni loading resulted in lower performance.[159] Dekker *et al.* reported high conversion of toluene of Ni-GDC anode,[160] which is a considerably better result than reported by Kim *et al.* for Ni-YSZ anode.[63] On the other hand, Dekker measured low kinetics for naphthalene, phenanthrene and pyrene reforming, which implies that these compounds require further adaptation of the anode formulation. Similar observations have been reported by Aravind *et al.*[161] and Shekhawat *et al.*[162]

5.2.3.1.2 Nickel-free Cermet Anodes.

The use of copper, gold and cobalt to replace nickel as electronic conducting compound in SOFC anodes can be found in literature. The Gorte group has reported results on the system Cu-ceria-YSZ.[163-170] Most interestingly, the amount of copper used in the anode microstructure was insufficient to establish a percolating structure. This was done to prevent serious sintering effects of copper during fuel cell operation. In order to obtain electronic conduction, the anodes required activation by carbon deposition prior to operation. The catalytic activity of copper was reported to be low, so the ceria was presumed responsible for the observed reforming reactions. The use of cobalt has been reported by Milt *et al.*[171] and Lee *et al.*[172] This however gave no improvement compared to the nickel cermet with respect to carbon tolerance. The use of gold was investigated by Marina *et al.* as a model compound for GDC anodes in developing a nickel free anode.[173] This revealed that GDC is active for hydrogen oxidation but only has low activity for methane oxidation. Similar observations were obtained by Lu *et al.*[174]

5.2.3.1.3 Nickel-free Oxide Anodes.

Research into oxide based anode systems has grown in recent years and this general development is covered by several fine reviews.[21,175,176] Much of this work has been motivated to finding an alternative material that would mitigate the well-documented issues arising from the volume changes exhibited by Ni-based anodes on redox cycling.[177,178] However, benefits relating to redox stability are not the only benefits that may be gained by successful development of an oxide-based system. There are also potential benefits with regards to fuel flexibility and impurity tolerance and it is these aspects of oxide anodes which will be discussed in this section. As has already been discussed, nickel, although being very active for hydrogen oxidation, is also very active for the cracking of hydrocarbons leading to a risk of coking if the steam-to-carbon ratio in the fuel stream drops below 2. This leads to increased complexity in fuel processing and control; the steam also dilutes the fuel, leading to reduced open circuit voltages and thus lower cell efficiencies.

The surfaces of oxide-based materials show a far lower activity for hydrocarbon cracking and therefore can be considered for the oxidation of dry hydrocarbon fuel streams as the oxidation of the fuel is favoured over cracking. However, for any oxide anode to be effective it must still meet the basic requirements for any anode material:

- Good catalytic activity for oxidation of fuel
- Good electrical conductivity for current distribution and minimising ohmic losses
- Some level of ionic conductivity for triple phase boundary functionality
- Chemical compatibility with adjacent layers
- Process compatibility with cell and stack design.

This is a tough set of requirements for any single material and so far no single phase material has been discovered that meets all of these; in reality it is most likely that any successful system will be a composite of two or more materials. The catalytic activity within many of the ceramic systems is often related to the variation in oxidation state of transition metal cations within the material when exposed to a reducing environment. Common transitions utilised are Mn^{3+}/Mn^{2+}, Ce^{4+}/Ce^{3+} and Ti^{4+}/Ti^{3+}. To be an effective catalyst the activation energy of the transition needs to be low enough to allow transfer of electrons from the fuel adsorbed onto the oxide surface so facilitating oxidation.[21]

One popular candidate material is the perovskite $La_{1-x}Sr_xCr_yMn_{1-y}O_3$ (LSCM). $La_{1-x}Sr_xCrO_3$ has long been studied as an interconnect material for higher temperature SOFC due to its good electrical conductivity and stability in both reducing and oxidising environments.[179] The substitution of Mn for Cr on the B-site introduces the Mn^{3+}/Mn^{2+} redox couple and so catalytic activity for fuel oxidation. Early investigations looked at smaller levels of Mn doping, generally below 20 at% and often with Ni additions.[180] However, subsequent studies showed that significant improvements in anodic behaviour could be attained when the ratio of Cr and Mn was increased to 50:50 with polarisation resistances in the order of $0.3 \, \Omega \, cm^2$ being reported at 900 °C in wet H_2.[181] This B-site ratio is now widely accepted and utilised in many investigations with a typical composition being $La_{0.75}Sr_{0.25}Cr_{0.5}Mn_{0.5}O_3$.[182–184] The Mn^{3+}/Mn^{2+} transition also results in a volume change of roughly 1% on the LSCM when transitioning between oxidising and reducing atmospheres.[185] However, by careful consideration of cell design and electrode microstructure it is possible to mitigate any detrimental effects of this change. Although this material has been shown to be reasonably stable with YSZ, an A-site deficiency of up to 5% may also be introduced to minimise the chance of any side reactions during processing.

LSCM is redox stable and its catalytic activity has been shown for both hydrogen and CH_4, with no susceptibility for coking in the latter under conditions of low humidification. The material exhibits p-type conductivity which decreases with reducing pO_2. This is one significant drawback of this material, while conductivities in air are of the order of $30–40 \, S \, cm^{-1}$, and suitable for electrode functionality. This value rapidly reduces in low pO_2 and may be in the range of $1–3 \, S \, cm^{-1}$ in an anode chamber environment. It was mentioned earlier that good conductivity is an important requirement of an anode material and while this is true it does not paint the complete picture. What is more important is that the ohmic drop across the anode layer is kept within acceptable limits, this is generally accepted as $< 0.1 \, \Omega \, cm^2$.[175] Taking this into account if the anode layer is kept thin, say $5–10 \, \mu m$, then acceptable losses may be attained with materials exhibiting less than optimal conductivity. However a major prerequisite to this approach is the presence of a highly conducting anode current collecting (ACC) layer in contact with the anode to ensure adequate current distribution from the interconnect contact points.[186] This strategy has been successfully demonstrated in the recent integration of an LSCM anode in the industrially relevant Rolls–Royce IP-SOFC design and shows the importance of considering cell design and geometry as a factor in materials selection.[187]

In most cases the LSCM is not used as a single phase but as a composite with GDC, this has been shown to improve the electrode performance by introducing further ionic conductivity and increasing triple phase boundary length, with polarisation resistances at 800 °C being reported in the region of $0.12 \, \Omega \, cm^2$ and $0.44 \, \Omega \, cm^2$ for wet hydrogen and wet methane, respectively.[188] Furthermore, as has already been discussed, GDC itself is also recognised as an oxidation catalyst for hydrocarbon fuels and this will also have a beneficial effect on the performance of the system. Further performance improvements have been realised by the addition of small amounts ($> 5\%$) of catalytic material, such as Pd, Rh and Ni which can improve the catalysis of the electrode without leading to coking or redox problems.[180,183,189,190] An additional advantage of the p-type conduction in the LSCM is that under conditions of high fuel utilisations where there is an increased pO_2 due to the presence of large quantities of water, the conductivity of the LSCM will improve. This may allow operation of the cell in more aggressive conditions where Ni-based systems may have been prone to re-oxidation either through anode gas composition, electrochemical conditions or a combination of both. LSCM has also been shown to be an effective cathode for SOFC and symmetrical cells with the same material as both anode and cathode have been demonstrated, which has possible advantages in the reduction of the bill of materials for cell manufacturing.[191]

Ceramic systems based on titanate perovskites have also received significant attention. These materials exhibit n-type conductivity with the reduction of the Ti^{4+} to Ti^{3+} providing donor electrons to the conduction band. Further doping with a M^{3+} cation such as La^{3+} or Y^{3+} can further enhance conductivity by the modification of the defect chemistry through the introduction of excess oxygen.[192] Generally based around $SrTiO_3$ with varying amounts of yttrium or lanthanum doping on the A-site,[193] although improvements in conductivity have also been reported when doping the B-site with cations such as niobium,[194,195] early studies focused on $La_xSr_{1-x}TiO_3$ and although high conductivities were reported for the reduced material ($80–360\,S\,cm^{-1}$), catalytic activity was sluggish and a composite structure with ceria was required to improve the fuel oxidation kinetics.[193]

Although described as straightforward cubic perovskites in many studies more recent work has revealed that the actual structure of the material can be rather more complex. Depending on the level of doping it can actually exist as a double perovskite with the general formula $La_4Sr_{n-4}Ti_nO_{6n+1}$, with the value of n playing an important role in defining the defect structure of the material.[196] Below $n = 12$ the material exhibits a layered structure. Long-range crystallographic shears separate oxygen rich crystal planes. However, this extended defect structure is detrimental to conductivity. As the value of n increases so this layered structure begins to break down as the oxygen excess decreases, until at $n = 12$ the layered structure is no longer tenable and randomly distributed point defects are observed within a perovskite framework. Maximum conductivity is observed at $n = 11$, where the layered structure has disintegrated but the point defect concentration is at a maximum. Conductivity slowly decreases as n continues to rise and the defect concentration drops.

All of the titanate systems must be pre-reduced to exhibit their maximum conductivities. Firing in air results in conductivity values in the range of $1–16\,S\,cm^{-1}$. However, sintering in a reducing atmosphere results in conductivities of $80–360\,S\,cm^{-1}$ depending on the system and the reducing conditions. Some of these systems can be slow to reduce and the sintering cycles may involve high temperatures over long time ($1500–1600\,°C$ over $10\,h$) to show the maximum conductivity.[192] Obviously this could create some difficulties in a practical SOFC manufacturing scenario and effective solutions to this issue are still being sought. Although these systems exhibit a high conductivity once reduced, the catalytic activity is still slow. It has been suggested that this may be due to the co-ordination chemistry of the Ti cations and recent research has been exploring B-site doping to improve the coordination flexibility on these sites and so catalytic activity. Promising dopants at present are Mn, Ga and Al.[196] Alternatively the addition of small amounts of Ni catalyst (<5 vol%) by infiltration of nitrates into LST/YSZ composites is also under investigation as a method of improving the catalytic activity of these anode structures and has shown some promising results with anode polarisation resistances of 0.21 cm^2 reported.[197]

Other emerging materials sets are again often based on the oxygen flexibility offered by the double perovskite based structures. One such material is $Sr_2MgMoO_{6-\delta}$. Initial reported properties have been promising with good redox stability, high conductivity and catalytic activity.[198,199] However, as with some of the titanate systems, high process temperatures and long sinter times are required to realise these properties by standard ceramic process routes which then results in non-optimal microstructures. Current work is focused on other process routes to improve on this.[199] Other structures which have also shown interesting properties are niobium titanates such as $Nb_{1-x}Ti_xO_2$. These possess a rutile type structure when reduced and have been shown to exhibit very high n type conductivities (up to a maximum of 300 S cm^{-1} in a pO_2 of 10^{-22}); however, catalytic activity is low and may limit application as an anode.[203] The high conductivity value is interesting and may be of use in the current collecting role; however, another problem with these materials exists in that the coefficient of thermal expansion (CTE) is very much smaller than other cell components being only around 7×10^{-6} K^{-1} and further research into doping or use in composite structures may be useful in mitigating this issue.

As with many of these types of review much of the text is focused on composition. While composition is important one must not forget the vital contribution played by microstructure in determining the performance of an anode. It has already been stated that a number of these oxide systems require very specific processing regimes to exhibit optimal chemical properties. However, these may also result in non-optimal microstructures for fuel cell operation such as particle coarsening and low porosities. While some of the detrimental effects of these can be mitigated in part by the use of composite structures, the need in some cases for pre-reduction can result in serious complications for other cell materials (such as subsequent application of cathodes). Recent research has been investigating ways in which oxide-based materials can be deposited using techniques which result in optimised functional properties but avoid the high temperatures and long times often required. One promising route is the infiltration of solution precursors into an existing porous ceramic skeleton, originally developed to avoid some of the issues encountered when trying to integrate copper-based materials with low melting points into anode structures.[201] The technique has subsequently been extended to deposit high-performance LSCM and ceria-based catalysts into zirconia skeletons using solutions of mixed nitrates.[190] Recent studies have also shown success depositing lanthanum-doped strontium titanate (LST)-based materials and in both the LSCM and LST materials fine structural development (on the nanometre scale) was observed upon reduction at SOFC operating temperatures (800 °C). In both cases initial sintering was carried out in a temperature range normally associated with SOFC fabrication (1100–1300 °C).[202] These types of techniques and approaches offer considerable promise for the control and optimisation of engineered microstructures in future anode systems.

As was mentioned earlier, one of the great advantages promised by oxide-based anodes is the ability to run on dry hydrocarbons and the study of this direct utilisation has been a topic of emerging interest in recent years. The attractiveness in the elimination of the requirement for excess steam required for reforming, and so the fuel dilution and efficiency implications (along with the simplifications in the balance of plant) are obvious. Many of the ceramic systems have already been demonstrated to have a low propensity for coking and some researchers have proposed direct oxidation of the hydrocarbon, however this is subject to some debate and the exact mechanisms through which the hydrocarbon is oxidised is not yet clear.[6,7] A greater understanding of this would be of great value in the design and engineering of materials and microstructures to optimise performance. It is unlikely that the reaction takes place in a single step and several steps of various adsorption and intermediate products are more likely to be involved and variable depending on the hydrocarbon mixture entering the anode chamber. In this context the term direct utilisation is probably better than direct oxidation as the former better reflects the fact that multiple mechanisms are involved. To add further complexity to this system one must remember that one of the significant products of the oxidation is water and in any operating system there will be a considerable amount of steam present in the anode chamber. Therefore some level of reformation may take place on any suitable catalytic surface. This gives further importance to understanding the various steps taking place on the anode surface and how the dominance of these change across differing anode conditions. This will be vital in optimising both microstructure and composition for optimum performance over long time scales.

5.2.3.2   Improved Sulfur Tolerance

As for improved tolerance to carbon deposition, replacement of YSZ with ceria has been shown to be effective in improving the tolerance to sulfur. Ouweltjes *et al.* investigated the impact of sulfur on Ni-GDC cermet anodes in hydrogen or methane–steam at 850 °C.[203] Interestingly, it was found that the electrochemical activity for hydrogen oxidation and the water gas shift reaction were much less affected (tolerance at least 9 ppm H_2S) while the methane reforming reaction was already inhibited at 1 ppm H_2S. This is thought to be related to electrochemical activity of doped ceria, which was already shown by Marina *et al.*[173] Cu-ceria-YSZ anodes have shown to exhibit considerable sulfur tolerance. He *et al.*[204] have investigated this anode in presence of 450 ppm H_2S at 800 °C without substantial deactivation. Results indicate that the impact of sulfur on the anode electrochemical activity and catalytic activity probably follows the same trends as Pd-Fe-ceria and the results reported by Ouweltjes.

As well as providing a surface with a low activity to hydrocarbon cracking, it is hoped that the surfaces of the oxide anode systems may also prove more resistant to sulfur poisoning than those of the Ni-based cermets. Therefore this aspect of the behaviour of the oxide anodes has become an area of growing interest. In general, both of the common systems, LSCM and LST, are not completely immune to degradation in sulfur containing gases. However, the level of degradation observed is very much reduced when compared to Ni-based systems, especially at lower levels of H_2S (< 50 ppm) and show this improved tolerance even when small amounts of other catalytic materials are introduced.[205-207]

The form of the degradation appears as two distinct portions; the first is a rapid drop in performance when H_2S is introduced, followed by a slower time dependant degradation. The magnitude of this first drop is related to the amount of H_2S introduced, being larger in higher concentrations of H_2S. At lower concentrations this initial drop appears to be recoverable although this appears to be less the case as the H_2S concentration increases. The severity of the second part of the degradation is also affected by the H_2S concentration, the performance appearing stable at lower concentrations but decreasing as concentrations increase into the 100 s of ppm level. Of these two systems the literature suggests that LST may be more tolerant to sulfur poisoning than LSCM.[205] With the latter there may be a link to the level of tolerance and the amount of Mn in the system, with tolerance decreasing as the Mn concentration increases with impurities such as MnS and αMnOS being detected.[208] However, what impact these impurities have on the degradation mechanism is still unclear and under investigation.[209] However, these results suggest that the B-site cation plays an important part in determining sulfur tolerance.

Other systems based around vanadium-containing oxide systems have also been investigated. These have shown good sulfur resistance; however, the other anode requirements are not sufficiently good to displace existing anode materials although work continues into these materials and they are of interest for H_2S SOFCs discussed later.[19,210,211] The double perovskite $Sr_2Mg_{1-x}Mn_x MoO_{6-\delta}$ has shown some early promise as a sulfur tolerant anode, but work on this material is still at an early stage; however, as discussed earlier, much work remains to be carried out on this material to optimise both processing and microstructure.[190]

As with many aspects of anode performance, anode morphology and microstructure are sure to play an important part in the sulfur tolerance of an anode structure. However, this has not yet been explored in any great depth. Also, tests of oxide systems to date, although promising, have still only been conducted for relatively short time periods, low 100 s of hours. It remains to be seen how the sulfur tolerance of these materials exhibited over these time scales extends across to tests lasting thousands and tens of thousands of hours. However, the initial results are promising and they are sure to be a high priority for future study and development.

5.3 Other Alternatives to Hydrogen

5.3.1 Ammonia

Ammonia might be a fine substitute for hydrogen and hydrocarbons. Compared to hydrogen, ammonia offers the advantage of easy liquefaction while the volumetric density of liquefied ammonia is higher than that of liquefied hydrogen, which is useful in transport and storage. Compared to carbonaceous liquid fuels ammonia offers a cheap and carbon-free alternative with low flammability and explosivity at a competitive energy density in terms of GJ per m^3.[212,213] Moreover, there already exists a global distribution network for ammonia; it is produced in large quantities by the chemical industry by means of the Haber–Bosch process to serve as a feedstock and for fertiliser production. It is also a significant biogas, mainly due to the hydrolysis of ureum. In fact, the ammonia concentration in biogas obtained from organic waste may be so intense that European legislation limits the amount of ammonia which can be discharged into the atmosphere. Utilisation of ammonia released by organic waste may therefore provide a means as an environmental clean-up device.[214] A drawback of ammonia is that it is toxic, and the use and handling of pure ammonia therefore could bring about some serious safety hazards. A mitigating safety factor in favour of ammonia, however, is that the specific smell, which is detectable far below the maximum allowable concentration, makes it easy to detect in cases of leaks so preventative measures can be taken. Another approach to mitigate ammonia's shortcomings is to complex it with other chemical compounds so that the resulting compound is stable but not toxic. Interesting materials that could offer a safe, reversible, high density and potentially low-cost solution to ammonia are so-called metal ammine salts, which in theory could store 0.67 norm litre ammonia gas in 1 g of salt.[215,216]

The thermodynamic properties of ammonia make it an attractive fuel for fuel cell applications. In 2003, Staniforth and Ormerod were the first to report about the conversion of ammonia in a solid oxide fuel cell with a conventional Ni-YSZ anode. It was found that a Ni-YSZ anode operated at 900 °C was not only tolerant to ammonia but could actually utilise the ammonia to produce electrical power. Experimental results indicated that ammonia was likely to be utilised after catalytic decomposition into hydrogen and nitrogen, and that the hydrogen was then electrochemically oxidised to steam.[214] Furthermore, the authors reported that the NO_x concentration formed by the SOFC was below the detection limit, which is a clear advantage over ammonia utilisation in internal combustion engines.[217] One year later, Dekker and Rietveld showed that long-term operation of conventional SOFCs with pure ammonia is possible, with a degradation rate below 1% per 1000 h. The detected NO_x emissions were lower than 0.5 ppm at temperatures up to 950 °C, while at 1000 °C 4 ppm NO_x was measured.[218] In the race to further reduce NO_x emissions, SOFCs operated at intermediate temperature were also investigated. Ma *et al.* investigated anode supported cells with a samaria-doped ceria electrolyte between 500 and 700 °C and indeed found no detectable NO_x formation.[219] It is important to note that at this operating temperature a pre-reduction of the anode with hydrogen is required because the reduction process of NiO in ammonia is too slow.[220]

Particular interest exists also in the use of SOFCs with a proton-conducting electrolyte. Thermodynamic analysis carried out by Ni *et al.* revealed that SOFC-H is superior to SOFC-O in terms of theoretical maximum efficiency, which becomes more pronounced with increasing fuel utilisation and temperature. This is because, for an SOFC-H, a higher hydrogen partial pressure and a lower steam partial pressure exists at the anode than for an SOFC-O.[221,222] Considering the level of maturity of SOFCs with proton-conducting electrolytes compared to those with ion-conducting electrolytes, promising experimental results have been reported.[223–227]

5.3.2 Hydrogen sulfide

The poisoning effect of hydrogen sulfide in SOFCs has already been discussed in this chapter. However, H_2S is capable of being oxidised and is therefore a fuel in its own right which, with the correct electrode material, could be used in an SOFC system. H_2S is a significant industrial by-product and an important source of elemental sulfur. While several processes can be used to process this gas and extract heat and elemental sulfur, use in an SOFC would arguably be a more efficient way of utilisation. The two possible main reactions of H_2S are:

$$H_2S + O^{2-} \rightarrow H_2O + \tfrac{1}{2}S_2 + 2^{e-} \tag{5.8}$$

$$H_2S + 3O^{2-} \rightarrow H_2O + SO_2 + 6^{e-} \tag{5.9}$$

The reversible potential for these reactions against an oxygen or air electrode are 0.786 V or 0.761 V, respectively, at 727 °C. At higher operating temperatures (> 700 °C) some of the H_2S may decompose to form H_2 and the presence of this hydrogen can push the open circuit potential of the cell towards 1 V.[228]

To date, many of the anodes for operation with high sulfur content fuels have been based on vanadium-based compositions. Receiving particular attention in this respect have been vanadium perovskites, in particular the Sr-doped $La_{1-x}Sr_xVO_3$ family.[19,210] With a strontium substitution of 20–30% the LSV material is observed to display an n-type semiconductor behaviour with high electrical conductivities in typical anode chamber environments of 800 °C and oxygen partial pressures of 10^{-20}.[229] SOFCs tested with these anodes have shown varying performances ranging from around 130 mW cm^{-2} with 5% H_2S in H_2 atmospheres at 1000 °C to around 280 mW cm^{-2} in mixtures of 5% H_2S in CH_4 at 950 °C.[19,210] However, these tests were carried out on small button cell with an electrode area of only 0.25 cm^2 and the anodes fired *in situ* during the test. The latter aspect does bring into question the long-term microstructural stability of such systems during long-term operation at or close to the effective electrode firing temperature while the requirement for this *in situ* firing approach points to another possible issue with these systems: that of the fundamental stability of these perovskites.

The vanadium-based perovskites rely on the vanadium cation remaining in the V^{3+} state. However, this oxidation state is not stable in oxidising atmospheres. The LSV material has the widest range of pO_2 stability, over a range of 10^{-20} to 10^{-14}; however, in higher pO_2 the formation of the poorly conducting $Sr_3V_2O_8$ phase is observed, leading to irreversible decomposition of the perovskite structure.[229] This must bring into question the redox stability of this material for practical SOFC operation as the irreversible nature of the change makes it potentially even more damaging than the NiO/Ni transition in conventional cermet anodes. The redox stability can also cause issues in ceramics processing as this material must be processed in a reducing atmosphere, increasing complexity and possibly costs. Secondly, this requirement also places stringent requirements on the stability of other materials in the processing chain as these must either be stable in reducing environments or be processed before the application of the LSV. Although issues such as these are not insurmountable they do reduce the process flexibility, which may be vital in achieving optimal microstructures and interfaces for all components in a real device. Also the formation of V^{5+} compounds such as V_2O_5 must be avoided during synthesis and processing as these have low melting points (around 600 °C) and high volatility. Other oxides based on other vanadium-containing oxides or perovskites have also been investigated; however, these will likely suffer similar stability issue to the LSV-based perovskites.[211,230]

Another approach has been the investigation of various metal sulfides. Nickel sulfide is generally accepted as the reaction product of sulfur poisoning, but these materials do possess good conductivity and some catalytic activity and therefore could be potential anodes for high sulfur content fuels. A number of compositions have been investigated to date and have generally been based on sulfides of Ni, Mo, V and mixtures of these.[230–232] So far these studies have been limited and no truly competitive SOFC performance has yet been reported. These materials will also face stability issues, especially in redox environments or at higher temperatures. This will present similar processing and operational issues to those of the vanadium-based perovskite, which will result in additional complexities and therefore costs over conventional SOFC materials.

While a true H_2S-fuelled SOFC may be very much a niche application and technically still some way off, studies into the behaviour of SOFCs in these environments may provide further data on the stability of different materials in sulfur-rich fuel streams. This in turn could prove useful in the development of sulfur tolerant anodes for other sulfur-containing hydrocarbon fuel sources such as coal gas, biogas and diesel.

References

1. P. Singh and N. Q. Minh, *Int. J. Appl. Ceram. Technol.*, 2004, **1**, 5.
2. S. Dunn, *Int. J. Hydrogen Energy*, 2002, **27**, 235.

3. B. C. H. Steele, *Nature*, 1999, **400**, 619.
4. R. Peters, E. Riensche and P. Cremer, *J. Power Sources*, 2000, **86**, 432.
5. E. P. Murray, T. Tsai and S. A. Barnett, *Nature*, 1999, **400**, 649.
6. S. McIntosh and R. J. Gorte, *Chem. Rev.*, 2004, **104**, 4845.
7. M. Mogensen and K. Kammer, *Annu. Rev. Mater. Res.*, 2003, **33**, 321.
8. T. Takeguchi, Y. Kania, T. Yano, R. Kikuchi, K. Eguchi, K. Tsujimoto, Y. Uchida, A. Ueno, K. Omoshiki and M. Aizawa, *J. Power Sources*, 2002, **112**, 588.
9. G. J. Saunders and K. Kendall, *J. Power Sources*, 2002, **106**, 258.
10. Y. Yi, A. D. Rao, J. Brouwer and G. S. Samuelsen, *J. Power Sources*, 2005, **144**, 67.
11. K. Sasaki and Y. Teraoka, *J. Electrochem. Soc.*, 2003, **150**, 885.
12. W. A. Surdoval, *ECS Trans.*, 2009, **25**, 21.
13. K. Sasaki, K. Susuki, A. Iyoshi, M. Uchimura, N. Imamura, H. Kusaba, Y. Teraoka, H. Fuchino, K. Tsujimoto, Y. Uchida and N. Jingo, *J. Electrochem. Soc.*, 2006, **153**, A2023.
14. S. Zha, Z. Cheng and M. Lui, *J. Electrochem. Soc.*, 2007, **154**, B201.
15. J. Van herle, Y. Membrez and O. Bucheli, *J. Power Sources*, 2004, **127**, 300.
16. S. Gair, A. Cruden, J. McDonald, T. Hegarty and M. Chesshire, *J. Power Sources*, 2006, **154**, 472.
17. N. Maffei, L. Pelletier, J. P. Charland and A. McFarlan, *J. Power Sources.*, 2005, **140**, 264.
18. P. He, M. Liu, J. L. Luo, A. R. Sanger and K. T. Chuang, *J. Electrochem. Soc.*, 2002, **149**, A808.
19. L. Aguilar, S. Zha, Z. Cheng, J. Winnick and M. Liu, *J. Power Sources*, 2004, **135**, 17.
20. A. Siddle, K. D. Pointon, R. W. Judd and S. L. Jones, ETSU F/03/00252/ REP URN 031644, DTI Sustainable Energy Programmes, 2003, www.dti.gov.uk/energy/renewables/publications/pubs_fuel_cells.shtml.
21. J. B. Goodenough and Y.-H. Huang, *J. Power Sources*, 2007, **173**, 1.
22. T. Setoguchi, K. Okamoto, K. Eguchi and H. Arai, *J. Electrochem. Soc.*, 1992, **139**, 2875.
23. H. S. Spacil, U.S. Patent 3,558,360, 1970.
24. B. de Boer, M. Gonzalez, H. J. M. Bouwmeester and H. Verweij, *Solid State Ionics*, 2000, **127**, 269.
25. M. Mogensen and S. Skaarup, *Solid State Ionics*, 1996, **86–88**, 1151.
26. P. Holtappels, I. C. Vinke, L. G. J. de Haart and U. Stimming, *J. Electrochem. Soc.*, 1999, **146**, 2976.
27. S. P. Jiang, *J. Electrochem. Soc.*, 2001, **148**, A887.
28. C. H. Bartholomew, *Appl. Catal. A: Gen.*, 2001, **212**, 17.
29. J. Rostrup-Nielsen and D. L. Trimm, *J. Catal.*, 1977, **48**, 155.
30. R. T. K. Baker, P. S. Harris, J. Henderson and R. B. Thomas, *Carbon*, 1975, **13**, 37.
31. R. T. K. Baker, P. S. Harris and S. Terry, *Nature*, 1975, **253**, 37.
32. C. W. Keep, R. T. K. Baker and J. A. France, *J. Catal.*, 1977, **47**, 232.

33. R. T. K. Baker, *Carbon*, 1989, **27**, 315.
34. B. Monnerat, L. Kiwi-Minsker and A. Renken, *Chem. Eng. Sci.*, 2001, **56**, 633.
35. C. H. Toh, P. R. Munroe, D. J. Young and K. Foger, *Mater. High Temp.*, 2003, **20**, 129.
36. J.-M. Klein, S. Georges and Y. Bultel, *J. Electrochem. Soc.*, 2008, **155**(4), B333.
37. I. Alstrup, M. T. Tavares, C. A. Bernardo, O. Sørensen and J. Rostrup-Nielsen, *Mater. Corros.*, 1998, **49**, 367.
38. H. S. Bengaard, J. K. Nørskov, J. Sehested, B. S. Clausen, L. P. Nielsen, A. M. Molenbroek and J. R. Rostrup-Nielsen, *J. Catal.*, 2002, **209**, 365.
39. W. L. Holstein, *J. Catal.*, 1995, **152**, 42.
40. A. L. Lee, R. F. Zabransky and W. J. Huber, *Ind. Eng. Chem. Res.*, 1990, **29**, 766.
41. S. Bebelis, S. Neophytides and C. G. Vayenas, in *Proceedings of the 1st European SOFC Forum, Lucerne, 1994*, ed. U. Bossel, European SOFC Forum Secratariat, Baden, Switzerland, vol. 1, p. 197.
42. I. V. Yentekakis, Y. Jiang, S. Neophytides, S. Bebelis and C. G. Vayenas, *Ionics*, 1995, **1**, 491.
43. M. Stoukides, *Catal. Rev. Sci. Eng.*, 2000, **42**, 1.
44. S. Bebelis and S. Neophytides, *Solid State Ionics*, 2002, **152–153**, 447.
45. A. Weber, B. Sauer, A. C. Müller, D. Herbstritt and E. Ivers-Tiffée, *Solid State Ionics*, 2002, **152–153**, 543.
46. G. Goula, V. Kiousis, L. Nalbandian and I. V. Yentekakis, *Solid State Ionics*, 2006, **177**, 2119.
47. A. M. Sukeshini, B. Habinzadeh, B. P. Becker, C. A. Stolz, B. W. Eichhorn and G. S. Jackson, *J. Electrochem. Soc.*, 2006, **153**(4), A705.
48. B. Habibzadeh, B. P. Becker, A. M. Sukeshini and G. S. Jackson, *ECS Trans.*, 2008, **11**(33), 53.
49. N. C. Triantafyllopoulos and S. G. Neophytides, *J. Catal.*, 2003, **217**, 324.
50. H. P. He, J. M. Vohs and R. J. Gorte, *J. Electrochem. Soc.*, 2003, **150**, A1470.
51. K. Nikooyeh, R. Clemmer, V. Alzate-Restrepo and J. M. Hill, *Appl. Catal. A*, 2008, **347**, 106.
52. Y. B. Lin, Z. I. Zhan, J. Liu and S. A. Barnett, *Solid State Ionics*, 2005, **176**, 1827.
53. V. Alzate-Restrepo and J. M. Hill, *Appl. Catal. A*, 2008, **342**, 49.
54. J. Liu and S. A. Barnett, *Solid State Ionics*, 2003, **158**, 11.
55. C. M. Finnerty, N. J. Coe, R. H. Cunningham and R. M. Ormerod, *Catal. Today*, 1998, **46**, 137.
56. K. Yamaji, H. Kishimoto, Y. Xiong, T. Horita, N. Sakai, M. E. Brito and H. Yokokawa, *ECS Trans.*, 2007, **7**(1), 1661.
57. K. Eguchi, K. Tanaka, T. Matsui and R. Kikuchi, *Catal. Today*, 2009, **146**, 154.
58. H. Timmermann, W. Sawady, D. Campbell, A. Weber, R. Reimert and E. Ivers-Tiffee, *ECS Trans.*, 2007, **7**(1), 1429.

59. G. J. Saunders, J. Preece and K. Kendall, *J. Power Sources*, 2004, **131**, 23.
60. H. Kishimoto, T. Horita, K. Yamaji, Y. Xiong, N. Sakai and H. Yokokawa, *Solid State Ionics*, 2004, **175**, 107.
61. H. Kishimoto, T. Horita, K. Yamaji, Y. Xiong, N. Sakai, M. E. Brito and H. Yokokawa, *J. Electrochem. Soc.*, 2005, **152**(3), A532.
62. H. Kishimoto, Y.-P. Xiong, K. Yamaji, T. Horita, N. Sakai, M. E. Brito and H. Yokokawa, *ECS Trans.*, 2007, **7**(1), 1669.
63. H. Kim, S. Park, J. M. Vohs and R. J. Gorte, *J. Electrochem. Soc.*, 2001, **148**(7), A693.
64. R. Coll, J. Salvado, X. Farriol and D. Montane, *Fuel Process. Technol.*, 2001, **74**, 19.
65. T. Kivisaari, P. Bjornbom, C. Sylwan, B. Jacquinot, D. Jansen and A. D. Groot, *Chem. Eng. J.*, 2004, **100**, 167.
66. Y. Yi, A. D. Rao, J. Brouwer and G. S. Samuelsen, *J. Power Sources*, 2005, **144**, 67.
67. A. Verma, A. D. Rao and G. S. Samuelsen, *J. Power Sources*, 2006, 158, 417.
68. F. N. Cayan, M. Zhi, S. R. Pakalapati, I. Celik, N. Wu and R. Gemmen, *J. Power Sources*, 2008, **185**, 595.
69. J. P. Trembly, R. S. Gemmen and D. J. Bayless, *J. Power Sources*, 2007, **163**, 986.
70. K. Haga, S. Adachi, Y. Shiratori, K. Itoh and K. Sasaki, *Solid State Ionics*, 2008, **179**, 1427.
71. N. Q. Minh and T. Takahashi, in *Science and Technology of Ceramic Fuel Cells*, Elsevier, Amsterdam, 1995, pp. 209–210.
72. L. L. Hegedus and R. W. McCabe, *Catalyst Poisoning*, Marcel Dekker, New York, 1984.
73. J. Dong, Z. Cheng, S. Zha and M. Liu, *J. Power Sources*, 2006, **156**, 461.
74. V. B. Tare and J. B. Wagner Jr., *J. Appl. Phys.*, 1983, **54**, 252.
75. S. C. Singhal, R. J. Ruka, J. E. Bauerle and C. J. Spengler, Report No. DOE/MC/22046-237 I, US Department of Energy, Washington, DC, 1986.
76. D. Dees, U. Balachandran, S. Doris, J. Heiberger, C. McPheeter and J. Picciolo, in *Solid Oxide Fuel Cells*, ed. S. Singhal, The Electrochemical Society, Pennington, NJ, 1989, PV 89-11, p. 317.
77. D. Stolten, R. Spah and R. Schamm, in *Proc. 5th Int. Symp. on Solid Oxide Fuel Cells (SOFC-V)*, ed. U. Stimming, S. C. Singhal, H. Tagawa and W. Lehnert, The Electrochemical Society, Pennington, NJ, 1997, PV 97-40, p. 88.
78. J. Geyer, H. Kohlmuller, H. Landes and R. Stubner, in *Proc. 5th Int. Symp. on Solid Oxide Fuel Cells (SOFC-V)*, U. Stimming, S. C. Singhal, H. Tagawa and W. Lennert, (ed.), The Electrochemical Society, Pennington, NJ, 1997, PV 97-40, p. 585.
79. S. Primdahl and M. Mogensen, in *Proc. 6th Int. Symp. on Solid Oxide Fuel Cells (SOFC-VI)*, ed. S. C. Singhal and M. Dokiya, The Electrochemical Society, Pennington, NJ, 1999, PV 99-190, p. 530.
80. Y. Matsuzaki and I. Yasuda, *Solid State Ionics*, 2000, **132**, 261.

81. J. F. B. Rasmussen and A. Hagen, *J. Power Sources*, 2009, **191**, 534.
82. Z. Cheng, S. Zha and M. Liu, *J. Power Sources*, 2007, **172**, 688.
83. P. Lohsontoorn, D. J. L. Brett and N. P. Brandon, *J. Power Sources*, 2007, **175**, 60.
84. M. Gong, X. Liu, J. Trembly and C. Johnson, *J. Power Sources*, 2007, **168**, 289.
85. J. B. Hansen, *Electrochem. Solid-State Lett.*, 2008, **11**, B178.
86. J. Dong, S. Zha and M. Liu, in *Proc. 9th Int. Symp. on Solid Oxide Fuel Cells (SOFC-IX)*, ed. S. C. Singhal and J. Mizusaki, Electrochemical Society, Pennington, NJ, 2005, PV 2005-07, p. 1284.
87. A. Braun, M. Janousch, J. Sfeir, J. Kiviaho, M. Noponen, F. E. Huggins, M. J. Smith, R. Steinberger-Wilckens, P. Holtappels and T. Graule, *J. Power Sources*, 2008, **183**, 564.
88. Z. Cheng and M. Liu, *Solid State Ionics*, 2007, **178**, 925.
89. J. L. Oliphant, R. W. Fower, R. B. Pannell and C. H. Bartholomew, *J. Catal.*, 1978, **51**, 229.
90. C. H. Bartholomew, P. K. Agrawal and J. R. Katzer, in *Sulphur Poisoning of Metals*, ed. D. D. Eley, H. Pines and P. B. Weisz, Academic Press, New York, 1982, pp. 135–242.
91. J. R. Rostrup-Nielsen, J. B. Hansen, S. Helveg, N. Christiansen and A.-K. Jannasch, *Appl. Phys. A*, 2006, **85**, 427.
92. T. R. Smith, A. Wood and V. L. Birss, *Appl. Catal. A*, 2009, **354**, 1.
93. J. E. Bao, G. N. Krishnan, P. Jayaweera, K. H. Lau and A. Sanjurjo, *J. Power Sources*, 2009, **193**, 617.
94. O. A. Marina, L. R. Pederson, D. J. Edwards, C. W. Coyle, J. Templeton, M. Engelhard and Z. Zhu, in *Proc. 8th Annual SECA Workshop, San Antonio, TX, 2007*, National Energy Technology Laboratory, NETL, Pittsburgh, PA, USA, 2007.
95. J. P. Trembly, R. S. Gemmen and D. J. Bayless, *J. Power Sources*, 2007, **163**, 986.
96. M. Zhi, H. Finklea, N. Madhiri, I. Celik, B. Kang, X. Liu and N. Q. Wu, in *Proc. 212th ECS Meeting, Washington, DC, 2007*, Electrochemical Society, Pennington, New Jersey, USA, 2007.
97. O. A. Marina, L. R. Pederson, J. W. Templeton, C. A. Coyle, D. J. Edwards and M. H. Engelhard, *ECS Trans.*, 2008, **11**(33), 63.
98. M. Zhi, X. Chen, H. Finklea, I. Celik and N. Q. Wu, *J. Power Sources*, 2008, **183**, 485.
99. J. Bao, G. N. Krishnan, P. Jayaweera, J. Perez-Mariano and A. Sanjurjo, *J. Power Sources 9*, **193**, 607.
100. C. Xu, J. W. Zondlo, H. O. Finklea, O. Demircan, M. Gong and X. B. Liu, *J. Power Sources*, 2009, **193**, 739.
101. K. Gerdes, J. Trembly and R. Gemmen, in *Proc. Coal Based Fuel Cell Technology: Status, Needs and Future Applications, Morgantown, WV, 2007*, National Energy Technology Laboratory, NETL, Pittsburgh, PA, USA, 2007.
102. B. Nielsen and J. Villaden, *Appl. Catal.*, 1984, **11**, 123.

103. J. P. Trembly, R. S. Gemmen and D. J. Bayless, *J. Power Sources*, 2007, **171**, 818.
104. G. N. Krishnan, in *Proc. 8th Annual SECA Workshop, San Antonio, TX, 2007*, National Energy Technology Laboratory, NETL, Pittsburgh, PA, USA, 2007.
105. C. A. Coyle, O. A. Marina, E. C. Thomsen, D. J. Edwards, C. D. Cramer, G. W. Coffey and L. R. Pederson, *J. Power Sources*, 2009, **193**, 730.
106. G. N. Krishnan, P. Jayaweera, K. Lau and A. Sanjurjo, *Technical progress report 2, 3 and 4*, SRI International, Morgantown, 2006.
107. H. J. Goldschmidt and M. J. Walker, *J. Appl. Crystallogr.*, 1969, **2**, 273.
108. J. Bao, G. Krishnan, P. Jayaweera, (Meet. Abstr.) *Electrochem. Soc.*, 2009, **901**, 421.
109. A. E. Hughes and S. P. S. Badwal, *Solid State Ionics*, 1991, **46**, 265.
110. A. Tsoga, A. Naoumidis and P. Nikopoulos, *Acta Mater.*, 1996, **44**(9), 3679.
111. M. Mogensen, S. Primdahl, K. Vels Jensen, M. Jorgensen and C. Bagger, in *Proc. 7th Int. Symp. on Solid Oxide Fuel Cells (SOFC-VII)*, ed. Y. Yokokawa and S.C. Singhal, Electrochemical Society, Pennington, NJ, PV 2001-16, p. 521.
112. K. Vels Jensen, S. Primdahl, I. Chockendorf and M. Mogensen, *Solid State Ionics*, 2001, **144**, 197.
113. M. Mogensen, K. Vels Jensen, M. J. Joergensen and S. Primdahl, *Solid State Ionics*, 2002, **150**, 123.
114. Y. L. Liu, S. Primdahl and M. Mogensen, *Solid State Ionics*, 2003, **161**, 1.
115. Y. L. Liu and C. Jiao, *Solid State Ionics*, 2005, **176**, 435.
116. M. S. Schmidt, K. Vels Hansen, K. Norrman and M. Mogensen, *Solid State Ionics*, 2008, **179**, 1436.
117. R. H. Cunningham, C. M. Finnerty and R. M. Ormerod, in *Proc. 5th Int. Symp. on Solid Oxide Fuel Cells (SOFC-V)*, ed. U. Stimming, S. C. Singhal, H. Tagawa and W. Lehnert, Electrochemical Society, Pennington, NJ, 1997, p. 973.
118. Z. Cheng, Q. Wu, J. Li and Q. Zhu, *Catal. Today*, 1996, **30**, 147.
119. D. L. King, J. J. Strohm, X. Wang, H.-S. Roh, C. Wang, Y.-H. Chin, Y. Wang, Y. Lin, R. Rozmiarek and P. Singh, *J. Catal.*, 2008, **258**, 356.
120. H. Praliaud, M. Primet and G. A. Martin, *Appl. Surf. Sci.*, 1983, **17**, 107.
121. H. Praliaud, J. A. Dalmon, C. Mirodatos and G. A. Martin, *J. Catal.*, 1986, **97**, 344.
122. J. R. Rostrup-Nielsen and L. J. Christiansen, *Appl. Catal. A*, 1995, **126**, 381.
123. T. Horiuchi, K. Sakuma, T. Fukui, Y. Kubo, T. Osaki and T. Mori, *Appl. Catal. A*, 1996, **144**, 111.
124. S. Alberttazzi, G. Busca, E. Finocchio, R. Glockler and A. Vaccai, *J. Catal.*, 2004, **223**, 372.
125. H. Morioka, Y. Shimizu, M. Sukenobu, K. Ito, E. Tanabe, T. Shishido and K. Tasehira, *Appl. Catal. A*, 2001, **215**, 11.

126. T. Takeguchi, T. Yano, Y. Kani, R. Kikuchi and K. Eguchi, in *Proc. 8th. Int.Symp. on Solid Oxide Fuel Cells (SOFC-XIII)*, ed. S. C. Singhal and M. Dokiya, Electrochemical Society, Pennington, NJ, 2003, PV 2003-07, p. 704.
127. J. Srinakruang, K. Sato, T. Vitidsant and K. Fujimoto, *Fuel*, 2006, **85**, 2419.
128. H. Kim, C. Lu, W. L. Worrell, J. M. Vohs and R. J. Gorte, *J. Electochem. Soc.*, 2002, **149**(3), A247.
129. K. Sato, Y. Ohmine, K. Ogasa and S. Tsuji, in *Proc. 8th. Int. Symp. on Solid Oxide Fuel Cells (SOFC-XIII)*, ed. S. C. Singhal and M. Dokiya, Electrochemical Society, Pennington, NJ, 2003, PV 2003-07, p. 695.
130. T. Hibino, A. Hashimoto, M. Yano, M. Suzuki and M. Sano, *Electrochim. Acta*, 2003, **48**, 2531.
131. M. Lo Faro, G. Monforte, V. Antonucci and A. S. Arico, in *Proc. 9th Int. Symp. on Solid Oxide Fuel Cells (SOFC-IX)*, ed. S. C. Singhal and J. Mizusaki, Electrochemical Society, Pennington, NJ, PV 2005-07, 2005, p. 1445.
132. Z. Zhan, Y. Lin and S. Barnett, in *Proc. 9th Int. Symp. on Solid Oxide Fuel Cells (SOFC-IX)*, ed. S. C. Singhal and J. Mizusaki, Electrochemical Society, Pennington, NJ, PV 2005-07, 2005, p. 1321.
133. A. Ishihara, E. W. Qian, I. N. Finahari, I. P. Sutrisna and T. Kabe, *Fuel*, 2005, **84**, 1462.
134. T. Iida, M. Kawano, T. Matsui, R. Kikuchi and K. Eguchi, *J. Electrochem. Soc.*, 2007, **154**, B234.
135. J.-M. Klein, M. Henault, P. Gelin, Y. Bultel and S. Georges, *Electrochem. Solid-State Lett.*, 2008, **11**(8), B144.
136. E. S. Putna, J. Stubenrauch, J. M. Vohs and R. J. Gorte, *Langmuir*, 1995, **11**, 4832.
137. T. Takeguchi, R. Kikuchi, T. Yano, K. Eguchi and K. Murata, *Catal. Today*, 2003, **84**, 217.
138. I. Gavrielatos, V. Drakopoulos and S. G. Neophytides, *J. Catal.*, 2008, **259**, 75.
139. Y. Nabae, I. Yamanaka, M. Hatano and K. Otsuka, *J. Phys. Chem. C*, 2008, **112**, 10308.
140. E. Nikolla, J. Schwank and S. Linic, *J. Catal.*, 2007, **250**, 85.
141. B. C. H. Steele, P. H. Middleton and R. A. Rudkin, *Solid State Ionics*, 1990, **40–41**, 388.
142. Z. L. Zhan and S. A. Barnett, *Science*, 2005, **308**, 844.
143. A. E. C. Palmqvist, E. M. Johansson, S. G. Jaras and M. Muhammed, *Catal. Lett.*, 1998, **56**, 69.
144. E. Ramirez-Cabrera, A. Atkinson and D. Chadwick, *Appl. Catal. B*, 2002, **36**, 193.
145. E. Ramirez-Cabrera, A. Atkinson and D. Chadwick, *Appl. Catal. B*, 2004, **47**, 127.
146. S. Zhao and R. J. Gorte, *Appl. Catal. A*, 2003, **248**, 9.

147. T. Kim, K. Ahn, J. M. Vohs and R. J. Gorte, *J. Power Sources*, 2007, **164**, 42.
148. P. Fornasiero, R. Dimonte, G. R. Rao, J. Kaspar, S. Meriani, A. Trovarelli and M. Graziani, *J. Catal.*, 1995, **151**, 168.
149. P. Fornasiero, G. Balducci, R. Di Monte, J. Kaspar, V. Sergo, G. Gubitosa, A. Ferrer and M. Graziani, *J. Catal.*, 1996, **164**, 173.
150. C. E. Hori, H. Permana, K. Y. S. Ng, A. Brenner, K. More, K. M. Rahmoeller and D. Belton, *Appl. Catal. B*, 1998, **16**, 105.
151. J. R. Gonzales-Velasco, M. A. Gutierrez-Ortiz, J.-L. Marc, J. A. Botas, M. P. Gonzales-Marcos and G. Blanchard, *Appl. Catal. B*, 1999, **22**, 167.
152. C. Bozo, N. Guilhaume, E. Garbowski and M. Primet, *Catal. Today*, 2000, **59**, 33.
153. N. Laosiripojana and S. Assabumrungrat, *Appl. Catal. A*, 2005, **290**, 200.
154. V. V. Kharton, A. A. Yaremchenko, A. A. Valente, E. V. Frovola, M. I. Ivanovskaya, J. R. Frade, F. M. B. Marques and J. Rocha, *Solid State Ionics*, 2006, **177**, 2179.
155. H. P. He and J. M. Hill, *Appl. Catal. A*, 2007, **317**, 284.
156. S. J. A. Livermore, J. W. Cotton and R. M. Ormerod, *J. Power Sources*, 2000, **86**, 411.
157. B. Rösch, H. Tu, A. O. Störmer, A. C. Müller and U. Stimming, in *Proc. 8th. Int. Symp. on Solid Oxide Fuel Cells (SOFC-XIII)*, ed. S. C. Singhal and M. Dokiya, Electrochemical Society, Pennington, NJ, 2003, PV 2003-07, p. 737.
158. J. B. Wang, J.-C. Jang and T.-J. Huang, *J. Power Sources*, 2003, **122**, 122.
159. M. Asamoto, S. Miyake, Y. Itagaki, Y. Sadaoka and H. Yahiro, *Catal. Today*, 2008, **139**, 77.
160. N. J. J. Dekker, J. P. Ouweltjes and G. Rietveld, *ECS Trans.*, 2007, **7**(1), 1465.
161. P. V. Aravind, J. P. Ouweltjes, N. Woudstra and G. Rietveld, *Electrochem. Solid-State Lett.*, 2008, **11**(2), B24.
162. D. Shekhawat, D. A. Berry, D. J. Haynes and J. J. Spivey, *Fuel*, 2009, **88**, 817.
163. S. Park, R. Craciun, J. M. Vohs and R. J. Gorte, *J. Electrochem. Soc.*, 1999, **146**, 3603.
164. R. Craciun, S. Park, R. J. Gorte, J. M. Vohs, C. Wang and W. L. Worrell, *J. Electrochem. Soc.*, 1999, **146**, 4019.
165. S. Park, J. M. Vohs and R. J. Gorte, *Nature*, 2000, **404**, 265.
166. C. Wang, W. L. Worrell, S. Park, J. M. Vohs and R. J. Gorte, *J. Electrochem. Soc.*, 2001, **148**(8), A864.
167. C. Lu, W. L. Worrell, C. Wang, S. Park, H. Kim, J. M. Vohs and R. J. Gorte, *Solid State Ionics*, 2002, **152–153**, 393.
168. S. McIntosh, J. M. Vohs and R. J. Gorte, *Electrechem. Solid-State Lett.*, 2003, **6**(11), A240.
169. S. McIntosh, H. P. He, S.-I. Lee, O. C. Nunes, V. V. Krishnan, J. M. Vohs and R. J. Gorte, *J. Electrochem. Soc.*, 2004, **151**, A604.
170. R. J. Gorte, J. M. Vohs and S. McIntosh, *Solid State Ionics*, 2004, **175**, 1.

171. V. G. Milt, E. A. Lombardo and M. A. Ulla, *Appl. Catal. B*, 2002, **37**, 63.
172. S.-I. Lee, J. M. Vohs and R. J. Gorte, *J. Electrochem. Soc.*, 2004, **151**(9), A1319.
173. O. A. Marina, C. Bagger, S. Primdahl and M. Mogensen, *Solid State Ionics*, 1999, **123**, 199.
174. C. Lu, W. L. Worrell, J. M. Vohs and R. J. Gorte, in *Proc. 8th. Int. Symp. on Solid Oxide Fuel Cells (SOFC-XIII)*, ed. S. C. Singhal and M. Dokiya, Electrochemical Society, Pennington, NJ, 2003, PV 2003-07, p. 773.
175. A. Atkinson, S. Barnett, R. J. Gorte, J. T. S. Irvine, A. J. McEvoy, M. Mogensen, S. C. Singhal and J. Vohs, *Nat. Mater.*, 2004, **3**, 17.
176. Q. X. Fu and F. Tietz, *Fuel Cells*, 2008, **8**, 283.
177. M. Cassidy, K. Kendall and G. Lindsay, *J. Power Sources*, 1996, **61**, 189.
178. J. Van herle, D. Perednis, K. Nakamura, S. Diethelm, M. Zahid, A. Aslanides, T. Somekawa, Y. Baba, K. Horiuchi, Y. Matsuzaki, M. Yoshimoto and O. Bucheli, *J. Power Sources*, 2008, **182**, 389.
179. N. Q. Minh, *J. Am. Ceram. Soc.*, 1993, **76**, 563.
180. J. Lui, B. D. Madsen, Z. Ji and S. A. Barnett, *Electrochem. Solid-State Lett.*, 2002, **5**, A122.
181. S. Tao and J. T. S. Irvine, *Nat. Mater.*, 2003, **2**, 320.
182. M. Van den Bossche and S. McIntosh, *J. Catal.*, 2008, **255**, 313.
183. J. Wan, J. H. Zhu and J. B. Goodenough, *Solid State Ionics*, 2006, **177**, 1211.
184. J. Peña-Matínez, D. Marrero-López, J. C. Ruiz-Morales, C. Savaniu, P. Núñez and J. T. S. Irvine, *Chem. Mater.*, 2006, **18**, 1001.
185. S. Tao and J. T. S. Irvine, *J. Electrochem. Soc.*, 2004, **151**, A252.
186. M. D. Gross, J. M. Vohs and R. J. Gorte, *Electrochem. Solid-State Lett.*, 2007, **10**, B65.
187. M. Cassidy, S. Boulfrad, J. Irvine, C. Chung, M. Jorger, C. Munnings and S. Pyke, *Fuel Cells*, 2009, DOI: 10.1002/fuce.200800188.
188. S. P. Jiang, X. J. Chen, S. H. Chan and J. T. Kwok, *J. Electrochem. Soc.*, 2006, **153**, A850.
189. Y. Ye, T. He, Y. Li, E. H. Tang, T. L. Reitz and S. P. Jiang, *J. Electrochem. Soc.*, 2008, **155**, B811.
190. G. Kim, G. Corre, J. T. S. Irvine, J. M. Vohs and R. J. Gorte, *Electrochem. Solid-State Lett.*, 2008, **11**, B16.
191. D. M. Bastidas, S. Tao and J. T. S. Irvine, *J. Mater. Chem.*, 2006, **16**, 1603.
192. O. A. Marina, N. L. Canfield and J. W. Stevenson, *Solid State Ionics*, 2002, **149**, 21.
193. O. A. Marina and L. R. Pederson, in *Proc. 5th Eur. SOFC Forum, Lucerne, 2002*, ed. J. Huijsmans. European SOFC Forum, Secretariat, Baden, Switzerland, p. 481.
194. J. T. S Irvine, P. R. Slater and P. A. Wright, *Ionics*, 1996, **2**, 213.
195. P. Blennow, A. Hagen, K. K. Hansen, L. Reine Wallenberg and M. Mogensen, *Solid State Ionics*, 2008, **179**, 2047.

196. J. C. Ruiz-Morales, J. Canales-Vázquez, C. Savaniu, D. Marrero-López, W. Zhou and J. T. S. Irvine, *Nature*, 2006, **439**, 568.
197. Q. Fu, F. Tietz, D. Sebold, S. Tao and J. T. S. Irvine, *J. Power Sources*, 2007, **171**, 663.
198. Y. Huang, R. I. Dass, Z. Xing and J. B. Goodenough, *Science*, 2006, **312**, 254.
199. D. Marrero-López, J. Peña-Matínez, J. C. Ruiz-Morales, D. Pérez-Coll, M. A. G. Aranda and P. Núñez, *Mater. Res. Bull.*, 2008, **43**, 2441.
200. C. M. Reich, A. Kaiser and J. T. S. Irvine, *Fuel Cells*, 2001, **1**, 249.
201. R. J. Gorte, S. Park, J. M. Vohs and C. Wang, *Adv. Mater.*, 2000, **12**, 1465.
202. G. Corre, G. Kim, M. Cassidy, J. M. Vohs, R. J. Gorte and J. T. S. Irvine, *Chem. Mater.*, 2009, **21**, 1077.
203. J. P. Ouweltjes, P. V. Aravind, N. Woudstra and G. Rietveld, *J. Fuel Cell Sci. Technol.*, 2006, **3**, 495.
204. H. P. He, R. J. Gorte and J. M. Vohs, *Electrochem. Solid-State Lett.*, 2005, **8**(6), A279.
205. R. Mukundan, E. L. Brosha and F. H. Garzon, *Electrochem. Solid-State Lett.*, 2004, **7**, A5.
206. X. C. Lu, J. H. Zhu, Z. Yang, Guanguang and J. W. Stevenson, *J. Power Sources*, 2009, **192**, 381.
207. M. R. Pillai, I. Kim, D. M. Bierschenk and S. A. Barnett, *J. Power Sources*, 2008, **185**, 1086.
208. S. Zha, P. Tsang, Z. Cheng and M. Liu, *J. Solid State Chem.*, 2005, **178**, 1844.
209. X. J. Chen, Q. L. Liu, S. H. Chan, N. P. Brandon and K. A. Khor, *J. Electrochem. Soc.*, 2007, **154**, B1210.
210. Z. Cheng, S. Zha, L. Aguilar, D. Wang, J. Winnick and M. Liu, *Electrochem. Solid-State Lett.*, 2006, **9**, A31.
211. Z. Xu, J. Luo and K. T. Chuang, *ECS Trans.*, 2008, **11**, 1.
212. Q. Ma, J. Ma, S. Zhou, R. Yan, J. Gao and G. Meng, *J. Power Sources*, 2006, **161**, 95.
213. C. Zamfirescu and I. Dincer, *Fuel. Process, Technol.*, 2009, **90**, 729.
214. J. Staniforth and R. M. Ormerod, *Green Chem.*, 2003, **5**, 606.
215. C. H. Christensen, T. Johannessen, R. Z. Sørensen and J. K. Nørskov, *Catal. Today*, 2006, **111**, 140.
216. T. Lipman and N. Shah, Research report UCB-ITS-TSRC-RR-2007-5, UC Berkeley Institute of Transportation Studies, Berkeley, 2007.
217. H. Brandhorst, B. Tatarchuk, D. Cahela, T. Barron and M. Baltazar-Lopez, in *4th Annual Conference on Ammonia – A Sustainable, Emission-free Fuel, 15–16 October, San Francisco, CA*, Iowa Energy Center, 2007.
218. N. Dekker and B. Rietveld, in *6th European Solid Oxide Fuel Forum, Lucerne*, ed. M. Mogensen, European SOFC Forum, Oberrohrdorf, Switzerland, 2004, p. 1524.
219. Q. Ma, R. Peng, L. Tian and G. Meng, *Electrochem. Comm.*, 2006, **8**, 1791.

220. Q. Ma, J. Ma, S. Zhou, R. Yan, J. Gao and G. Meng, *J. Power Sources*, 2007, **164**, 86.
221. M. Ni, D. Y. C. Leung and M. K. H. Leung, *J. Power Sources*, 2008, **183**, 687.
222. M. Ni, D. Y. C. Leung and M. K. H. Leung, *J. Power Sources*, 2008, **185**, 233.
223. A. McFarlan, L. Pelletier and N. Maffei, *J. Electrochem. Soc.*, 2004, **151**, A930.
224. L. Pelletier, A. McFarlan and N. Maffei, *J. Power Sources*, 2005, **145**, 262.
225. Q. Ma, R. Peng, Y. Lin and J. Gao, *J. Power Sources*, 2006, **161**, 95.
226. K. Xie, Q. Ma, B. Lin, Y. Jiang, J. Gao, X. Liu and G. Meng, *J. Power Sources*, 2007, **170**, 38.
227. L. Zhang and W. Yang, *J. Power Sources*, 2008, **179**, 92.
228. C. Yates and J. Winnick, *J. Electrochem. Soc.*, 1999, **146**, 2841.
229. S. Hui and A. Petric, *Solid State Ionics*, 2001, **143**, 275.
230. A. Aguadreo, C. de la Calle, J. A. Alonso, D. Pez-Coll, M. J. Escudero and L. Daza, *J. Power Sources*, 2009, **192**, 78.
231. V. Vorontsov, W. An, J. L. Luo, A. R. Sanger and K. T. Chuang, *J. Power Sources*, 2008, **179**, 9.
232. V. Vorontsov, J. L. Luo, A. R. Sanger and K. T. Chuang, *J. Power Sources*, 2008, **183**, 76.

CHAPTER 6

Direct Carbon Fuel Cells

P. DESCLAUX,[a] S. NÜRNBERGER[a] AND
U. STIMMING[a, b]

[a] ZAE Bayern, Division 1, Walther-Meiner-Str. 6, 85748 Garching, Germany;
[b] TU München, Department of Physics E19, James-Franck-Str. 1, 85748
Garching, Germany

6.1 Electrochemical Oxidation of Carbon

The direct carbon fuel cell (DCFC) is a high-temperature fuel cell which allows direct conversion of the chemical energy of carbon materials into electricity in an efficient way. DCFC systems are potentially significantly more efficient than conventional coal-fired power plants due to fewer conversion steps needed for electric power generation. In addition, the maximum efficiency of a fuel cell, thus also of a DCFC, is not limited by the Carnot efficiency, in contrast to thermal-electricity generators. The carbon fuel can be in the form of, for example, coal, lignite, char, carbonised biomass or carbon containing waste (such as plastic). These materials are easy to store and transport and the released energy per unit volume of carbon in oxidation with oxygen is very high ($20.0\,kWh\,L^{-1}$). In comparison with various other fuels, this value is the highest: liquid hydrogen ($2.4\,kWh\,L^{-1}$), gasoline ($9.0\,kWh\,L^{-1}$) and diesel ($9.8\,kWh\,L^{-1}$). The anodic exhaust gas of a DCFC consists of almost pure CO_2 that can be captured and sequestered with less difficulty than with conventional thermal power plants, for example, where CO_2 capture and management solutions reduce efficiency, increase complexity and add cost. If fuel materials for DCFCs are derived from biomass, a negative carbon dioxide balance is possible in case of sequestration.[1–5]

RSC Energy and Environment Series No. 2
Innovations in Fuel Cell Technologies
Edited by Robert Steinberger-Wilckens and Werner Lehnert
© Royal Society of Chemistry 2010
Published by the Royal Society of Chemistry, www.rsc.org

6.1.1 Thermodynamics

The overall cell reaction in a DCFC is based on the complete electrochemical oxidation of carbon to carbon dioxide (CO_2) in a four-electron process:[4]

$$C + O_2 \rightarrow CO_2 \qquad \Delta_r H = -394.0 \, kJ \, mol^{-1} \qquad (6.1)$$

The thermodynamic efficiency η_{th} of a fuel cell is defined as the ratio between the Gibbs free energy of the reaction, $\Delta_r G$, and the reaction enthalpy, $\Delta_r H$:

$$\eta_{th} = \frac{\Delta_r G}{\Delta_r H} = 1 - T \times \frac{\Delta_r S}{\Delta_r H} \qquad (6.2)$$

with $\Delta_r G = \Delta_r H - T\Delta_r S$, where $\Delta_r S$ is the reaction entropy and T the operating temperature. The thermodynamic efficiency η_{th} of direct carbon conversion slightly exceeds 100% (Figure 6.1) due to a positive reaction entropy: $\Delta_r S = 1.6 \, J \, K^{-1} \, mol^{-1}$. The value of thermodynamic equilibrium voltage for standard conditions is $E_0 = -\Delta_r G(800 \, °C)/(4F) = 1.026 \, V$ at 800 °C, with F the Faraday constant. The actual cell voltage strongly depends on the activity of the carbon material and partial pressure of CO_2 in the anode compartment, and on the partial pressure of oxygen in the cathode compartment, according to the Nernst equation. As $\Delta_r S$ is very small compared to the reaction enthalpy $|\Delta_r H|$, the dependence on temperature variation of thermodynamic efficiency and equilibrium voltage is negligible. The thermodynamic efficiency of the electro-

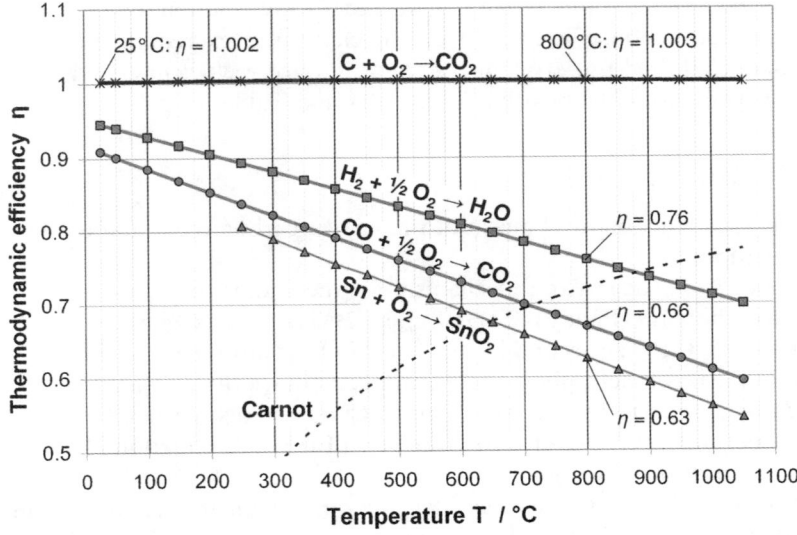

Figure 6.1 Comparison of thermodynamic efficiencies. Efficiency of direct carbon conversion exceeds 100%, almost independent of conversion temperature.

Table 6.1 Conversion efficiencies of DCFCs compared to hydrogen fuel cells at 800 °C[6]

Fuel	Theoretical limit ($\Delta G/\Delta H$)	Fuel utilisation	Voltage efficiency	Maximal actual efficiency
C	1.00	1.0	0.8	0.8
H_2	0.76	0.8	0.8	0.49

oxidation of carbon, in comparison to other fuels, is given in Figure 6.1 in the temperature range 25–1050 °C.

The fuel utilisation in a DCFC system can reach up to 100%, due to the fact that the reaction product, CO_2, exists in a separate gas phase and thus does not influence the activity of the solid carbon. This is in contrast to hydrogen fuel cells, for example, where water produced can affect the activity of the fuel. Therefore, fuel cell concepts working on gaseous fuels have to be performed at a lambda value greater 1 to guarantee sufficient and homogeneous fuel supply over the whole electrode surface. Thus, a typical fuel utilisation of 80% can be assumed for such systems (Table 6.1). For pure hydrogen in a dead-end configuration also 100% is possible. Another factor of the overall efficiency of a fuel cell is the voltage efficiency, the ratio between operating voltage and equilibrium voltage (given by the Nernst equation). The operating point of a fuel cell is somewhat arbitrary; power density and electrical efficiency show opposite behaviour: the higher the power output, the lower the operating voltage, which means lower efficiency. In Table 6.1 a value of 80% (typically 60–80%) is assumed for voltage efficiency. As thermodynamic efficiency of carbon electro-oxidation is very high and fuel utilisation is almost 100%, practical conversion efficiencies (coal to electricity) exceeding those of conventional coal-fired power plants (limited by Carnot efficiency) and other fuel cell concepts should be possible (Figure 6.1 and Table 6.1).[2,4]

6.1.2 Mechanism

The detailed reaction mechanism of the anodic carbon oxidation is still under investigation.[4,7]

Cooper *et al.* suggested a mechanism for the anodic oxidation of carbon in molten carbonates (Figure 6.2 and Table 6.2): The first oxygen ion adsorbs at a reactive site, C_{rs}. The second adsorption of O^{2-} onto the $C_{rs}O$ site (Table 6.2 Equation (3)) is kinetically hindered and constitutes the rate-determining step (RDS) for complete carbon oxidation to CO_2. The discharge of the adsorbed species occurs in two one-electron transfers (Table 6.2, Equation (4)), followed by the desorption of CO_2.

To achieve significant reaction rates, operating temperatures in the range of 500–1000 °C are required. As the activation energy of electrochemical oxidation of carbon is very high, appropriate catalyst materials are needed in order to enhance kinetics.

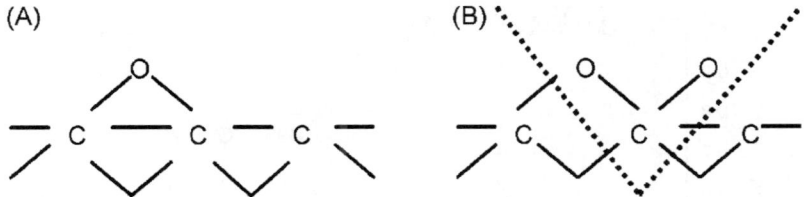

(A)   (B)

Figure 6.2 Pictorial description of the carbon electrochemical oxidation: (A) initial oxygen ion adsorption and (B) second oxygen ion adsorption and CO_2 release.[7] Reprinted, with permission, from Cao *et al.*[4] © Copyright 2007 Elsevier.

Table 6.2 Mechanism of the carbon electrochemical oxidation.[4,5,7]

$C_{rs} + O^{2-} \rightarrow C_{rs} - O^{2-}$	First adsorption of O^{2-} on carbon	(1)
$C_{rs} - O^{2-} \rightarrow C_{rs}O + 2e^-$	Fast discharge, two $1e^-$ discharges	(2)
$C_{rs}O + O^{2-} \rightarrow C_{rs}O - O^{2-}$	Slow O^{2-} adsorption: RSD	(3)
$C_{rs}O - O^{2-} \rightarrow CO_{2,ad} + 2e^-$	Fast discharge, two $1e^-$ discharges	(4)
$CO_{2,ad} \rightarrow CO_2$	Desorption	(5)

6.1.3 Boudouard Reaction

At high temperatures CO can be produced by a subsequent reaction of the product gas CO_2 with the solid carbon, according to the Boudouard reaction equilibrium (Equation (6.3), Figure 6.3). The CO can possibly be electro-oxidised to CO_2 in a second step, which leads to a fuel loss. CO formation could be avoided at high current densities and lower operating temperatures.[4,5,8,9]

$$C_{(s)} + CO_{2(g)} \leftrightarrow 2CO_{(g)} \qquad \Delta_r H^\circ = 172.4\,\text{kJ}\,\text{mol}^{-1} \qquad (6.3)$$

6.2 Different Types of Direct Carbon Fuel Cells

The first concept of direct carbon conversion in fuel cells was already patented in 1896 by William W. Jacques. Researchers had trouble duplicating Jacques' results until the mid 1970s, when R. Weaver and co-workers proved that generating electricity by complete electrochemical oxidation of carbon is feasible. Since the 1990s, with the significant development of fuel cell technology and materials research and the need for a clean and efficient technology to convert coal into electricity, research on this topic has been more intense. Recently, three different concepts of a DCFC based on different electrolytes have been discussed: molten carbonate electrolyte (liquid salt), molten hydroxide electrolyte (liquid salt) and solid oxide electrolyte (solid ceramic layer).[2,4,8,10–12]

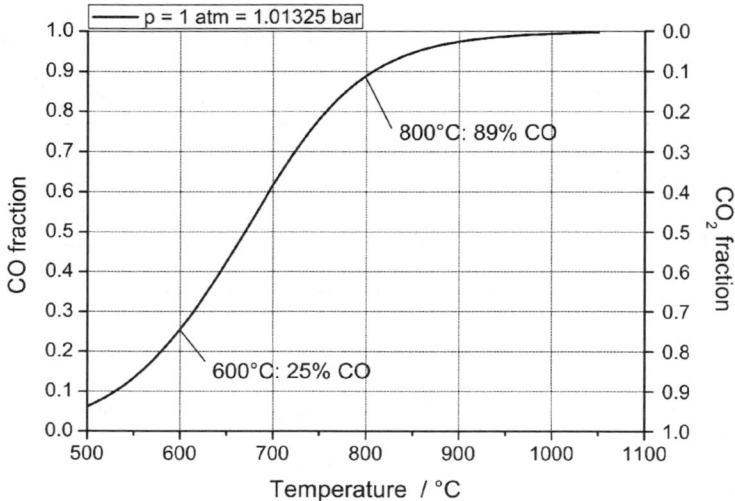

Figure 6.3 Equilibrium gas composition for the Boudouard reaction, Equation (6.3).

Most of the concepts will require some preprocessing of the fuel. In addition to pulverisation, impurities (*e.g.* sulfur, mercury and ash) need to be removed. For instance, ash accumulating in liquid electrolytes might lead to its degradation and solid electrolytes could be blocked. Removing impurities may result in parasitic power requirements and higher costs.[1]

In this section different concepts of a direct carbon fuel cell will be presented. A direct comparison of cell performances is not possible, because these strongly depend on, for instance, carbon fuel material used, operating temperature and the concept itself (anode/electrolyte material).

6.2.1 Molten Carbonate Electrolyte

One important concept of a DCFC is based on a molten carbonate fuel cell (MCFC). A mixture of carbonates, *e.g.* Li_2CO_3, K_2CO_3 and/or Na_2CO_3 is advantageous for use as an electrolyte in DCFCs for various reasons. Firstly, employing molten carbonates in the anode compartment allows the contact area between carbon and electrolyte to be extended. Secondly, eutectic mixtures of carbonates possess suitable melting temperatures (500–900 °C). In addition, carbonate melts show high ionic conductivity and good stability in the presence of CO_2.[4,13–16]

The half-cell reactions are given in Table 6.3. Carbon dioxide has to be fed with oxygen to the cathode compartment, in order to form carbonate ions.

At present, this technique is investigated for instance by John F. Cooper and co-workers at Lawrence Livermore National Laboratory (LLNL, USA). The concept (Figure 6.4) is based on an alumina tube, wherein carbon particles are

Table 6.3 Half-cell reactions for the electrochemical carbon oxidation in molten carbonate.[4,17]

Cathode reaction	$O_2 + 2CO_2 + 4e^- \rightarrow 2CO_3^{2-}$	(1)
Anode reaction	$C + 2CO_3^{2-} \rightarrow 3CO_2 + 4e^-$	(2)
Overall reaction	$C + O_2 \rightarrow CO_2$	(3)

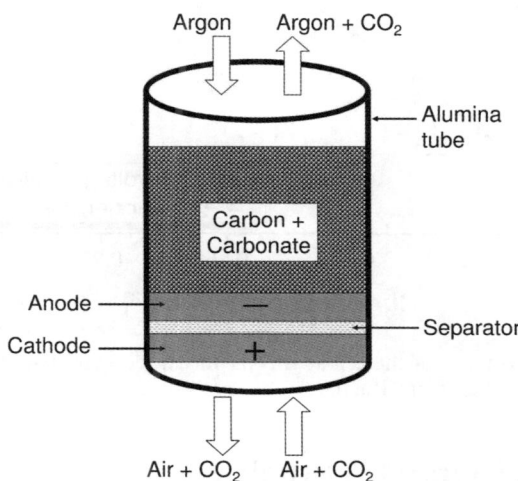

Figure 6.4 Schematic of LLNL's direct carbon fuel cell.[4,7]

dispersed in the molten carbonate mixture on the anode side. The electrolyte used is a liquid carbonate-based salt in the proportion 32 mol% of Li_2CO_3 and 68 mol% of K_2CO_3. The operating temperatures are in the range of 750–850 °C. The main problem of this technique is the high corrosivity of molten carbonates, which limits the choice of materials for cell housing and current collectors.[3,4,11,18]

Between anode and cathode, an inert separator of porous ceramic is used, which is saturated with molten carbonate and allows the carbonate ions to migrate between the two compartments. The cathode contains lithium as the catalytically active material and the anode is composed of carbon particles ($<100\,\mu m$) in the melt with an open-porous nickel layer as current collector.[3,4,7,18,19]

The up-to-date results of Cooper *et al.* are shown in Figure 6.5: power densities up to $210\,mW\,cm^{-2}$ have been obtained at 0.8 V (voltage efficiency, 72%) at an operating temperature of 750 °C. De-ashed coal was used for this measurement (ash content <0.1–0.2%). In addition, the sulfur content was reduced by half in the same process of coal pretreatment. In Figure 6.5 the voltage efficiency is also shown to illustrate the conflicting correlation between power output and efficiency.[6]

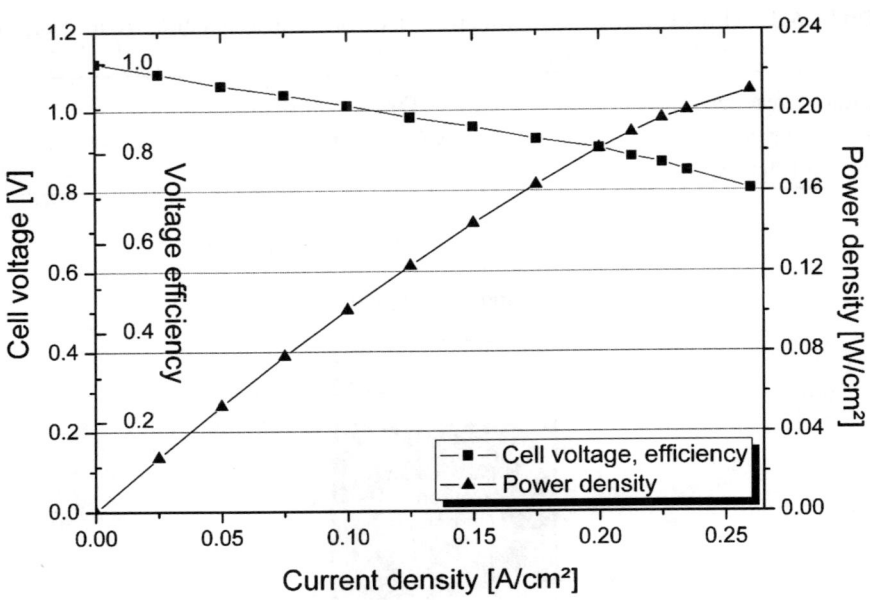

Figure 6.5 Performance of the cell at LLNL, measured with de-ashed coal at 750 °C. Data taken from Rastler.[6]

6.2.2 Molten Hydroxide Electrolyte

Similarly to the molten carbonate electrolyte concept, molten hydroxides such as LiOH, KOH or NaOH can be used as electrolyte in carbon–air fuel cells. In contrast to molten carbonates, molten hydroxides show sufficient ionic conductivities even in the temperature range 400–750 °C. Thus cheaper materials, *e.g.* for containers and cathodes, can be used. Additionally, at lower temperatures the corrosivity of hydroxide melts is reduced. Moreover, according to the Boudouard equilibrium, operating temperatures below 700 °C avoid CO formation. Using molten hydroxides, a higher activity of the electrochemical oxidation of carbon is achieved. The latter is equivalent to a higher carbon oxidation rate at a given over-potential.[4,14,20,21] The half-cell reactions are given in Table 6.4.

Insufficient water content leads to consumption of the liquid electrolyte by reaction of OH^- ions with produced CO_2 which forms carbonates:

$$2OH^- + CO_2 \leftrightarrow CO_3^{2-} + H_2O \tag{6.5}$$

In order to avoid this effect, humidified air has to be used. This increases the water content of the electrolyte, which minimises the production of carbonates, so that the equilibrium reaction (Equation (6.5)) is shifted to the left. Moreover, water vapour brought into the electrolyte increases the ionic conductivity of the melt by adding polar molecules.[4,5,21–23]

Table 6.4   Half-cell reactions for the electrochemical carbon oxidation in molten hydroxide.[4,22,23]

Cathode reaction	$O_2 + 2H_2O + 4e^- \rightarrow 4OH^-$	(1)
Anode reaction	$C + 4OH^- \rightarrow CO_2 + 2H_2O + 4e^-$	(2)
Overall reaction	$C + O_2 \rightarrow CO_2$	(3)

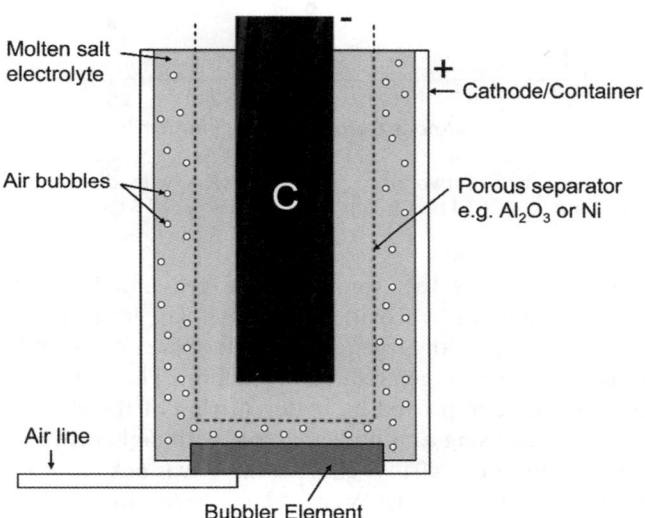

Figure 6.6   Schematic description of the molten hydroxide electrolyte fuel cell of SARA.[1,20,21]

This concept is, for instance, investigated at Scientific Application & Research Associates (SARA) in Cypress (CA, USA). The SARA cell (Figure 6.6) consists of a metal basket, wherein a carbon rod anode is immersed in the molten hydroxide electrolyte. SARA use an eutectic mixture of LiOH, KOH and/or NaOH as electrolyte. The operating temperatures are typically around 600 °C. The cell container, which is made of a non-porous Fe-Ti alloy, is used as cathode and current collector. Iron was doped with titanium in order to enhance the corrosion stability of the material. Moreover, this cathode presents excellent catalytic activity for oxygen reduction. Humidified air is brought into the electrolyte by bubbling. To segregate anode and cathode compartments, a porous membrane of metal or ceramic (*e.g.* Al_2O_3 or Ti-doped Ni) is used. This separator is impermeable to gas bubbles and is able to transport hydroxyl and metal ions between both compartments.[4,5,20,21]

At SARA, Zecevic *et al.* have measured different cell prototypes with different anode surface areas at 630 °C. Representative current–voltage curves are presented in Figure 6.7. The carbon anode (pure graphite rod without

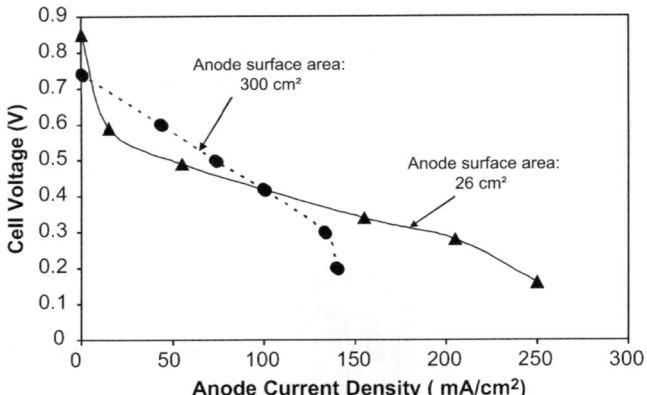

Figure 6.7   Performance of two prototypes at SARA with different anode surface areas at 630 °C. Data taken from Zecevic *et al.*[5]

pretreatment) was used as fuel source in the cell and humidified air was introduced into the sodium hydroxide electrolyte. The lower observed current densities of the prototype with larger anode surface area can be related to mass transport problems of the oxygen reduction reaction. The mass transport to the electrode surface can be improved by better stirring of the electrolyte melt by gas bubbling in the cell using a cathode material with higher surface roughness. An average power output of $40 \, mW \, cm^{-2}$ at 0.3 V was achieved over 540 h and peak power densities of 120–$180 \, mW \, cm^{-2}$ have been observed.[5]

6.2.3   Solid Oxide Electrolyte

Solid oxide fuel cells (SOFCs) show a high fuel flexibility. As oxygen ions are transported through the electrolyte, SOFCs can theoretically be operated on any combustible fuel, thus also on carbon-containing materials. A solid oxide electrolyte does not suffer from corrosion or degradation and is therefore a strong candidate to be used in a DCFC. The conventional material used in SOFCs is yttria-stabilised zirconia (YSZ) which requires operating temperatures above 700 °C for sufficient ionic conductivity. At those high temperatures, carbon monoxide production, according to Boudouard reaction, might lead to a fuel loss. As O^{2-} is the migrating ion, only oxygen is required at the cathode side and carbon at the anode side, so that no recycle loops are needed.[12,24] The half-cell reactions are given in Table 6.5.

An important aspect is to maintain good physical contact between the two solid materials. Carbon particles have to be in close contact with the charge transfer reaction sites at the anode/electrolyte interface.[25]

Direct conversion of carbon in an SOFC system is investigated, for example, at the Bavarian Center for Applied Energy Research (ZAE Bayern) in Garching (Germany). A scheme of the set-up is shown in Figure 6.8. The solid

Table 6.5 Half-cell reactions for the electrochemical carbon oxidation on solid oxide electrolyte.[1]

Cathode reaction	$O_2 + 4e^- \rightarrow 2O^{2-}$	(1)
Anode reaction	$C + 2O^{2-} \rightarrow CO_2 + 4e^-$	(2)
Overall reaction	$C + O_2 \rightarrow CO_2$	(3)

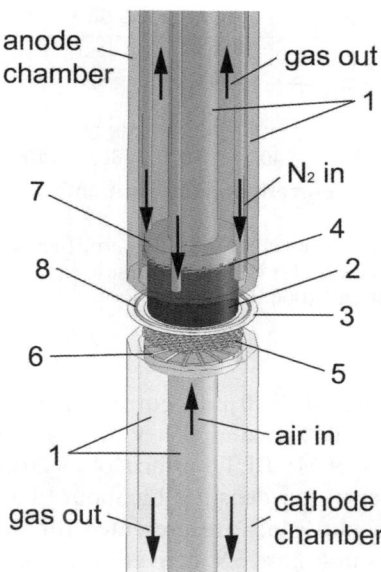

Figure 6.8 Schematic drawing of the cell: (1) aluminium oxide tubes; (2) carbon pellet; (3) YSZ electrolyte; (4 and 5) current collector mesh; (6) cathode flow field; (7) anode flow field; (8) gold wire used as sealing.[26]

electrolyte (YSZ) is placed between two aluminium oxide tubes and the carbon is introduced in form of a pellet on the anode side. The pellet is porous, allowing the product gas CO_2 to exit the cell without having a negative effect on the contact between carbon and anode/electrolyte interface. Commercial electrolyte supported cells with a standard cathode are used. As conventional anode layers are not stable under the actual conditions during carbon electro-oxidation, experiments are made without or with modified anode layers. The contact between carbon pellet and anode/electrolyte is controlled by applying a defined pressure at the back side of the carbon pellet. In order to avoid CO formation, alternative electrolyte materials have to be used to lower operating temperatures. For instance, gadolinium-doped ceria (CGO) offers sufficient ionic conductivity already at 600 °C.[26]

The current–voltage curves of two different carbon materials are presented in Figure 6.9. For the amorphous material (Vulcan XC72: BET surface of

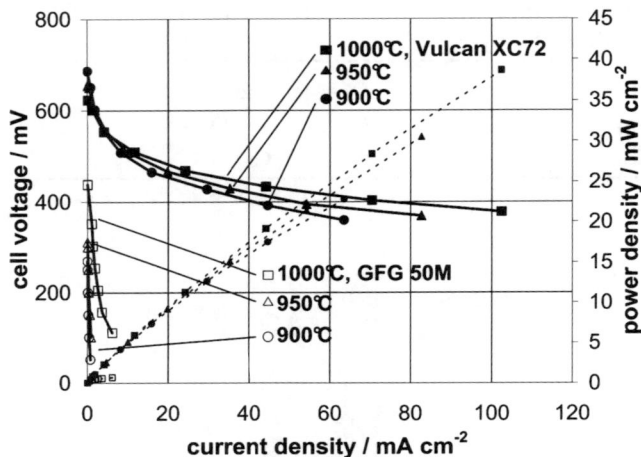

Figure 6.9   Performance of the cell at ZAE Bayern, fuelled by Vulcan XC72 (amorphous carbon) and GFG 50M (graphitic carbon). Operating temperatures between 900 and 1000 °C.[26]

$\sim 250 \, \text{m}^2 \, \text{g}^{-1}$, particle size of 5–10 μm), current densities up to $100 \, \text{mA cm}^{-2}$ have been observed at 1000 °C and 0.4 V (*i.e.* $40 \, \text{mW cm}^{-2}$). In contrast, graphitic carbon (GFG 50M: BET surface of $\sim 20 \, \text{m}^2 \, \text{g}^{-1}$, particle size of 40–60 μm) showed lower power densities by about two orders of magnitude.

In Figure 6.10, Arrhenius plots are presented for the both materials used. According to the results in Figure 6.9, activation energies of Vulcan XC72 and GFG 50M are about $100 \, \text{kJ mol}^{-1}$ and $295 \, \text{kJ mol}^{-1}$, respectively. The activation energy of the amorphous material is about three times lower than those of the graphitic material. This might indicate that the reactivity of the amorphous material is much higher.[26]

6.2.4   Other Concepts

6.2.4.1   *Combining SOFC with a Liquid Carbonate-based Anode*

This concept combines electrolyte and cathode of an SOFC with an anode consisting of carbon particles dispersed in a molten salt. The carbonate melt acts as electrochemical mediator, overcoming the difficulty of good physical contact between carbon and solid oxide electrolyte, and additionally extends the reaction zone into the anode volume. As molten carbonate is never in contact with the cathode layer, cathode corrosion is avoided. Thus the same materials as employed in SOFCs can be used for this component. In contrast to MCFCs, CO_2 recirculation is dispensable, because no CO_2 has to be supplied to the cathode compartment.[12,27]

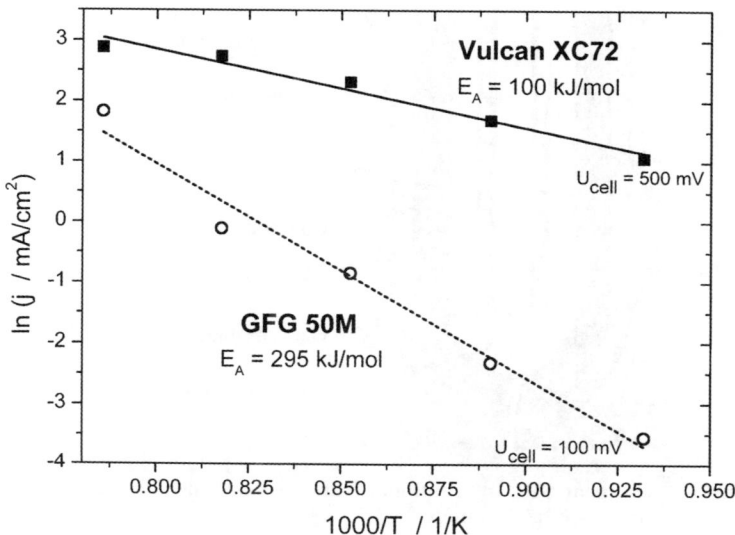

Figure 6.10 Arrhenius plots for two different carbon materials used at ZAE Bayern: The activation energy of Vulcan XC72 is about $100 \, \text{kJ} \, \text{mol}^{-1}$ and for GFG 50M $295 \, \text{kJ} \, \text{mol}^{-1}$.

Table 6.6 Electrode reactions of an SOFC cell combined with a molten carbonate anode; carbonate ions are formed before carbon is electrochemically consumed.[1]

Cathode reaction	$O_2 + 4e^- \rightarrow 2O^{2-}$	(1)
Anode reaction	$2O^{2-} + 2CO_2 \rightarrow 2CO_3^{2-}$	(2)
	$C + 2CO_3^{2-} \rightarrow 3CO_2 + 4e^-$	(3)
Overall reaction	$C + O_2 \rightarrow CO_2$	(4)

On the anode side, the reaction $C + 2O^{2-} \leftrightarrow CO_2 + 4e^-$ occurs in two steps (Table 6.6). First, at the interface solid electrolyte/liquid anode, carbonate ions are formed in the molten salt (Table 6.6, Equation (2)). Then, the electrochemical conversion of carbon takes place through carbonate ions to form carbon dioxide (Table 6.6, Equation (3)).

SOFCs with molten carbonate layer at the electrode/electrolyte interface are studied with different cell geometries, *e.g.* by Irvine *et al.* at St. Andrews University in Scotland and at Stanford Research Institute (SRI International) in Menlo Park (CA, USA).[1,4,12,27–30]

A schematic drawing of the set-up at SRI International is shown in Figure 6.11. A tubular standard SOFC half-cell (YSZ electrolyte and standard cathode) is immersed into the molten salt anode, which incorporates the carbon fuel, typically 30 vol% or more. YSZ is used as electrolyte material, which leads

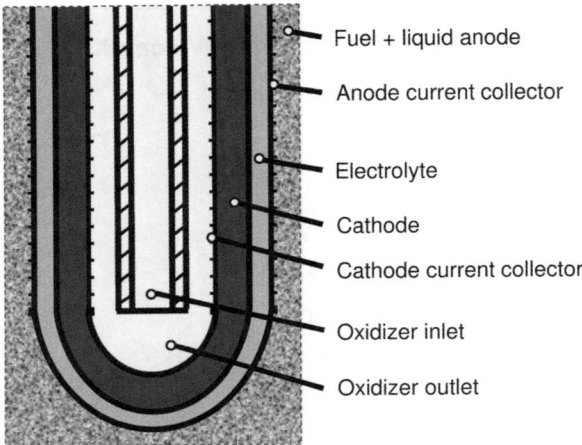

Fuel + liquid anode

Anode current collector

Electrolyte

Cathode

Cathode current collector

Oxidizer inlet

Oxidizer outlet

Figure 6.11   Set-up at SRI International (tubular design): SOFC cell with a liquid carbonate-based anode.[30]

to operating temperatures in the range of 700–950 °C. At these temperatures, carbonates are very corrosive and it was shown that standard YSZ is not stable in lithium-containing carbonate melts. Alternative electrolyte material, for instance CGO, seems to be more resistant and additionally would allow lower operating temperatures.[1,4,12,30]

Up-to-date results from SRI International are shown in Figure 6.12. A peak power density of 300 mW cm^{-2} at 0.5 V has been observed at 700 °C. A power output of 170 mW cm^{-2} has been achieved at 0.84 V with a voltage efficiency of 82%. Pulverised raw coal was used as fuel in the cell (powder of 150 μm size or finer).[6]

6.2.4.2   Carbon Conversion with a Preceding Chemical Reaction Step

Concepts of a DCFC also exist where carbon is not involved in the electrochemical reaction step but in a preceding endothermic chemical reaction. Thus, thermal energy is converted into chemical energy by means of formation of an intermediate product. The latter is subsequently oxidised in an electrochemical reaction, but with lower thermodynamic efficiency compared to direct carbon electro-oxidation. This can be seen, for example, for CO or Sn oxidation in Figure 6.1, but as a part of the generated heat is converted into chemical energy of the intermediate, overall efficiency might be comparable to the direct carbon conversion process. For high reaction rates, providing a sufficient amount of educt for electro-oxidation is essential. The electrochemical reaction is always coupled to the preceding chemical reaction. This fact might lead to an inefficiency regarding kinetics. The advantage of such a concept is that there are

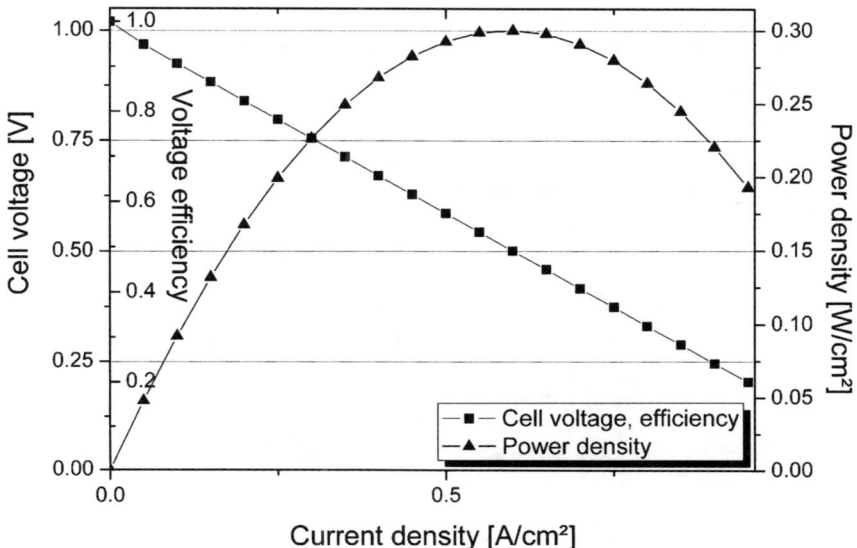

Figure 6.12   Performance of the cell at SRI International, measured with pulverized raw coal at 700 °C. Data taken from Rastler.[6]

fewer problems caused by impurities because only the pure carbon of the fuel material is involved in the preceding chemical reaction. As only the intermediate product is electrochemically oxidised at the anode/electrolyte interface, the contact between possible impurities, contained in carbon material, and sensitive parts of the anode/electrolyte are hindered or can be fully prevented. In the following, two different concepts are described that use a preceding chemical reaction step within conversion of carbon into electricity.

6.2.4.2.1   Solid Oxide Electrolyte Combined with Fluidised-bed Technologies.

At Clean Coal Energy (CCE) in Menlo Park (CA, USA) an SOFC is combined with fluidised-bed technologies. As CO_2 is used to fluidise carbon particles, the gaseous fuel CO is produced *in situ via* the Boudouard reaction (Table 6.7, Equation (6.2)), which is subsequently electro-oxidised at the SOFC anode to CO_2.[25,31]

A schematic drawing of the set-up is shown in Figure 6.13. The electrolyte of the SOFC is made of YSZ requiring typical operating temperatures of 900 °C. The anode materials have to be stable during CO electro-oxidation.

Lee *et al.* from CCE performed experiments with synthetic carbon and biomass from almond shells in CO_2 atmosphere at 900 °C (Figure 6.14). For comparison, the cell performance without carbon, but with pure CO as fuel, has been determined. The observed power output of synthetic carbon in a CO_2 fluidised bed is only a bit lower compared to pure CO, while biomass from

Table 6.7   Electrode reactions of a DCFC combining SOFC with fluidised-bed technologies.[25,31]

Cathode reaction	$O_2 + 4e^- \rightarrow 2O^{2-}$	(1)
Anode reaction	$C + CO_2 \rightarrow 2CO$	(2)
	$2CO + 2O^{2-} \rightarrow 2CO_2 + 4e^-$	(3)
Overall reaction	$C + O_2 \rightarrow CO_2$	(4)

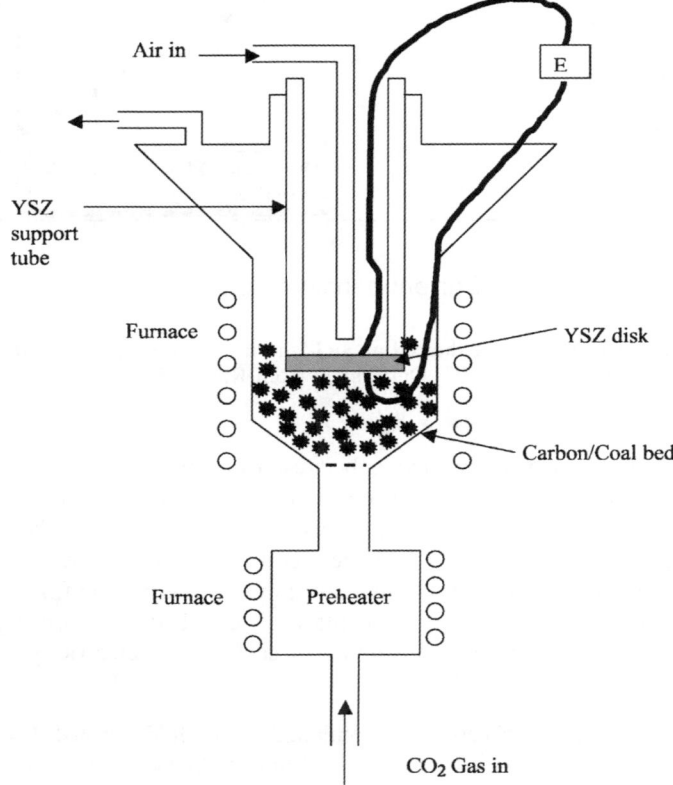

Figure 6.13   Schematic of a DCFC combining SOFC with fluidised-bed technologies. Reproduced, with permission, from Li *et al.*[25] © 2008 The Electrochemical Society.

almond shells only offers low performance. The peak power density of synthetic carbon is located at about $140\,\mathrm{mW\,cm^{-2}}$ at a cell voltage of approximately $0.5\,\mathrm{V}$.[31]

6.2.4.2.2   Solid Oxide Electrolyte Combined with a Liquid Tin-based Anode. The company CellTech Power in Westborough (MA, USA) developed a

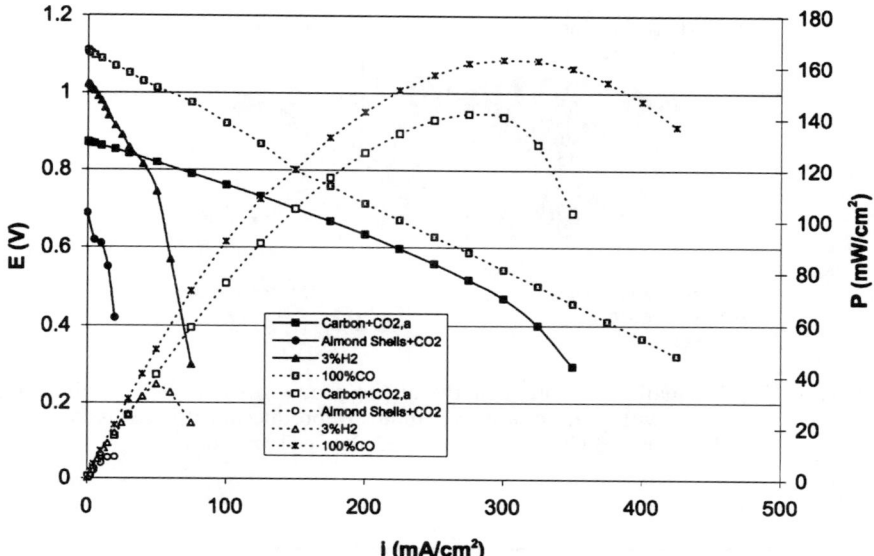

Figure 6.14 Comparison of the (I–V–P) curves at 900 °C of synthetic carbon and biomass from almond shells in CO_2 atmosphere. Measurements with 100% CO and 3% H_2 (balance N_2) are used as benchmarking. Reproduced, with permission, from Lee *et al.*[31] © 2008 The Electrochemical Society.

Table 6.8 Electrode reactions of a DCFC combining an SOFC with a liquid tin-based anode.[12,32]

Cathode reaction	$O_2 + 4e^- \rightarrow 2O^{2-}$	(1)
Anode reaction	$C + SnO_2 \rightarrow CO_2 + Sn$	(2)
	$Sn + 2O^{2-} \rightarrow SnO_2 + 4e^-$	(3)
Overall reaction	$C + O_2 \rightarrow CO_2$	(4)

concept which combines an SOFC with a liquid tin-based anode. In this case, Sn is electrochemically oxidised to SnO_2 at the SOFC anode. Sn is regained in a chemical reaction step (Table 6.8, Equation 6(2)), where carbon and SnO_2 form Sn under CO_2 release.[32,33]

A schematic of the process is given in Figure 6.15. The ceramic electrolyte used is typically made of YSZ and the operating temperature is 1000 °C.

Performance data of the cell at CellTech Power are shown in Figure 6.16. At an operating temperature of 1000 °C a power density of 213 mW cm^{-2} has been observed at 0.64 V (voltage efficiency of 82%) and the peak power density at 0.57 V is about 290 mW cm^{-2}. Pulverised raw coal was used as fuel for the measurements.[6]

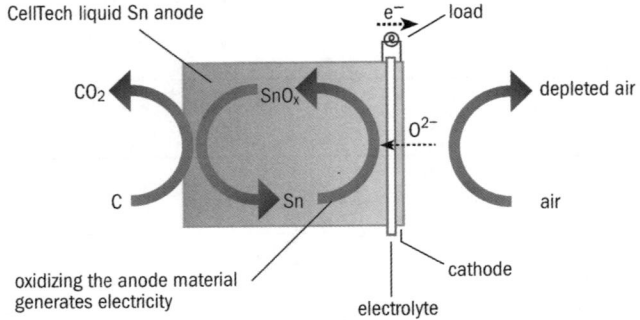

Figure 6.15 Schematic description of the anode process of a DCFC combining a SOFC with a liquid tin-based anode. Reproduced, with permission, from Heydorn and Crouch-Baker.[2] © 2006 CellTech Power.

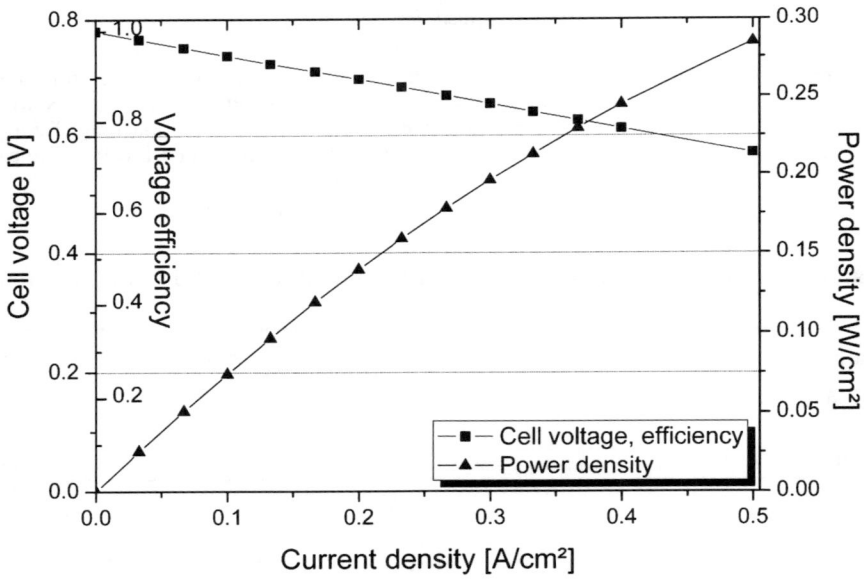

Figure 6.16 Performance of the cell at CellTech Power, measured with pulverised raw coal at 1000 °C. Data taken from Rastler.[6]

Moreover, a liquid tin anode SOFC has been demonstrated for direct conversion of any carbonaceous fuel including solid fuels (carbon, coal, biomass), liquid fuels (*e.g.* gasoline, diesel) and gaseous fuels, such as natural gas. In addition, this anode has the advantage of showing a good tolerance against fuel impurities.

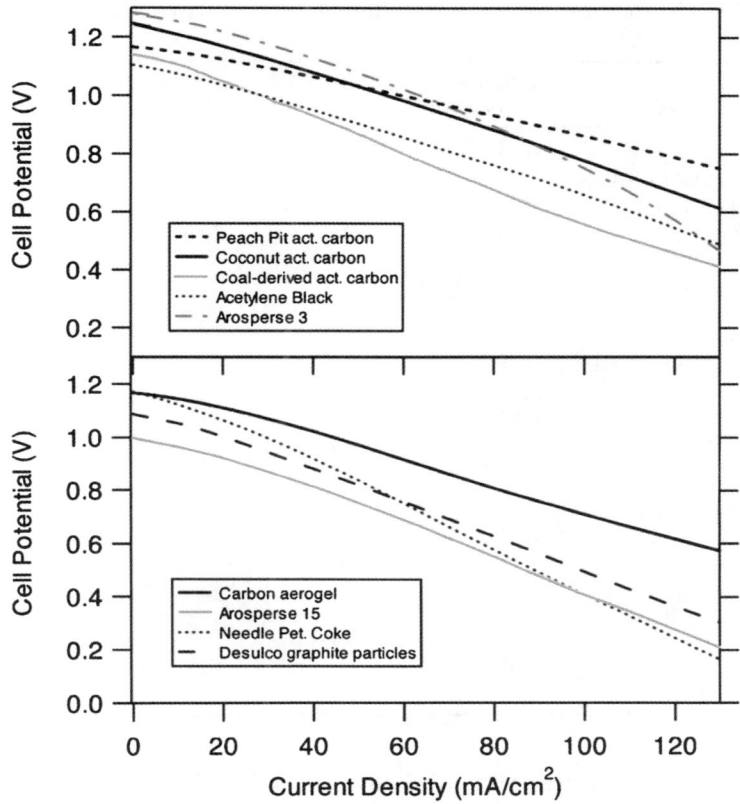

Figure 6.17 Performance of the cell at LLNL for nine different carbon-containing materials at 800 °C. Reproduced, with permission, from Cherepy *et al.*[7] © 2005 The Electrochemical Society.

6.3 Comparison of Different Carbon Fuels

As already mentioned, direct comparison of cell performances of different DCFC concepts is difficult, because it depends on the used carbon fuel material, operating temperature and the concept itself (anode/electrolyte material). In this chapter, measurements performed on two selected systems, but with various carbon fuels are shown, in order to illustrate the influence of different carbon properties on cell performance.[4]

Carbon properties, such as crystallisation, particle size, specific surface area and electrical conductivity, have more or less an effect on the reactivity of the carbon fuel. Cooper and co-workers at LLNL have measured nine different types of carbon used as fuel in the DCFC based on molten carbonate electrolyte (Figure 6.17). Carbon properties, such as particle size and degree of crystallinity were investigated with electron microscopy and X-ray diffraction, respectively. In Table 6.9, the main properties of the carbon materials are presented. They are

Table 6.9 Properties of the nine different carbon-based materials used by LLNL.[7]

Carbon material	Crystal-lisation rate	Surface area $(m^2 g^{-1})$	Primary parti-cles size (nm)	Type of particles
Peach pit activated carbon, AW (biochar-derived carbon)	0.558	>1000	30–3000	–
Coconut act carbon, AW, milled (biochar-derived carbon)	0.565	1050	20	Spheres
Coal-derived act carbon, AW, milled	0.649	950	60–10 000	–
Acetylene black, from acetylene	0.696	75	40	Spheres
Aerosperse 3, from furnace oil	0.652	75	30	Spheres
Carbon aerogel microspheres, pyr-olysed at 1050 °C	0.545	1225	20 000–100 000	Porous spheres
Aeroperse 15, thermal black, from methane	0.665	15	290	Spheres
Needle petroleum coke calcined at 1400 °C, milled	0.821	0.4	3000–100 000	Needles
Desulco graphite particles	0.972	9	5000–30 000	Stacked sheets

very variable, *e.g.* surface area varies between 9 and 1225 $m^2 g^{-1}$ and crystallinity index between 0.545 and 0.972. Amorphous carbons (*e.g.* biochar, aerogel carbon), which are characterised by poor crystallisation and large surface area, are in general more reactive compared to graphitic materials. But it can be seen that neither variations among the samples in primary particle size, surface area, nor degree of crystallinity have a cumulative effect greater than a factor of about 2 on carbon fuel performance for their measurements. This is in contrast to the results of ZAE Bayern (DCFC with solid oxide electrolyte). It has been found that the reactivity of amorphous carbon is two orders of magnitude larger than those of graphitic carbon (Figure 6.9). Therefore, the cumulative effect of carbon properties on cell performance varies strongly with the used concept.[1,4,7]

At SRI International, a variety of carbon-containing fuels (*e.g.* plastic) have been tested between 700 and 950 °C with a DCFC, which combines an SOFC and a molten carbonate electrolyte (Figure 6.18). These results again show the variety of carbon fuels that can be used in a DCFC system.

Besides carbon morphology, impurities in the carbon fuel may influence cell performance and long term stability. For instance, Cooper *et al.* have performed experiments, focused on the impurities of the sample in the cell (*e.g.* sulfur and ash). A raw petroleum coke contains on average 2.9 wt% of sulfur and 0.3 wt% of ash and shows decreased discharge rates by a factor of 2 compared to those of purified petroleum coke. The presence of sulfur might

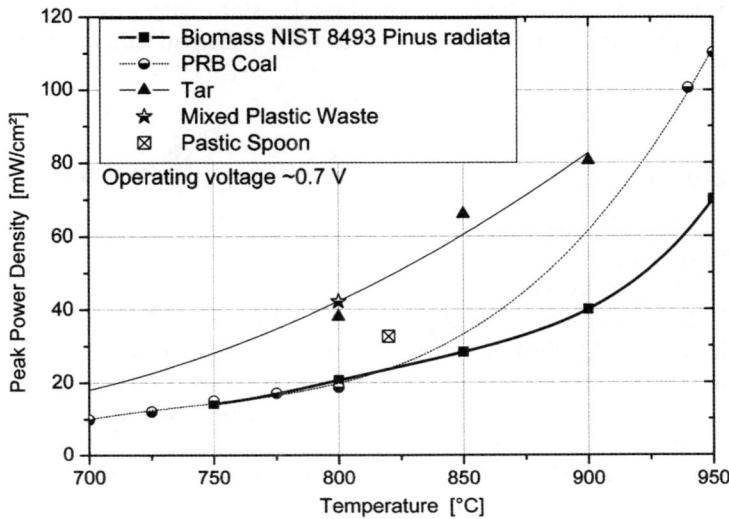

Figure 6.18 Comparison of peak power densities attained for different carbon-based materials in a DCFC between 700 and 950 °C at SRI International. Data taken from Balachov *et al.*[30]

degrade the cell because of corrosion of anode materials. Ash accumulating in liquid electrolytes might lead to its degradation and solid electrolytes could be blocked. The effect of carbon impurities (such as sulfur, mercury, ash) and moisture on the cell performance and the system's life time varies strongly with the used concept.[4,7]

6.4 Conclusions

Direct carbon fuel cells allow the direct conversion of the chemical energy of carbon into electricity with a high thermodynamic efficiency (slightly exceeding 100%) and a fuel utilisation of up to 100%. Various carbon-containing materials can be used, such as coal or renewable materials (*e.g.* biomass), which are directly available fuels. However, because of the utilisation of a solid fuel, DCFC systems have to meet some special requirements, leading to modified fuel cell concepts with optimised materials and design configurations. In this chapter, different DCFC techniques have been presented and the feasibility of such systems has been shown. The reactivity of different carbon materials is strongly dependent on their properties, such as particle size, BET surface area, degree of crystallinity and impurities. Some of the DCFC technologies require a pretreatment of the raw coal beyond pulverisation to remove impurities. In order to avoid preprocessing steps, DCFC systems should be optimised to operate on raw carbon materials. With the use of biofuels, a negative CO_2 effect is possible. Moreover, both long-term stability of the reviewed DCFCs and power output have to be optimised using efficient catalyst materials.[4,7]

In the future, the main goal will be to realise stationary power plants using DCFCs in the range of some megawatts with electrical efficiency exceeding 60%.[6,19,33]

References

1. D. Rastler, *Technical report 1013362*, EPRI, 2006.
2. B. Heydorn and S. Crouch-Baker, *Fuel Cell Rev.*, 2006, **2**(6), 15.
3. J. F. Cooper, US patent 2006/0057443 A1, 2006.
4. D. Cao, Y. Sun and G. Wang, *J. Power Sources*, 2007, **167**(2), 250.
5. S. Zecevic, E. M. Patton and P. Parhami, *Carbon*, 2004, **42**(10), 1983.
6. D. Rastler, Technical report 1016170, EPRI, 2008.
7. N. J. Cherepy, R. Krueger, K. J. Fiet, A. F. Jankowski and J. F. Cooper, *J. Electrochem. Soc.*, 2005, **152**(1), A80.
8. K. Hemmes and M. Cassir, *Proceeding of the 2nd International Conference on Fuel Cell Science, Engineering and Technology, Rochester, NY, USA*, June 14–16, 2004, pp. 395.
9. J. S. J. Van Devanter and P. R. Visser, *Thermochim. Acta*, 1987, **111**, 89.
10. W. W. Jacques, Patent US 555511 A, 1896.
11. A. L. Dicks, *J. Power Sources*, 2006, **156**(2), 128.
12. K. Pointon, B. Lakeman, J. Irvine, J. Bradley and S. Jain, *J. Power Sources*, 2006, **162**(2), 750.
13. J. F. Cooper, URCL-TR-210346, LLNL, 2005.
14. P. V. Pesavento, US patent 2001/6200697 B1, 2001.
15. W. H. A. Peelen, M. Olivry, S. F. Au, J. D. Fehribach and K. Hemmes, *J. Appl. Electrochem.*, 2000, **30**(12), 1389.
16. W. H. A. Peelen, K. Hemmes and J. H. W. de Wit, *High Temp. Mater. Processes*, 1998, **2**(4), 471.
17. J. F. Cooper, Presented in *Direct Carbon Fuel Cell Workshop*, NETL, Pittsburg, PA, USA, 30th July, 2003.
18. J. F. Cooper, R. Krueger and N. Cherepy, US patent 2002/0106549 A1, 2002.
19. J. F. Cooper, Presented in *International Conference on Fuel Cell Science, Engineering and Technology, Rochester, NY, USA*, June 14–16, 2004.
20. S. Zecevic, E. M. Patton and P. Parhami, US patent 2005/0282063 A1, 2005.
21. S. Zecevic, E. M. Patton, P. Parhami, Presented in *Direct Carbon Fuel Cell Workshop*, NETL, Pittsburg, PA, USA, 30th July, 2003.
22. SARA, Presented in *Fuel Cell Seminar, Direct Carbon Fuel Cell Workshop*, Palm Springs, CA, USA, 14th November, 2005.
23. G. A. Hackett, J. W. Zondlo and R. Svensson, *J. Power Sources*, 2007, **168**(1), 111.
24. A. Weber and E. Ivers-Tiffée, *J. Power Sources*, 2004, **127**, 273.
25. S. Li, A. C. Lee, R. E. Mitchell and T. M. Gür, *Solid State Ionics*, 2008, **179**, 1549.

26. S. Nürnberger, R. Bußar, P. Desclaux, B. Franke and U. Stimming, *Energy Environ. Sci.*, 2010, **3**, 150.
27. S. L. Jain, Y. Nabae, B. J. Lakeman, K. D. Pointon and J. T. S. Irvine, *Solid State Ionics*, 2008, **179**, 1417.
28. S. L. Jain, B. J. Lakeman, K. D. Pointon and J. T. S. Irvine, *J. Fuel Cell Sci. Technol.*, 2007, **4**(3), 280.
29. Y. Nabae, K. D. Pointon and J. T. S. Irvine, *J. Electrochem. Soc.*, 2009, **156**(6), B716.
30. I. I. Balachov, L. H. Dubois, M. D. Hornbostel and A. S. Lipilin, Presented in *Fuel Cell Seminar, Direct Carbon Fuel Cell Workshop*, Palm Springs, CA, USA, 14th November, 2005.
31. A. C. Lee, S. Li, R. E. Mitchell and T. M. Gür, *Electrochem. Solid-State Letts.*, 2008, **11**(2), B20.
32. W. McPhee and T. Tao, Presented in *Fuel Cell Seminar & Exposition*, Phoenix, AZ, USA, October 28-30, 2008.
33. T. Tao, Presented in *Fuel Cell Seminar, Direct Carbon Fuel Cell Workshop*, Palm Springs, CA, USA, 14th November, 2005.

Part 4
Modelling and Lifetime Prediction

Introduction

In stationary applications lifetimes of up to 100 000 h are expected. This is nothing that can be validated in laboratories anymore, and necessitates accelerated testing methods, as in many engineering areas. Unfortunately, the predominately used parameters of increased temperature, humidity and pressure influence the operation of fuel cells directly. Therefore the search for accelerated testing methods is paramount in giving guarantees for fuel cell products and opening up markets. Part of this exercise is the integration of degradation phenomena into fuel cell models in order to be able to predict the lifetime and effects of degradation.

CHAPTER 7

Integrating Degradation into Fuel Cell Models and Lifetime Prediction

ANDREAS GUBNER

Enerday GmbH, Speicherstr. 3, 17033 Neubrandenburg, Germany

7.1 Introduction

Modelling of fuel cells takes various approaches. Which approach to choose is determined by the intended scope. For example, at stack level, very often computational fluid dynamics (CFD) models are employed that deliver three-dimensional (3D) temperature and current density fields occurring inside a stack. Fuel cell systems contain many more components than just the stack, so the individual components including the fuel cell stack must be geometrically simplified. Hence spatially one-dimensional (1D) models describing the change of local quantities, *e.g.* along the flow path (gas channel), are used or the spatial resolution is dropped completely. The latter results in models whose thermal behaviour is characterised by a single temperature instead of a temperature field. Models describing a stationary state tend to be mathematically simpler than models for dynamic simulations which need the time as an additional independent variable. Exceptions arise if the stationary model is comprised of a set of nonlinear equations: An iterative procedure for finding the stationary solution must be used anyway. Then the effort of creating a dynamic model that finds the stationary solution by stepping forward in time may turn out about equal. The aim of this chapter is to suggest a way to combine simple time dependent degradation models with a simple fuel cell (FC) model All properties

RSC Energy and Environment Series No. 2
Innovations in Fuel Cell Technologies
Edited by Robert Steinberger-Wilckens and Werner Lehnert
© Royal Society of Chemistry 2010
Published by the Royal Society of Chemistry, www.rsc.org

of the FC remain constant in time. Degradation is introduced as a concept of deterioration of properties over time based on rate equations. This chapter will limit itself to solid oxide fuel cell (SOFC) modelling as an example because SOFC models can be based on ideal gas behaviour eliminating the need to deal with an additional liquid phase. However, it is possible to extend the approach to other FC types as well because of the modular concept. The degradation and the FC model may be replaced by different models whereby it is possible to reuse existing stationary FC models. The description will limit itself to isothermal, thus rather small stacks as used at Forschungszentrum Jülich (in the following just abbreviated as Jülich) in their SOFC development programme.[1] The author worked in Jülich for 5 years where a practical approach is pursued. This article follows this approach. The model is kept simple so it can be easily implemented by the reader and used for evaluating own test results. The discussed experimental results were generated in Jülich. This article now appends the change of properties over time to an isothermal 1D model considering the change of the local current density along the gas channel due to locally different compositions. The model is isothermal because it focuses on small short stacks (one or two planes) in a large furnace. This is a typical laboratory set-up that is typically used for performance (current–voltage (I–V) curves) measurements long-term degradation monitoring (endurance tests). In Jülich degradation rates are measured in terms of the overall area specific resistance (ASR) as an alternative in addition just recording the stack/cell voltage over time.[2] The cell voltage depends on particular operation conditions whereas the overall ASR is depends on material properties and the quality of their interfaces. However, an endurance test must be interrupted a considerable number of times for recording an I–V curve which is a considerable disadvantage. The model developed in this article can be used to determine ASR-based degradation rates without I–V curves. The ASR appears as a parameter in the model derivable by a given stack current (or current density) and stack (or cell) voltage.

7.2   Background

Since technically relevant SOFC stacks can only be cooled by the air stream flowing through the cathode, simulating isothermal SOFC operation certainly is a quite idealised approach to real-world conditions that is not suitable for thermally self-sustaining (ideally adiabatic) stacks running at high fuel utilisation.

However, assuming isothermal operation is reasonable when dealing with small-scale laboratory cells and short stacks (in the watt range) placed in comparatively large furnaces especially when operated at low (relative) fuel utilisation when the heat production rate in the cell or the stack is small compared with the heat transfer rate between the gas streams and the SOFCs solid body.

Typically, these small scale SOFCs are used to characterise or to evaluate the electrochemical performance of new cells, materials, concepts *etc.*[3–6] The stack performance is usually described by plots of its electrical current–voltage (I–V) behaviour recorded at constant operating temperature, fuel and air flow rates. The operating temperature is often assumed to be constant everywhere in the SOFC. It is measured directly by thermocouples inside a cell (of a stack) or, alternatively, simply the furnace temperature is taken. The flow rates of fuel and air are often so high that the overall fuel utilisation in the cell or the stack is quite low (approx. 8–15%).

Another area in which an isothermal 1D model describing the local change of the current density and the composition along a gas channel can be a sensible approach is the investigation of the long-term degradation behaviour as introduced above. Eliminating irreversible performance degradation, or at least its reduction, is still one of the key challenges at least for planar SOFCs with metallic interconnector plates.

When investigating long-term behaviour, typically the cell voltage is recorded as a function of the operating time, t, while the electric current output, the operating temperature and all other parameters are kept constant. From this, a degradation rate in volts or % per 1000 h can be calculated.

Usually, the SOFC performance is characterised by the ASR, *e.g.* in Ω cm^2, which represents the SOFCs current–voltage behaviour in the neighbourhood of the chosen operating point quite well if the cell voltage is a linear function of the electric current density. In other words, using a constant ASR is a reasonable simplification if the SOFC shows an overall ohmic current–voltage behaviour. This is usually the case. Nonlinear effects due to electron transfer polarisation manifest themselves in the vicinity of the open circuit voltage (OCV) and a progressively increasing mass transfer-related resistance is usually not seen in Jülich at reasonable SOFC operating currents.

The ASR is usually calculated by taking the slope at the measured I–V curve around the operating point. This method produces an error since the Nernst voltage is a function of the fuel and air composition. The current at a given cell voltage is a local quantity which changes along the channel.

The 1D model developed in this chapter offers a possibility of determining the ASR from I–V points so that the described gas composition effect is properly accounted for by numerical integration of the local I–V characteristics.

Since only one single I–V pair at the operating point is required, it is not necessary to measure an I–V curve to obtain sufficiently accurate ASR values. The ASR is independent of the operating current whereas the cell voltage depends on the operating current. Since the operating current generates the voltage loss, $- i \cdot \hat{R}$, it is difficult to compare cell voltage based degradation rates of two fuel cells if they are operated at different current densities. Recording the ASR instead of the cell voltage is recommended because the operating current is a factor that amplifies the voltage loss. Recording the ASR as it can be calculated by the 1D model eliminates the described superficial influence of the operating current and allows a better comparability of the results. This allows a clear quantification of the true impact the operating current has on the degradation rate. Of course, assuming a current-independent polarisation resistance is a crude simplification that must be justified. The possibility of expressing the degradation rate in terms of the ASR was already discussed.[7] The 1D model offers an *in situ* ASR calculation and it can be used for post-calculating the ASR(t) information from endurance tests that have been completed already. This is possible because it is not necessary to determine the slope at the I–V curve which requires interrupting the endurance test and to measure an I–V curve or at least a second point, respectively.

7.3 Basic Solid Oxide Fuel Cell Modelling Theory

An overview of the current approaches and techniques discussing 1D, two-dimensional (2D) and 3D models is found in the literature.[8] A list of the symbols used in this chapter is given in section 7.10 (p. 245).

The result of the following derivation is a basic 1D model allowing the simulation of stationary SOFC operation with (humidified) hydrogen and air at constant temperature and pressure in co-flow mode. It is straightforward to extend this isothermal model by an energy balance for the fluid streams and the heat conduction equation (Fourier's law) for including non-isothermal behaviour as well. The pressure loss in the gas channels only has a minor impact on the Nernst voltage so it can be neglected:

$$\frac{\partial V_{\text{Nernst}}}{\partial p} = \frac{1}{4F}\frac{RT}{p} \tag{7.1a}$$

At $T = 1073.15\,\text{K}$ and $p = 100\,\text{kPa}$, the Nernst voltage increases by just $0.23\,\text{V}\,\text{MPa}^{-1}$ or $0.023\,\text{V}\,(100\,\text{kPa})^{-1}$.

A fuel cell's I–V behaviour is generally governed by a sum of individual voltage losses stemming from different sources that reduce the OCV:[9]

$$V_{\text{cell}} = V_{\text{Nernst}} - \eta^{(A)}(i) - i\hat{R}_{\Sigma}^{(\Omega)} + \eta^{(C)}(i) \tag{7.1b}$$

In this equation, V_{cell} is the operating cell voltage in V, V_{Nernst} is the Nernst voltage in V, the functions $\eta^{(A)}(i)$ and $\eta^{(C)}(i)$ are the voltage losses in V due to charge transfer polarisation and $\hat{R}_{\Sigma}^{(\Omega)}$ in $\Omega\,\text{cm}^2$ is the sum of all ohmic resistance losses. This includes contact resistances and also the ionic resistance of the electrolyte. Although the latter is not an electronic resistance, it still exhibits ohmic behaviour for SOFC electrolytes. The Nernst voltage, V_{Nernst}, is based on thermodynamic considerations, *i.e.* it is valid for the thermodynamic equilibrium state. If internal losses (*e.g.* due to mixed potential formation or noticeable electronic conductivity of the electrolyte) occur, then the OCV is different from the Nernst voltage. Hence identifying the OCV with the Nernst voltage is not valid automatically and must be checked from case to case.

Using a constant, V_{cell}, implies that the whole cell (plane of a stack) exhibits the same cell voltage everywhere. That allows a significant simplification of the theory and is allowable as long as the electrical conductivity of the interconnector plate, or the bipolar plate, respectively, is sufficiently high. This is a reasonable simplification at least for metallic interconnector plates. It seems taking a constant, V_{cell}, is a widely accepted assumption.[9]

Voltage losses due to mass transfer limitation are neglected. This simplification is motivated by the technical scope of this chapter. It is not intended to simulate SOFC behaviour at current densities much greater than the envisaged operating current at which the cell voltage is high enough to yield acceptable efficiency values.

In order to calculate the charge transfer over-potential at both electrodes as a function of the current density, the functions $\eta^{(A)}(i)$ and $\eta^{(C)}(i)$ must be known for each electrode. The well-known Butler–Volmer (BV) equation describes the charge transfer polarisation:[10]

$$i = i_0 \left[\exp\left(\frac{\beta z F \eta}{RT} \right) - \exp\left(-\frac{(1-\beta)zF\eta}{RT} \right) \right] \qquad (7.2)$$

Here i_0 is the exchange current density in $A\,m^{-2}$, β is the transfer coefficient, z is the number or electrons transferred in the reaction and η is the over-potential in volts.

One BV equation must be specified for the anode and one for the cathode so there are four unknown parameters. It must also be kept in mind that i_0 at the anode depends on the gas phase's H_2 and the electrode's H^+ concentration and at the cathode it depends on the gas phase's O_2 and the electrode's O^{2-} concentration. Hence additional parameters for describing this concentration dependence occur as well. Finally, it also depends on the temperature in an Arrhenius-type fashion so additional unknown parameters for the anode and the cathode activation energies appear on top of all that. A far more thorough discussion about how to obtain usable data for all those unknown BV parameters is given in the literature.[11] It seems that applying the BV equation to practical calculations that have a technical relevance is a sophisticated challenge. A recent review of the current SOFC modelling indicates that work in this area is still needed.[12] One paper, for example, states data for the full BV mechanism but without a discussion on the choice of the charge transfer coefficient β.[13] That data, originating from the late 1980s to the mid 1990s, is also used for non-isothermal 1D models introduced by Li and Chyu.[14] A second paper using the full BV equation was presented by Costamagna and Honegger.[15] Those equations were also used by Bove *et al.* and currently went into an updated review on SOFC modelling.[16,17] These literature findings may emphasise how difficult it is to obtain satisfactory data for employing the BV equation. Hence, it is quite understandable that there is a long history in SOFC modelling of simplifying the BV equation: Achenbach used linearised expressions and an equivalent electric circuit model.[18] Those equations were still used by *e.g.* Roos *et al.* and even more recently by Pfafferodt *et al.*[19,20] A slight simplification without a linearisation is achieved by prescribing $\beta = 0.5$ which obviously eliminates one unknown parameter and allows inverting the BV equation analytically.[14,21–23]

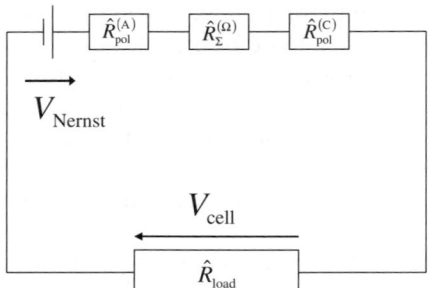

Figure 7.1 Electrical resistance network model for the I–V behaviour.

The simplest and most practical solution to the problem of the lack of BV parameters is found when the BV equations for the two electrodes can be linearised. Then the individual charge transfer over-potentials can be treated as ohmic voltage losses as well so the entity of all voltage losses can be expressed in terms of ohmic resistances and the SOFC model can be expressed in terms of an equivalent electric circuit as shown in Figure 7.1. It will be shown that at least for the cells from Jülich this seems possible for operating temperatures equal and greater than 700 °C.

All electric resistances shown in Figure 7.1 can be replaced by the single value:

$$\hat{R}_\Sigma = \hat{R}_{pol}^{(A)} + \hat{R}_\Sigma^{(\Omega)} + \hat{R}_{pol}^{(A)} \tag{7.3}$$

This quantity is often referred to as the ASR expressible in $\Omega\,cm^2$ as introduced above. The ASR can be determined quite easily by taking the slope of measured I–V plots if a few prerequisites are met which are discussed in this chapter. It is a strong function of the temperature because the charge transfer polarisation at each single electrode and the ionic conductivity of the electrolyte exponentially depend on the temperature. This would certainly require three individual activation energy values but an attempt can be made to replace these by a single lump-sum value according to:

$$\hat{R}_\Sigma = \hat{R}_\Sigma^\infty \exp(E_A/(RT)) \tag{7.4}$$

A tolerable temperature range for this rather crude simplification is found as long as a fairly straight line can be identified in the Arrhenius plot. Using this simplification, the following equation is obtained for the I–V behaviour that now replaces Equation (7.16):

$$V_{cell} = V_{Nernst} - i\hat{R}_\Sigma \tag{7.5}$$

Equation (7.5) is easily solved for i:

$$i = (V_{\text{Nernst}} - V_{\text{cell}})/\hat{R}_\Sigma \qquad (7.6)$$

It is important to note that i must be understood as a local current density expressible in A cm^{-2} because V_{Nernst} is not constant throughout the cell. V_{Nernst} depends on the concentrations of all contributing species H_2, H_2O at the anode side and O_2 at the cathode side. Since the H_2 is turned into H_2O by consuming O_2 as the gases flow along both electrodes, the local concentrations are different at every location in the SOFC. The concentrations do not vary independently; they are linked by chemical stoichiometry and Faraday's law as shown in Figure 7.2 for a co-flow configuration of H_2 and air or oxygen, respectively.

While H_2 and O_2 are flowing along their gas channels at the anode side and the cathode side, a fraction, $d\dot{n}_{H_2}$, of the H_2 mole flow rate, $\dot{n}_{H_2}$, and a fraction, $d\dot{n}_{O_2}$, of the O_2 mole flow rate, $\dot{n}_{O_2}$, in mol s^{-1} is converted into H_2O at every catalytically active area element of both electrodes dA of the membrane electrode assembly (MEA) which produces the electric current $i\,dA$ in amps. Hence the current density, i, is a local quantity that changes with the gas channel length denoted by x (measured, for example, in cm):

$$i = dI/dA \qquad (7.7)$$

With this, Equation (7.5) becomes:

$$\frac{dI}{dA} = \frac{V_{\text{Nernst}}(I) - V_{\text{cell}}}{\hat{R}_\Sigma} \qquad (7.8)$$

Equation (7.8) is an ordinary differential equation describing the evolution of the electric current produced in an SOFC as a function of the active area A. It

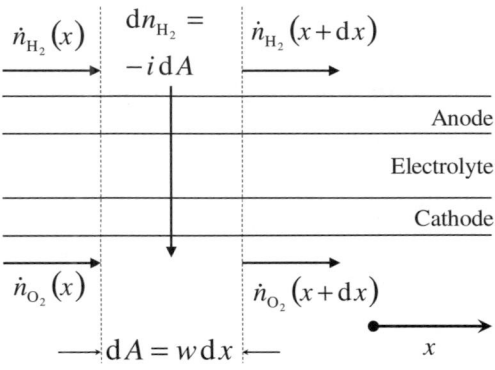

Figure 7.2 Local character of the electric current production.

can be linked to the length of the gas channel by the width of the MEA $dA = w\,dx$ so that Equation (7.8) can be expressed in terms of the gas channel length, x:

$$\frac{dI}{dx} = \frac{w}{R_\Sigma}\left(V_{\text{Nernst}}(I) - V_{\text{cell}}\right) \tag{7.9}$$

This approach was already introduced by Standaert *et al.* for molten carbonate fuel cells (MCFCs)[24] and it was recently applied to SOFCs.[25] The goal of the two papers was to derive analytical models with minimum CPU demands but not to neglect the local nature of the electrochemical current production. In order to solve *e.g.* Equation (7.9) analytically, the logarithmic term in the Nernst equation (which will be discussed below) had to be linearised. Since a numerical solution of the 1D model's ODEs presented in this chapter is preferred over an analytical one, the linearisation is unsuitable here. For determining $V_{\text{Nernst}}(I)$ the gas concentrations must be found as a function of the electric current I. The Nernst voltage is given by:

$$V_{\text{Nernst}} = V^0(T) - \frac{RT}{2F}\ln\frac{x_{\text{H}_2\text{O}}}{x_{\text{H}_2}} + \frac{RT}{4F}\ln\frac{x_{\text{O}_2}p}{p^0} \tag{7.10}$$

In Equation (7.10), $x_{\text{H}_2\text{O}}$ is the mole fraction of steam at the anode side, x_{H_2} is the mole fraction of hydrogen there, x_{O_2} is the mole fraction of oxygen at the cathode side and p is the total absolute operating pressure (Pa). $V^0(T)$ is the Nernst voltage at standard pressure p^0. It is calculated from free enthalpy data as described by textbooks on physical chemistry.[26] The mole fractions of all the involved species are given by the ratio of their individual mole flow rates and the total mole flow rate of fuel gas flowing at the anode side and air flowing at the cathode side:

$$x_i = \dot{n}_i/\dot{n}_t = \dot{n}_i \bigg/ \sum_j \dot{n}_j \tag{7.11}$$

so:

$$x_{\text{H}_2\text{O}} = \dot{n}_{\text{H}_2\text{O}}/\left(\dot{n}_{\text{H}_2} + \dot{n}_{\text{H}_2\text{O}}\right) \tag{7.12}$$

and:

$$x_{\text{H}_2} = \dot{n}_{\text{H}_2}/\left(\dot{n}_{\text{H}_2} + \dot{n}_{\text{H}_2\text{O}}\right) \tag{7.13}$$

and for the oxygen in the air at the cathode side:

$$x_{\text{O}_2} = \dot{n}_{\text{O}_2}/\left(\dot{n}_{\text{O}_2} + \dot{n}_{\text{N}_2}\right) \tag{7.14}$$

if air is simplified as a binary mixture that is just composed of O_2 and N_2.

The next step is to apply Faraday's law for expressing the electric current in terms of the electrochemically converted H_2:

$$\dot{n}_{H_2} = \dot{n}_{H_2}^{in} - I/(2F) \tag{7.15}$$

and:

$$\dot{n}_{H_2O} = \dot{n}_{H_2O}^{in} + I/(2F) \tag{7.16}$$

Finally, for the mole flow rate of oxygen at the cathode side (co-flow case) it is:

$$\dot{n}_{O_2} = \dot{n}_{O_2}^{in} - I/(4F) \tag{7.17}$$

while the flow rate of N_2 remains constant, $\dot{n}_{N_2} = \dot{n}_{N_2}^{in}$, because N_2 is not involved in the considered electrochemical reaction.

Inserting all this into Equation (7.10) yields:

$$
\begin{aligned}
V_{\text{Nernst}}(I) = V^0(T) &- \frac{RT}{2F} \ln \frac{\dot{n}_{H_2O}^{in} + I/(2F)}{\dot{n}_{H_2}^{in} - I/(2F)} \\
&+ \frac{RT}{4F} \ln \frac{\left(\dot{n}_{O_2}^{in} - I/(4F)\right)p}{\left(\dot{n}_{N_2}^{in} + \dot{n}_{O_2}^{in} - I/(4F)\right)p^0}
\end{aligned}
\tag{7.18}
$$

Now the Nernst voltage depends only on the electric current produced by the SOFC and the supplied gas flow rates, *i.e.* the initial composition. However, it is convenient to introduce a steam load defined by the ratio:

$$\alpha = \dot{n}_{H_2O}^{in} \big/ \dot{n}_{H_2}^{in} \tag{7.19}$$

and to define further:

$$I_{\text{Fuel}} = 2F\dot{n}_{H_2}^{in} \tag{7.20}$$

as the maximum current that could be drawn from the fuel if it was possible to operate the SOFC at 100% fuel utilisation. At the cathode side, it is convenient to express the supplied O_2 and N_2 flow rates in terms of an air excess number defined by

$$\lambda = 2x_{O_2}^{in} \dot{n}_{air}^{in} \big/ \dot{n}_{H_2}^{in} \tag{7.21}$$

with

$$\dot{n}_{O_2}^{in} = x_{O_2}^{in}\dot{n}_{air}^{in} = x_{O_2}^{in}\left(\dot{n}_{O_2}^{in} + \dot{n}_{N_2}\right) \tag{7.22}$$

Then Equation (7.18) finally becomes

$$V_{Nernst}(I) = V^0(T) - \frac{RT}{2F}\ln\frac{\alpha I_{Fuel} + I}{I_{Fuel} - I} + \frac{RT}{4F}\ln\frac{x_{O_2}^0(\lambda I_{Fuel} - I)p}{\left(\lambda I_{Fuel} - x_{O_2}^0 I\right)p^0} \tag{7.23}$$

After inserting all this into Equation (7.9), the desired model can eventually be represented by the single nonlinear first-order ordinary differential equation (ODE)

$$\frac{dI}{dx} = \frac{w}{\hat{R}_\Sigma}\left(V^0 - \frac{RT}{2F}\ln\frac{\alpha I_{Fuel} + I}{I_{Fuel} - I} + \frac{RT}{4F}\ln\frac{x_{O_2}^0(\lambda I_{Fuel} - I)p}{\left(\lambda I_{Fuel} - x_{O_2}^0 I\right)p^0} - V_{cell}\right) \tag{7.24}$$

Its solution for the initial condition

$$I(x = 0) = 0 \tag{7.25}$$

and given values for the cell voltage V_{cell} delivers the produced electric current as a function of the gas channel length, x. At the outlet of the SOFC at $x = L$, $I(x = L)$ is the current obtained at the SOFC's terminals. This would be equivalent to simulating potentiostatic operation.

As mentioned above, a solution is to be found by numerical approximation using readily available numerical initial value problem (IVP) software for solving Equations (7.24) and (7.25). The situation becomes slightly more complicated if the electric current at the terminals is prescribed and the corresponding cell voltage has to be determined instead. This corresponds to galvanostatic operation. A straightforward solution to this problem is to extend Equation (7.24) by introducing a second differential equation

$$\frac{dV_{cell}}{dx} = 0 \tag{7.26}$$

which is just another way to state mathematically that the cell voltage is treated as a constant. So Equations (7.24) and (7.26) now constitute a system of two first-order differential equations. The benefit of this is that it is now possible to specify another initial or boundary condition which obviously is:

$$I(x = L) = I_L \tag{7.27}$$

The mathematical task of simulating galvanostatic SOFC operation has now become a two-point boundary value problem (BVP).[27] The constant cell voltage V_{cell} in volts is now a part of the solution. Since the ODE system is not stiff, there are several ways to obtain a numerical solution. For example, the so-called shooting method could be used which is implemented by wrapping a root finder around an available IVP solver.[28] The task then simply becomes finding the root of:

$$I(x = L, V_{cell}^k) - I_L = 0 \qquad (7.28)$$

The superscript k is the iteration counter, $k = 1,2,3,\ldots, N$ where N is the number of iterations, *i.e.* numerical solutions of the corresponding initial value problem $I(x = 0) = 0$ using V_{cell} as the independent variable, required to achieve $I(x = L, V_{cell}^N) - I_L \leq |tol|$.

7.4 Calculations and Results

I–V curve modelling using Equations (7.24) and (7.26) is carried out in practice by constructing the I–V curve point by point. In other words, a table of as many values of I_L and V_{cell} as wished to cover the whole envisaged I–V range is calculated by solving Equations (7.24), (7.25), (7.26) and (7.27) for each individual point. For practical calculations, it is often more convenient to prescribe V_{cell} at each point and solve the simpler IVP introduced by Equations (7.24) and (7.25).

7.4.1 Determining the Area Specific Resistance

In order to be able to calculate an I–V curve, the ASR, or $\hat{R}_\Sigma$, must be determined first from a given set of experimental data. This can be accomplished quite easily by reformulating Equation (7.26) which may be replaced by:

$$\frac{d\hat{R}_\Sigma}{dx} = 0 \qquad (7.29)$$

again subject to the two boundary conditions, Equations (7.25) and (7.27). Now $\hat{R}_\Sigma$ has the role of the unknown parameter to be determined in place of V_{cell} so that both boundary conditions can be satisfied again.

Then the ASR can be calculated by a 1D least square minimisation routine wrapped around the IVP or BVP solver similar to the procedure described above:

$$\min \sum_i \left[I(x = L, V_{cell}^i, \hat{R}_\Sigma) - I_L^i \right]^2 \qquad (7.30)$$

However, for practical reasons it may be sufficient to just pick the operating point (I_L^{op}, V_{cell}^{op}) and to simply obtain the ASR from:

$$I\left(x = L, V_{cell}^{op}, \hat{R}_\Sigma\right) - I_L^{op} = 0 \qquad (7.31)$$

This shortcut is justifiable if the primary interest is to describe the SOFC's I–V behaviour accurately only in the vicinity of a given the operating point which is sufficient for many technical applications or system studies, respectively.

7.4.2 Experimental Validation

Single-cell I–V curve measurements were carried out in Jülich using small laboratory-type single-cell SOFCs with a cell area of $A_{cell} = 4\,cm \times 4\,cm = 16$ cm^2. Humidified H$_2$ at varying steam content was used as the fuel. These measurements were carried out at cell temperatures $T_{cell} = 800\,°C$, $750\,°C$ and $700\,°C$. The fuel humidity was expressed in terms of the steam contents in vol.%.

The measurements were then used to evaluate the model for recalculating the measured I–V curves. The results are shown in Figures 7.3, 7.4 and 7.5. There is a curve for a fuel humidity of 10%, 25% and 50% in each figure.

The three figures all show a comparison between measured and modelled I–V curves, denoted by a different mark for each one of the three fuel humidity values and by straight lines, respectively. The fuel utilisation, U_F, was 14% at a H$_2$ flow rate of $1000\,mL\,min^{-1}$ and at $\langle i \rangle = 1.25\,A\,cm^{-2}$ in all three cases. The

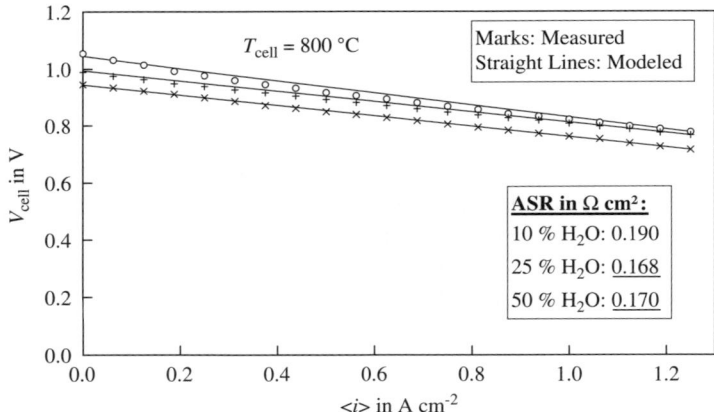

Figure 7.3 Measured and modelled I–V curves at different fuel humidities for 800 °C.

quantity $\langle i \rangle$ is the average current density in $A\,cm^{-2}$ calculated by the equation:

$$\langle i \rangle = I_L / A \qquad (7.32)$$

This value should not be confused with the local current density $i!$. It can be calculated based on Equation (7.20). This yields $I_{Fuel} = 143.5\,A$. With that, $\langle i_{Fuel} \rangle = 143.5\,A/16\,cm^2 = 8.97\,A\,cm^{-2}$ is then found which confirms $U_F = 1.25/8.97 = 0.14$. The H_2O flow rate was adjusted each time to match the target humidity.

It can be seen from Figure 7.3 that the modelled values match the experimental results very nicely at 800 °C. Although it is not surprising to note that the curves in Figure 7.3 seem fairly linear because of the constant ASR value, there is still a difference among the slopes of the curves. The reason is found in Equations (7.10) and (7.23), respectively, which show the dependency of the Nernst voltage on the local concentrations of H_2 and H_2O. It can also be seen in the box at the lower right corner in Figure 7.3 that the ASR values determined from the slopes (values given in the box in Figure 7.3) of the curves around $1.2\,A\,cm^{-2}$ seem to converge towards the value at 50% humidity. The I–V curves at 25% (marked with $+$) and 50% (marked with $\times$) humidity run almost in parallel while the one at 10% humidity (marked with circles) obviously has a steeper slope. The ASR shown in Figure 7.3 was calculated by Equation (7.31) yielding $\hat{R}_\Sigma = 0.16\,\Omega\,cm^2$ which is lower than the ASR obtained from slope information.

The conclusion from this is that the ASR as obtained from the slope of measured I–V curves is not the underlying intrinsic quantity. Because of the impact of the fuel humidity on the Nernst voltage, taking the slope of a measured I–V curve tends to lead to over-estimated values.

It is clear from Equation (7.10) that the influence of the steam content is minimal at 50% humidity.

The trends just described also apply to Figure 7.4 although the deviations between model and experiment are more pronounced. Using the same procedure as in the case of Figure 7.3, an ASR of $\hat{R}_\Sigma = 0.22\,\Omega\,cm^2$ was calculated which is higher than $0.21\,\Omega cm^2$ at 50% humidity but still lower than $0.27\,\Omega cm^2$ at 10% humidity.

Figure 7.5 shows the results for 700 °C. The trends and effects described above can be observed again. The calculated ASR for 700 °C was $\hat{R}_\Sigma = 0.31\,\Omega\,cm^2$.

However, the deviations between model and experiments are rather obvious at 700 °C. There may be a nonlinear behaviour that cannot be explained by a constant ASR. A speculative guess is to explain this effect by nonlinear polarisation according to Equation (7.2). If this can be confirmed, then the low-field approximation loses its validity at temperatures lower than 700 °C. As a result, more sophisticated models for describing the electrochemical oxidation than the simple one presented in this chapter would be

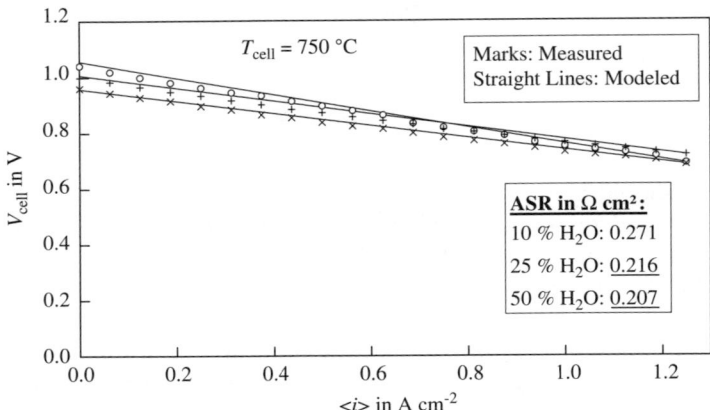

Figure 7.4   Measured and modelled I–V curves at different fuel humidities for 750 °C.

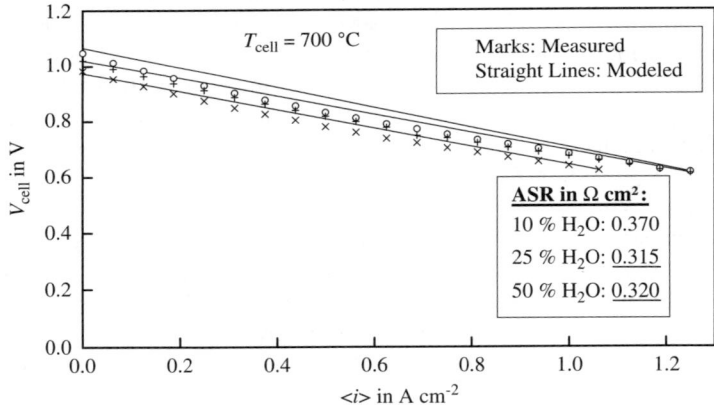

Figure 7.5   Measured and modelled I–V curves at different fuel humidities for 700 °C.

needed in order to model intermediate temperature SOFC behaviour, *e.g.* between 600 °C and 700 °C.

On the other hand, since the best fit was obtained for 800 °C, there is reason to believe that an assumption of a constant ASR is reasonable at temperatures around 800 °C. Quite likely the linear ASR behaviour stays valid above 800 °C because the electrochemical reactions become more rapid so electrochemical polarisation should be less relevant at high temperatures. Ideally, they could almost disappear at some point and mostly only the ohmic resistances would remain. If this can be confirmed, this will be very beneficial for modelling electrolyte-supported SOFCs operating between 850 °C and 1000 °C because it would then be possible to simulate their performance by this very simple ASR model.

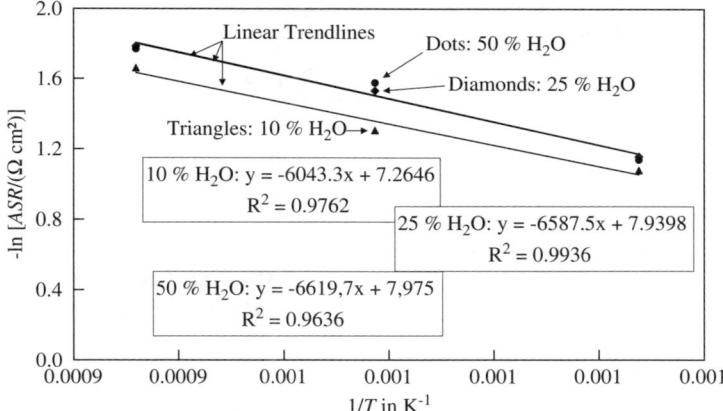

Figure 7.6 Estimation of an apparent activation energy, data measured at 50 vol.% humidity.

The temperature dependence of the ASR is displayed by an Arrhenius-type plot of the available experimental data (Figure 7.6). This procedure delivers an approximation of an apparent activation energy. It must be stated that the so-called apparent activation energy has no physical meaning because at best it is a lump sum average of all activation energies of all thermally activated reactions contributing to the overall electrochemical performance. These reactions are at least the cathode polarisation, the anode polarisation and the O^{2-} conduction in the electrolyte.

A lump sum activation energy of $E_A \sim 55 \, kJ \, mol^{-1}$ was estimated using this crude approximation.

A result of modelling the ASR temperature dependence by the simple Equation (7.4) is shown in Figure 7.7.

It shows measured I–V curves (marks) and modelled ones (straight lines) for $U_F = 14\%$ and 50% H_2O for 700 °C, 750 °C and 800 °C. It may be concluded that the discussed simple model approach is sufficient for the presented parameter range. The simple model can reproduce the measured SOFC behaviour satisfactorily although the quality of the results starts deteriorating at 700 °C, where the measured I–V curves start showing deviations from linearity not observed at higher temperatures.

7.4.3 Simulation of the Current–Voltage Behaviour

In this section some more details of the points discussed so far are investigated further by model calculations. In addition, the influence of the fuel utilisation is discussed.

Figure 7.8 shows the local current density, i, as a function of the cell area, A, for $U_F = 14\%$ and $U_F = 80\%$ as it develops while the fuel mixture (and the air,

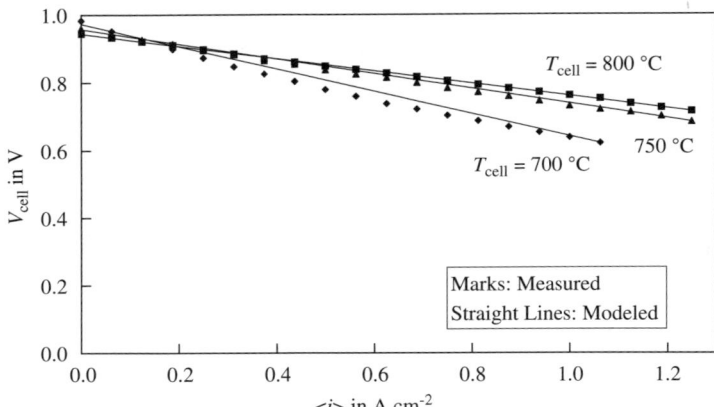

Figure 7.7 Measured and modelled I–V curves at 50 vol.% steam for different temperatures.

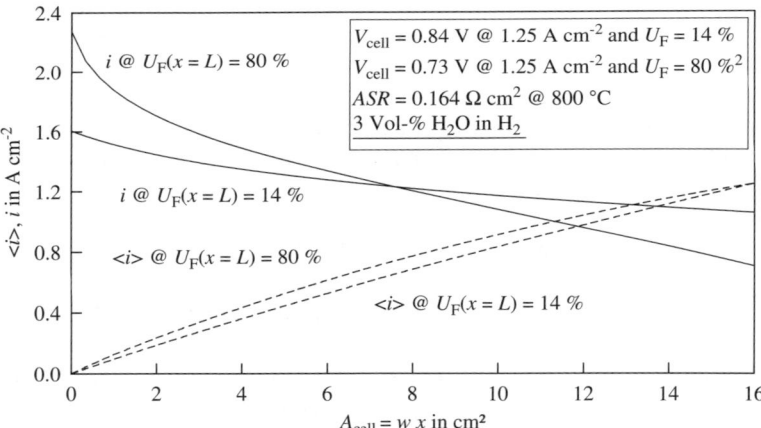

Figure 7.8 Local and average current density at 3 vol.% steam for 14% and 80% fuel utilisation.

respectively) flows along the gas channels, passing over the active area of the cell's MEA from the entrance at $A = 0$ cm^2 ($x = 0$ cm) to the exit at $A = 16$ cm^2 ($x = L = 4$ cm). These are two particular solutions of Equations (7.24), (7.25), (7.26) and (7.27). In addition, the corresponding average current densities $\langle i \rangle = I(x)/A$ are displayed. They all end up at the same point at $\langle i \rangle = \langle i_L \rangle = 1.25$ A cm^{-2} because this is required by the boundary condition, Equation (7.27).

It follows from Equations (7.10) and (7.24) that the Nernst voltage, and for that reason dI/dx, is most humidity sensitive at low steam contents. Therefore a humidity of 3% was chosen for emphasising the influence of the fuel humidity.

The inlet humidity is not the only factor influencing the steam concentration in the cell. In addition, the impact of U_F must be investigated, so $U_F = 14\%$ from the experiment was compared with a high value of $U_F = 80\%$ to highlight the differences. The impact of U_F can be made very clear by slightly rewriting Equation (7.23):[25]

$$V_{\text{Nernst}}(U_F) = V^0(T) - \frac{RT}{2F} \ln \frac{\alpha + U_F}{1 - U_F} + \frac{RT}{4F} \ln \frac{x_{O_2}^0 (\lambda - U_F)p}{\left(\lambda - x_{O_2}^0 U_F\right)p^0} \qquad (7.33)$$

It is obvious that U_F has the lowest impact on V_{Nernst}, and therefore on dI/dx, if $U_F \ll \alpha$. This is an important result for the proper selection of a suitable steam content for I–V curve experiments that are intended for ASR determination from slope information. Those experiments require the observation of a fairly linear I–V curve in the considered I–V range and thus $\alpha = 1$ or 50% humidity should be chosen. If the H_2 flow rate is then chosen, so that $U_F < 10\%$ is maintained to ensure a difference of one order of magnitude between both quantities (and between 1 and U_F), the local current density i should remain fairly constant throughout the cell and sufficiently linear I–V curves may be obtained allowing a proper determination of the true, intrinsic ASR. On the other hand, since the best fit was obtained for 800 °C, there is reason to believe that an assumption of a constant ASR is reasonable at temperatures around 800 °C. Quite likely the linear ASR behaviour stays valid above 800 °C because the electrochemical reactions become more rapid so electrochemical polarisation should be less relevant at high temperatures. Ideally, they could disappear at some point and only the ohmic resistances would remain. If this can be confirmed, this will be very beneficial for modelling electrolyte-supported SOFCs operating between 850 °C and 1000 °C because it would then be possible to simulate their performance by this very simple ASR model.

Operating the cell at higher U_F and lower α tends to over-estimate the ASR and may also cause a nonlinear I–V behaviour which makes it difficult to find a representative ASR for the entire I–V curve as displayed by Figure 7.9. It shows modelled I–V curves for 3% humidity.

Figure 7.10 may further support the conclusions drawn above. It shows corresponding results for 50% humidity. As expected, the least variation of i is obtained at $U_F = 14\%$ and $\alpha = 1$.

Corresponding I–V curves are shown in Figure 7.11. They now show a fairly linear behaviour at 50% humidity. They also show that the curves are converging against each other for decreasing U_F which supports the conclusion that a true, intrinsic ASR can be obtained from I–V measurements if $\alpha = 1$ and $U_F < 10\%$ are chosen as long as the electrochemical polarisation can be described properly by assuming ohmic behaviour.

The observation of nonlinear I–V curves despite choosing $\alpha = 1$ and $U_F < 10\%$ may be a good indication for the presence of nonlinear activation

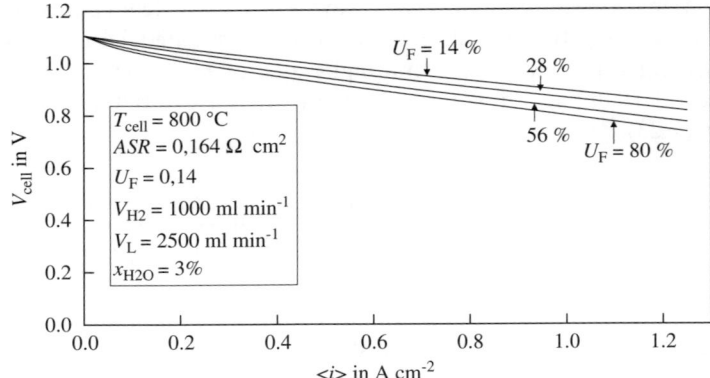

Figure 7.9   Modelled I–V curves for different fuel utilisations.

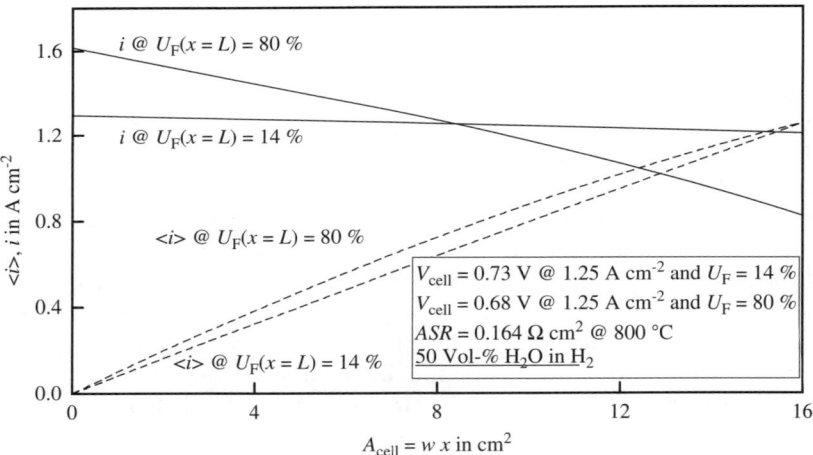

Figure 7.10   Measured and modelled I–V curves at 50 vol.% steam for different temperatures.

polarisation and/or mass transport polarisation. In that case, a proper description by a lump sum ASR is not valid any more and more sophisticated models must be applied if available.

7.5   Degradation Monitoring by Area Specific Resistance Simulation

The next core aspect to be discussed here is related to degradation behaviour monitoring and lifetime modelling. Degradation monitoring is usually carried out by operating the fuel cell as long as possible at constant current density

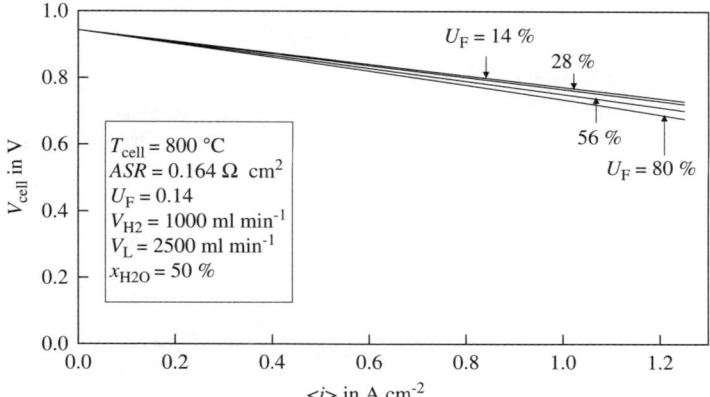

Figure 7.11 Measured and modelled I–V curves at 50 vol.% steam for different temperatures.

while recording the (usually decreasing) cell voltage over time. The degradation rate then is often expressed in ΔV per 1000 h.

Since the voltage drop diminishing the Nernst voltage during operation depends on the electric current, it seems that increasing the electric current also leads to an increased degradation rate. This may very well be true; however, the impact of the current may be over-estimated because the voltage drop is amplified by the current density. This can be understood by recalling the equation for the local I–V curve, Equation (7.5), which may be used for describing the overall SOFC behaviour if $\alpha \approx 1$ and $U_F < 10\%$ are chosen as discussed above.

The hypothesis is that the degradation is caused by an increase of the ASR over time which causes a corresponding time dependent voltage loss ΔV_{cell}:

$$\Delta V_{cell}(t) = -i\hat{R}_\Sigma(t) \tag{7.34}$$

so that

$$V_{cell}(t) = V_{Nernst} + \Delta V_{cell}(t) \tag{7.35}$$

In these equations, the current density i (or $\langle i \rangle$ for that matter) is an amplification factor that amplifies the voltage loss.

Thus a true, or intrinsic, current dependent degradation behaviour can only be detected if that amplification is eliminated first. That can be accomplished by using the 1D model, introduced above, for calculating the ASR as a function of time from V_{cell} and t information which can be recorded by an endurance test as usual. No changes need to be introduced to the experiment. If the Equations (7.24), (7.29), (7.27) and (7.31) are solved for each V_{cell} and t pair of a measured endurance test data table, then the ASR can be calculated by only one pair of

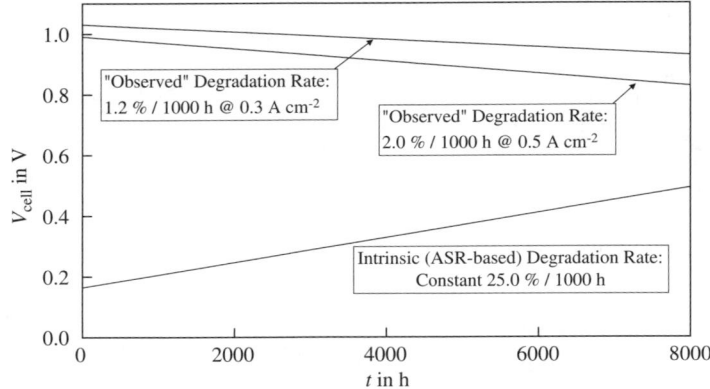

Figure 7.12   Comparison of cell voltage and ASR-based degradation rates.

V_{cell} and t. There is no need for interrupting the endurance test for measuring a second operating point at a different V_{cell} and $\langle i \rangle$ which is necessary for calculating the slope at the I–V curve. It is even possible to re-evaluate old endurance test data because the model's application is just a post-processing step that is independent of the endurance test experiments.

An example of how the calculation of the ASR as a function of the time could look like is given by Figure 7.12.

In order to illustrate how the observed voltage-based degradation translates into ASR-based degradation, a constant ASR increase of 25% per 1000 h was assumed. Then the corresponding voltage drop over time was calculated. It can be seen that the observed degradation rate at $0.3\,\mathrm{A\,cm^{-2}}$ is 1.2% per hour. It increases to 2% per 1000 h at $0.5\,\mathrm{A\,cm^{-2}}$ although the ASR increase over time is the same in both cases. The conclusion is that it is recommendable to express degradation information from endurance tests in terms of the ASR instead of using the cell voltage. As mentioned above, care must be taken to choose the operating conditions properly. This can be done by using the model introduced above or by ensuring $\alpha = 1$ and $U_F < 10\%$ and manual ASR determination by interrupting the endurance test and running an additional I–V curve. The charm of using the model is that the endurance test does not need to be interrupted and also already available test information is not lost because it can be re-evaluated: No additional I–V measurements are needed.

7.6   Extensions of the Model

The isothermal 1D model derived above can be used as a first building block for creating a set of more sophisticated 1D models. The detailed derivation of these extensions is beyond the scope of this chapter. As a next step, equations for describing the methane–steam reforming kinetics were included. Then, equations describing the energy balance of the gas streams and the heat conduction in the solid body (interconnector plates and membrane electrode assembly) were also built in. The resulting model is capable of simulating the electrochemical and thermal stationary behaviour of planar SOFCs in co-flow and counter-flow mode.[29] It was then further expanded to allow transient simulation as well.[30] It also proved to be rapid enough for inclusion in systems simulation tools.[31,32] This addresses the model scale issues discussed by Bove *et al.*[16], for example. The following section utilises the isothermal 1D model with internal methane reforming.

7.7   A Simple Lifetime Prediction Model

The previously developed linear ASR model shows very well how the increase of the ASR over time influences the voltage based degradation behaviour. As explained above it can be concluded that the degradation behaviour of different SOFCs can be made comparable by expressing it in terms of the ASR.

However, the current experience with SOFC degradation behaviour leads to the conclusion that there are further underlying factors depending on operating conditions that may accelerate the degradation behaviour. Hence the linear ASR degradation model developed so far has some educational value but does not describe the observed behaviour sufficiently.

This section is therefore dedicated to developing ideas on incorporating the observed effects into more realistic ASR degradation models. The ideas developed below are intended as ideas contributing to an ongoing discussion and by no means as a comprehensive list of all known degradation root causes.

The approach for developing improved ASR degradation models is to extend the previously described linear ASR behaviour in time by including some of the nonlinear degradation effects already described in literature.[2,33,34]

7.7.1   Some Literature Results

It was found that the cathode polarisation resistance of the LSM material has a time dependence correlating with the exponential equation:[33]

$$R_\eta = R_\infty - A \exp(-t/b) \qquad (7.36)$$

Where A and b are parameters fitted to the experimental findings. A similar expression was found for the conductivity decrease, denoted by σ, in Ni/8YSZ electrodes after changing from oxidising to reducing conditions:[34]

$$\sigma(t) = \sigma_\infty + A_1 \exp(-t/\tau) \qquad (7.37)$$

These two examples may indicate that at least some degradation root causes may exhibit an exponential decay behaviour.

A very comprehensive study on parameters that may have an impact on the degradation behaviour was presented by de Haart *et al.*[2] It was found that the degradation behaviour was not influenced by the fuel choice (*i.e.* there was no observable difference between CH_4 and H_2 operation) and it was not influenced by the fuel utilisation (*i.e.* the water or CO_2 content). The influence of Cr cathode poisoning remained unclear. However, three different time domains for three seemingly different degradation effects could be distinguished: The first time domain extended from 0 to 200 h in which an initial exponential decay took place. A second time domain was identified between 200 and 2000 h in which linear degradation was predominant. Finally, a third time domain took over from 2000 to 4500 h which exhibited progressive degradation and which clearly was lifetime limiting.

This work will compile the three quoted degradation effects into a simple lifetime model that will replace the purely linear ASR degradation model mentioned above by a more detailed approach also including exponential terms.

7.7.2   Development of a Non-linear Area Specific Resistance Over Time Behaviour

All three papers mentioned above work on the assumption that exponential degradation is predominant, at least for some periods in a stack's lifetime. Hence exponential degradation is used as a working hypothesis for the following derivation.

De Haart *et al.* described that the degradation behaviour of an SOFC may be categorised into three distinct time domains.[2] Time domain 1 (TD1) is a rapid initial degradation; in the example case, tapering off at about 200 h. Time domain 2 (TD2) exhibits a linear decrease of the cell voltage that can last up to several thousand hours; and the third time domain (TD3) shows an increasingly progressive decay of the cell voltage eventually leading to total failure (all at constant current). TD3 is clearly lifetime limiting and indicates there is a second nonlinear, maybe exponential, degradation process with a different time constant besides the initial rapid decay that comes to a complete stop at some hundred hours.

Assuming exponential degradation for TD3 can also explain a linear ASR increase over time in TD2 as well if the time constant τ matches up with the experimental observations. This becomes clear when performing the Taylor expansion of an exponential function; for example:

$$\hat{R}(t) = \hat{R}_\infty \exp(t/\tau) \tag{7.38}$$

which is

$$\hat{R}(t) = \hat{R}_\infty \left(1 + \frac{1}{\tau} t + \frac{1}{2\tau^2} t^2 + \ldots \right) \tag{7.39}$$

Hence the ASR function shows a linear increase over time as long as the linear approximation

$$\hat{R}(t) = \hat{R}_\infty \left(1 + \frac{1}{\tau}t\right) \tag{7.40}$$

dominates the overall ASR behaviour.

This leads to the assumption that time domains 2 and 3 mentioned by de Haart *et al.* could belong to the same root cause and that they are only different stages of Equation (7.38).[2] In TD2, the linear part of the Taylor expansion governs the degradation behaviour, while in TD3 the second and the higher order terms take over which finally leads to a progressive acceleration of the degradation.

Whatever the root cause actually is, it seems sensible to base the lifetime model on a simple first-order kinetics that obviously leads to an exponential time dependence. A crude physical justification is to assume that the electric (or ionic) conductivity σ is proportional to the concentration of contact bridges, C_{CB}, in the electrodes, contact layers and interfaces:

$$\sigma \propto C_{CB} \tag{7.41}$$

A very simple idea for the underlying physical process of the degradation process now is to assume that the degradation is caused by a loss of contact bridges, *i.e.* by a decrease of C_{CB} by which σ is reduced.

There also seems to be a dependence of the degradation rate on the current density. It may be reasonable to assume as a first hypothesis that the degradation rate increases or decreases linearly with the current density, *i.e.* to postulate a first-order kinetics as well. The following rate equation can then be suggested:

$$\frac{d\sigma}{dt} = -k_2 i_{\text{loc}} \sigma \tag{7.42}$$

In Equation (7.42) the current density is more precisely denoted as i_{loc} because i it is a local quantity that changes inside the cell or the stack.

The rate constant was denoted as k_2 because this kind of decay behaviour leads to a catastrophic failure occurring in TD3 (see above) as it will be described below.

TD1 can be described mathematically as well if a threshold value σ_8 is introduced:

$$\frac{d\sigma}{dt} = -k_1 i_{\text{loc}} (\sigma - \sigma_\infty) \tag{7.43}$$

These two rate equations (for time TD1 and TD2) can now be integrated using the initial condition

$$\sigma(t = 0) = \sigma_0 \tag{7.44}$$

The result is the individual time dependences of the degradation behaviour. The local current density is treated as a constant. Although the overall current is usually kept constant during an endurance test, i_{loc} may nonetheless change over time. This will be discussed below. In the meantime, for TD1 it is obtained that:

$$\sigma(t) = \sigma_\infty + \sigma_0 \exp(-k_1 i_{loc} t) \qquad (7.45)$$

Relating the conductivity to the electrode area yields the area specific conductivity which is inverse to the ASR:

$$\hat{R}_1(t) = \frac{A}{\delta}\left(\frac{1}{\sigma_\infty + \sigma_0 \exp(-k_1 i_{loc} t)}\right) = \frac{1}{\hat{\sigma}_\infty + \hat{\sigma}_0 \exp(-k_1 i_{loc} t)} \qquad (7.46)$$

A is the active electrode surface area and δ is the thickness of the conducting layer. The quantity $\hat{\sigma} = \sigma\delta/A$ is the area specific conductivity. For time domain 2 it is obtained again by using initial the condition, Equation (7.25):

$$\hat{R}_2(t) = \frac{1}{\hat{\sigma}(t)} = \hat{R}_0 \exp(k_2 i_{loc} t) \qquad (7.47)$$

These two results are now finally compiled into a time dependent local I–V curve:

$$V_{cell}(t) = V_{Nernst} - i_{loc}\left(\hat{R}_c + \frac{\hat{R}_\infty}{1 + f_0 \exp(-k_1 i_{loc} t)} + \hat{R}_0 \exp(k_2 i_{loc} t)\right) \qquad (7.48)$$

In Equation (7.48) the modifications $\hat{R}_\infty = 1/\hat{\sigma}_\infty$ and $f_0 = \sigma_0/\sigma_\infty$ were introduced for convenience. The equation is a simple lifetime model that can describe the degradation behaviour of an SOFC as a function of the time and forecast when its lifetime limit may be reached. Although it is a very simple model with little physical information, it already contains six parameters that must be fitted to experimental results. These are the constant offset $\hat{R}_c$, the amplification factors $\hat{R}_\infty, f_0$ and $\hat{R}_0$ as well as the rate constants k_1 and k_2. This may give an impression already how challenging it might be to find detailed physical models for individual degradation phenomena. Equation (7.48) is valid if the fuel utilisation is low enough to keep i_{loc} nearly constant throughout the SOFC and if V_{Nernst} can be kept nearly constant. This is the case at 50% humidity for H_2 operation as discussed in the first section of this chapter. The constraint of constant i_{loc} (or low fuel utilisation, respectively) can be dropped if Equation (7.48) is entered into a model that resolves the spatial distribution of i_{loc}. One suitable model is the isothermal 1D model introduced above. This will also be discussed in the following.

7.8 Calculations

This chapter concludes with a discussion of some simulations using the isothermal 1D model derived above and the simple lifetime model that was just developed. First, Equation (7.48), which strictly speaking is only valid for vanishing fuel utilisation, will be discussed and compared with spatially resolved 1D model results at 10% and at 75% fuel utilisation. Then it will be shown how the local current density distribution and the ASR change over time covering a service time range up to 5000 h.

7.8.1 Constant Current Density

Figure 7.13 shows the result of a simple evaluation of Equation (7.48). The result is the cell voltage-based degradation behaviour based on a rough approximation of the available data for three different current densities ($0.3\,\mathrm{A\,cm^{-2}}$, $0.5\,\mathrm{A\,cm^{-2}}$ and $0.7\,\mathrm{A\,cm^{-2}}$) obtained by fitting the six parameters of Equation (7.48).[2] Fortunately, it was possible to eliminate one parameter by choosing $f_0 = 1$. Figure 7.13 shows a surprisingly good agreement with the experimental findings already.[2]

The rapidly diminishing effect of time TD1 can be seen as well as TD2 and TD3. The actual duration of TD2 and TD3 depends heavily on the current density: While the degradation rate remains comparatively low and linear (TD3 not visible yet) at $0.3\,\mathrm{A\,cm^{-2}}$, it is clearly progressive and lifetime limiting at $0.7\,\mathrm{A\,cm^{-2}}$ (almost no TD2 behaviour any more). At $0.5\,\mathrm{A\,cm^{-2}}$, all three time domains can be identified. These calculations seem to confirm that there may only be two time domains. The linear behaviour may, in fact, be the initial stage of progressive degradation leading to catastrophic failure later. At what instant

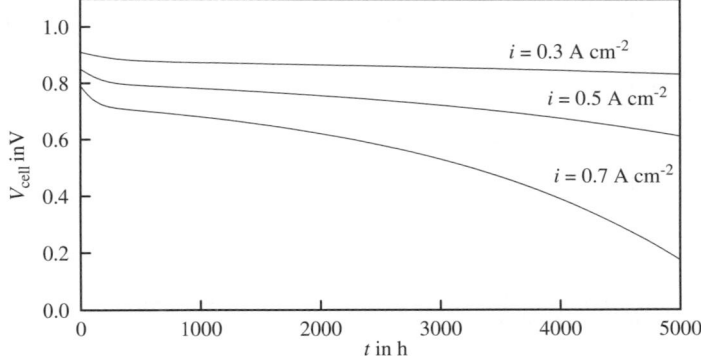

Figure 7.13 Degradation modelling at constant local current density.

progressive degradation starts to show may differ from fuel cell to fuel cell depending on the individual initial ASR values and the envisaged operating point. However, one indication may be given by the product of the rate constant k_2 and the current density as suggested by Equation (7.47). It would certainly be highly speculative at this time to make statements on the temperature dependence of k_2. However, if the obviously most critical progressive degradation can be described by a rate equation leading to an exponential conductivity decay then it may be reasonable to assume that k_2 has an Arrhenius-type temperature dependence of the well known type:

$$k_2(T) = k_2^{\infty} \exp(-E_A/(RT)) \tag{7.49}$$

The background would be that a loss of C_{CB} would progress more rapidly at greater temperatures, which is not an unreasonable thought. Hence it is suggested that the operating temperature is investigated as an additional parameter as well. The tests discussed here were all carried out at the same temperature.

7.8.2 Varying Local Current Density and Fuel Utilisation Influence

The next step was to quantify the influence of the fuel utilisation on the degradation model. This was carried out by incorporating Equation (7.48) into Equation (7.24) which yields an implicitly given nonlinear ODE for the spatial resolution of a variable local current density that must be numerically solved

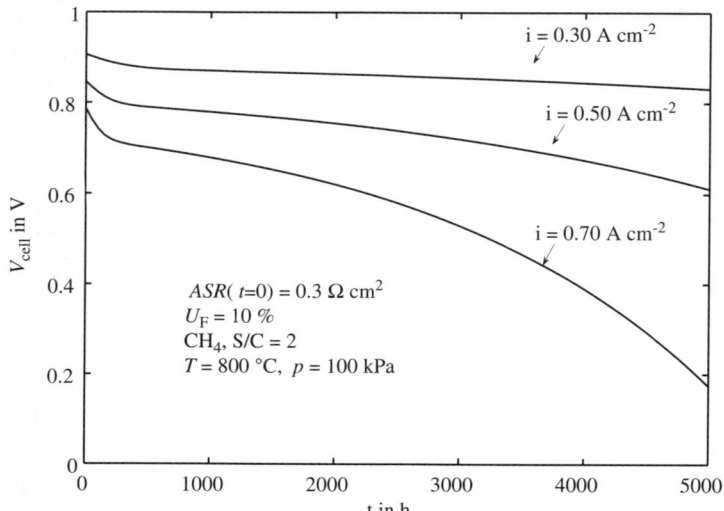

Figure 7.14   Degradation modelling at varying local current density and $U_F = 10\%$.

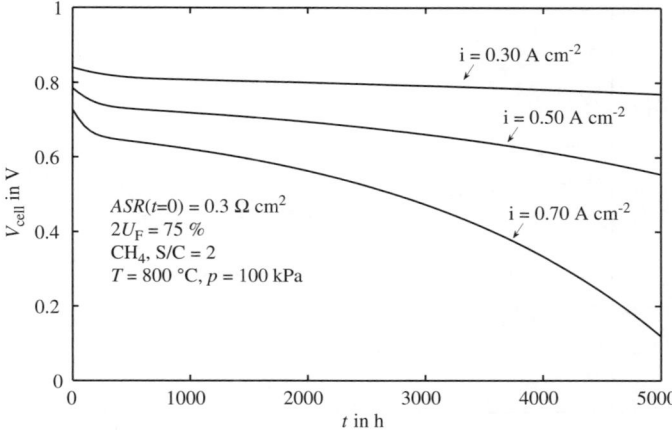

Figure 7.15 Degradation modelling at varying local current density and $U_F = 75\%$.

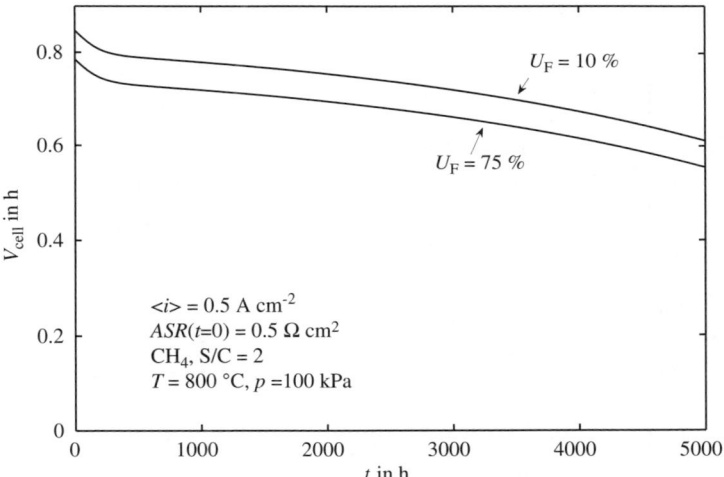

Figure 7.16 Influence of the fuel utilisation, U_F.

for $i_{loc} = dI/dA$ at each integration step. This is shown in Figure 7.14 for $U_F = 10\%$ and in Figure 7.15 for $U_F = 75\%$.

The difference to the result shown in Figure 7.13 (vanishing fuel utilisation) is almost not noticeable. Hence, Equation (7.48) is a reasonable approximation for low fuel utilisation as expected. If the fuel utilisation is increased further, the cell voltage drops by approx. 70 mV on average (compare Figure 7.14 and Figure 7.15).

The difference between $U_F = 10\%$ and $U_F = 75\%$ for 0.5 A cm^{-2} is finally illustrated in Figure 7.16. The difference of 70 mV is an almost constant offset by which the two curves run apart from each other.

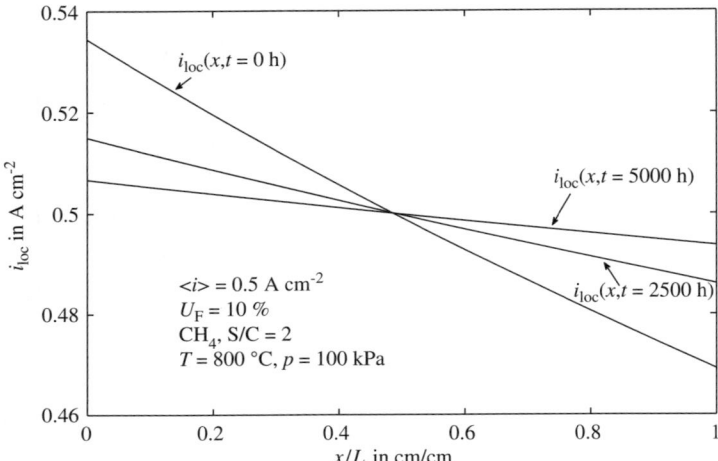

Figure 7.17   Local current density distribution.

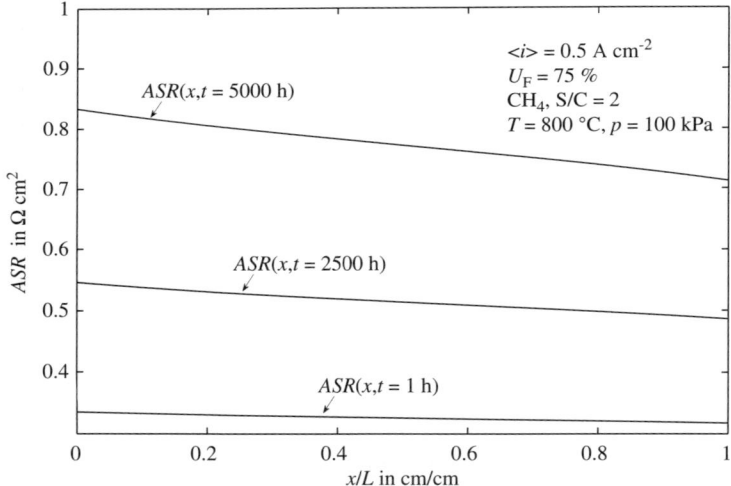

Figure 7.18   Local ASR distribution.

7.8.3   Change of Local Current Density and Area Specific Resistance Profile over Time

Figure 7.17 shows the simulated local current density profile as a function of the gas channel position after 0 h (new cell), 2500 h and 5000 h for an average current density of $0.5\,A\,cm^{-2}$ and $U_F = 10$. All curves go through the same point at $0.5\,cm^{-2}$ because the overall current generated by the cell is given by the area under the curves.

The local current density profile becomes increasingly uniform (*i.e.* its gradient decreases) over time. This is explained by the increasing ASR which is shown in Figure 7.18 for $U_F = 75\%$. The curve is quite similar for $U_F = 10\%$ and 75% so not much else can be learned and thus it is deemed unnecessary to discuss the difference again.

7.9 Conclusions

The 1D model approach is sufficient at least for the parameter range presented. For H_2 operation, it was demonstrated that U_F has the lowest impact on V_{Nernst}, and therefore on dI/dx, if $U_F \ll \alpha$. This is an important result for the proper selection of a steam contents suitable for I–V curve experiments intended for ASR determination from slope information: Those experiments require the observation of a fairly linear I–V curve in the considered I–V range. Thus $\alpha \approx 1$ or 50% humidity should be chosen. If in addition the H_2 flow rate is chosen so that $U_F < 10\%$ is maintained to ensure a difference of one order of magnitude between α and U_F (and between 1 and U_F), the local current density, i, should stay fairly constant throughout the cell and sufficiently linear I–V curves may be obtained allowing a proper determination of the true, intrinsic ASR.

It is recommended that degradation information from endurance tests is expressed in terms of the ASR instead of using the cell voltage.

The possibility to express the degradation process as an increase of the ASR over time leads to a simple lifetime model. It is based on the idea that SOFC degradation is mainly caused by a loss of electrical contacts. Although a broader validation of the model is still missing, it seems to allow some SOFC lifetime prediction already. The current density clearly came out as one lifetime limiting factor. This suggests that SOFCs should simply be operated at greater current densities than specified as the normal operating point for accelerated degradation testing.

However, so far, the influence of temperature on SOFC degradation has not been investigated. Since the loss of contacts could be described by a rate equation, it seems reasonable to assume that the rate constants follow an Arrhenius-type exponential temperature dependence. This indicates there might be a strong temperature dependence on SOFC degradation as well. This temperature dependency can be described by the model as well. Unfortunately, the experimental investigation must be left to future work.

7.10   List of Symbols Used in this Chapter

A	electrochemically active cell area (cm^2)
E	energy (J)
L	gas path length of the active area (m)

$\hat{R}$ area specific resistance (ASR) (Ω cm^2)

T absolute temperature (K)

U utilisation (1)

V voltage (V)

I current (A)

F   Factor, ratio of two quantities (1)

i current density (A cm^{-2})

$\langle i \rangle$ average current density (A cm^{-2})

k rate constant (cm^2 A^{-1} s^{-1})

$\dot{n}$ mole flow rate (mol s^{-1})

p pressure (Pa)

t time (s)

tol   numerical precision threshold (A)

w width of the active area (cm)

x mole fraction (1)

x length coordinate of the active area (m)

z number of electrons transferred per elementary reaction (1)

Δ difference (prefix)

α steam load (1)

β charge transfer coefficient (1)

η polarisation over-potential (V)

λ air excess number (1)

σ conductivity (S cm^{-1})

$\hat{\sigma}$ area specific conductivity (S cm^{-2})

Subscripts

A   activation energy

F   fuel

Fuel   fuel

L   load

loc   local

Nernst   Nernst (potential, voltage)

cell   cell

i, j species

T   total

$\sum$ sum

0   exchange current density

0   initial state

pol   polarisation

∞ limit

Superscripts

(A)   anode side

(C)   cathode side

(Ω)	ohmic
i, k	iteration counters
in	state at cell inlet
op	operating
0	standard conditions
∞	limit

Acknowledgements

The author would like to thank Robert Steinberger and Bert de Haart, Forschungszentrum Jülich, very much for their support and their careful proofreading. Special thanks are extended to Stefanie Berns and Jeniffer Kuhl, Forschungszentrum Jülich, for writing the programs and to Dieter Froning, Uwe Reimer, Martin Spiller and Andrei Kulikovsky, Forschungszentrum Jülich, for all the fruitful discussions.

References

1. R. Steinberger-Wilckens, L. Blum, H.-P. Buchkremer, L. G. J. de Haart, M. Pap, R. W. Steinbrech, S. Uhlenbruck and F. Tietz, in *Proceedings of the 11th ECS International Symposium on Solid Oxide Fuel Cells, 2009*, ECS Transactions, Pennington N.J., 2009, p. 213.
2. L. G. J. de Haart, L. Mougin, O. Posdziech, J. Kiviaho and N. H. Menzler, in *Proceedings of the 8th European SOFC conference, 2008*, EFCF, Oberrohrdorf, Switzerland, 2008, Paper no. COBO9O7.
3. V. A. C. Haanappel, N. Jordan, A. Mai, J. Mertens, J. M. Serra, F. Tietz, S. Uhlenbruck, I. C. Vinke, M. J. Smith and L. G. J. de Haart, *J. Fuel Cell Sci. Technol.*, 2009, **6**, 021302.
4. C. Lalanne, F. Mauvy, E. Siebert, M. L. Fontaine, J. M. Bassat, F. Ansart, P. Stevens and J. C. Grenier, *J. Eur. Ceram. Soc.*, 2007, **27**, 4195.
5. H. Moon, S. D. Kim, E. W. Park, S. H. Hyun and H. S. Kim, *Int. J. Hydrogen Energy*, 2008, **33**, 2826.
6. R. Steinberger-Wilckens, O. Bucheli, L. G. J. de Haart, A. Hagen, J. Kiviaho, J. Larsen, S. Pyke, B. Rietveld, J. Sfeir, F. Tietz and M. Zahid, in *Proceedings of the 11th ECS International Symposium on Solid Oxide Fuel Cells*, ECS Transactions, Pennington, N.J., 2009, p. 43.
7. D. Larrain, J. van Herle and D. Favrat, *J. Power Sources*, 2006, **161**, 392.
8. R. Bove and S. Ubertini, *J. Power Sources.*, 2006, **159**, 543.
9. N. Sammes, ed., *Fuel Cell Technology: Reaching Towards Commercialization*, Springer Verlag, London, 2006.
10. C. Hamann and W. Vielstich, *Electrochemistry*, 3rd edn. Wiley-VCH, Weinheim, 1998.
11. H. Zhu and R. J. Kee, *J. Power Sources*, 2003, **117**, 61.

12. J. D. J. Vandersteen, B. Kenney and J. G. Pharaoah, K. Karan, in *Proceedings of hydrogen and fuel cells. Toronto, Canada, 2004*, Fuel Cell Research Centre, Canada.
13. M. Iwata, T. Hikosaka, M. Morita, T. Iwanari, K. Ito, K. Onda, Y. Esaki, Y. Sakaki and S. Nagata, *Solid State Ionics*, 2000, **132**, 297.
14. P. W. Li and M. K. Chyu, *Trans. ASME*, 2005, **127**, 1344.
15. P. Costamagna and K. Honegger, *J. Electrochem. Soc.*, 1998, **145**, 3995.
16. R. Bove, P. Lunghi and N. M. Sammes, *Int. J. Hydrogen Energy*, 2005, **30**, 181.
17. S. Kakaç, A. Pramuanjaroenkij and X. Y. Zhou, *Int. J. Hydrogen Energy*, 2007, **32**, 761.
18. E. Achenbach, *J. Power Sources*, 1994, **49**, 333–348.
19. M. Roos, E. Batawi, U. Harnisch and Th. Hocker, *J. Power Sources*, 2003, **118**, 86.
20. M. Pfafferodt, P. Heidebrecht, M. Stelter and K. Sundmacher, *J. Power Sources*, 2005, **149**, 53.
21. P. Costamagna, P. Costa and V. Antonucci, *Electrochim. Acta*, 1998, **43**, 375.
22. S. H. Chan, K. A. Khor and Z. T. Xia, *J. Power Sources*, 2000, **93**, 130.
23. M. A. Khaleel, Z. Lin, P. Singh, W. Surdoval and D. Collin, *J. Power Sources*, 2004, **130**, 136.
24. F. Standaert, K. Hemmes and N. Woudstra, *J. Power Sources*, 1996, **63**, 221.
25. R. Bove, P. Lunghi and N. M. Sammes, *Int. J. Hydrogen Energy*, 2005, **30**, 189.
26. P. Atkins and J. Paula, *Physical Chemistry*, Oxford University Press, Oxford, 2006.
27. A. C. Hindmarsh, *SIGNUM Newsletter*, 1980, **15**, 10.
28. W. H. Press, S. A. Teukolsky and W. T. Vetterling, Numerical Recipes in FORTRAN 77, in *The Art of Scientific Computing: Fortran Numerical Recipes*, Cambridge, Cambridge University Press, 1992, 2nd edn, vol. 1, p. 749.
29. A. Gubner, D. Froning, B. de Haart and D. Stolten, in *Proceedings of the 8th ECS International Symposium on Solid Oxide Fuel Cells, 2003*, Proceedings of the Electrochemical Society 8, Pennington N.J., 2003, p. 1436.
30. A. Gubner, in *Proceedings of the 9th ECS International Symposium on Solid Oxide Fuel Cells*, Proceedings of the Electrochemical Society 9, Pennington, N.J., 2005, p. 814.
31. M. Finkenrath, *Simulation und Analyse des dynamischen Verhaltens von Kraftwerken mit oxidkeramischer Brennstoffzelle (SOFC)*, Aachen, Techn. Hochsch., Diss., 2005, Schriften des Forschungszentrums Jülich, Reihe Energietechnik/Energy Technology, 2005.
32. A. Gubner, J. Saarinen, J. Ylijoki, D. Froning, A. Kind, M. Halinen, M. Noponen and J. Kiviaho, in *Proceedings of the 7th European SOFC Conference, Lucerne, 2006*, EFCF, Oberrohrdorf, Switzerland, 2006 CD-ROM.
33. D. Bronin, B. Kuzin, D. Osinkin and I. Yaroslavtsev in *Proceedings of the 8th European SOFC Conference, Lucerne*, EFCF, Oberrohrdorf, Switzerland, 2008, paper no. B1001.
34. V. Sonn and E. Ivers-Tiffee, in *Proceedings of the 8th European SOFC conference, Lucerne*, EFCF, Oberrohrdorf, Switzerland, 2008, paper no. B1005.

CHAPTER 8

Accelerated Lifetime Testing for Phosphoric Acid Fuel Cells

JOHN DONAHUE,[a] NED CIPOLLINI[b] AND ROBERT FREDLEY[b]

[a] 469 Hickory Street, Suffield, Connecticut 06078, USA; [b] UTC Power, 195 Governors Highway, South Windsor, Connecticut 06074, USA

8.1 Introduction

The path to market acceptance of commercial products is fraught with many warranty risks; one of the quandaries for developers is the long lifetimes expected for commercial fuel cells. Prudent investors demand that these risks be mitigated before committing their capital. However, obtaining real-time verification and validation data on fully developed systems with 10-year lifetimes is cost and time prohibitive. Therefore, accelerated lifetime testing is required to adequately build confidence for developers and investors. Accelerated lifetime testing has evolved across the industry and testing conditions have been tailored for each failure mechanism for each component. The focus in this chapter is on the fundamental understanding of the degradation mechanisms of bi-polar plates, their rates and how representative data is gathered in times consistent with rapid product introduction.

8.2 Background

Carbon and graphite are fundamental building blocks for phosphoric acid fuel cells (PAFCs) and polymer electrolyte membrane fuel cells (PEMFCs) and, thus, are a major focus for cost reduction. Although carbon is not thermodynamically stable in all PAFC environments, the degradation kinetics are slow

RSC Energy and Environment Series No. 2
Innovations in Fuel Cell Technologies
Edited by Robert Steinberger-Wilckens and Werner Lehnert
© Royal Society of Chemistry 2010
Published by the Royal Society of Chemistry, www.rsc.org

enough for carbon materials to serve in components with a useful lifetime.[1]
PAFCs and PEMFCs are operated over a range of pressures, where the kinetic
rates vary significantly. The work reported here will be at ambient pressure with
temperatures in the range of 150 °C to 200 °C.

The carbon corrosion reaction,[2] Equation (8.1), describes the electrochemical
conversion of carbon and water to CO_2. The reaction proceeds in two steps:
carbon oxidation refers to the formation of metastable surface oxides on carbon
which are then oxidised to CO_2 in the corrosion reaction. Carbon corrosion and
carbon oxidation are the major sources of component-property decay, substrates lose mechanical crush strength and separators lose acid barrier properties. The driving force for Equation (8.1) is the oxygen reduction reaction:

$$C + 2H_2O \rightarrow CO_2 + 4H^+ + 4e^- \quad \varepsilon(V) = -0.207 - \frac{RT}{4F} \log \frac{P_{CO_2} a_{H^+}^4}{a_C a_{H_2O}^2} \quad (8.1)$$

where ε is the thermodynamic electrochemical potential for the reaction in volts
versus the standard hydrogen electrode, R is the gas constant, T is the absolute
temperature, F is the Faraday's constant, P_{CO_2} is the pressure of carbon dioxide,
and a is the activity of acid, carbon and water, as designated. Normally, the
activity of a solid phase like carbon is not included in the equation. We include it
here to show specifically that the stability of carbon impacts the corrosion
potential. If the activity of carbon is decreased by graphitisation, the electrochemical potential becomes more negative and the oxidation reaction is reduced.
The activity of water not only increases the thermodynamics of oxidation, but
also increases its kinetics. In the context of section 8.8.4 and Tables 8.1 and 8.2, ε
has been called the oxidation potential and is the potential where the corrosion
reaction is initiated; below the oxidation potential, the material is stable.

Even when thermodynamically favourable, the carbon corrosion rate is
kinetically controlled. The process is highly irreversible, meaning the reverse of
Equation (8.1) is never observed in fuel cells because carbon dioxide escapes.
Because of this irreversibility, the Butler–Volmer kinetic equation simplifies to
a Tafel-like relationship:

$$i_{cc} = i_{cc,0}(A) \left(P_{H_2O}^n \right) \exp \left[\frac{\alpha_a F}{RT} \left(E_{app} - E_C \right) \right] \quad (8.2)$$

where i_{cc} is the carbon corrosion current in $A\,cm^{-2}$, $i_{cc,0}$ is the exchange current
density, A is the the real area of the sample which will be discussed further
below, P_{H_2O} water vapor pressure raised to the n-th power, α_a is the anodic
transfer coefficient for the reaction of Equation (8.1), E_{app} is the applied
potential versus standard hydrogen electrode (SHE), and E_C the corrosion
potential of section 8.8.4 and Tables 8.1 and 8.2. Above the corrosion potential,
corrosion increases exponentially with applied potential. F, R and T have the
same meanings as in Equation (8.1). The value of n is determined

Table 8.1 Oxidation and corrosion potentials of PAFC separator plates with various heat treatments.

Conditions	$E_{oxidation}$ (mV)	$E_{corrosion}$ (mV)
3000 °C, Vendor A, no purge gas	720	1150
2100 °C, Vendor B, H_2 vacuum	800	1175
2100 °C, Vendor C, H_2 vacuum	750	1140
2100 °C, Vendor C, 40% H_2/Ar, top of furnace	725	1115
2100 °C, Vendor C, 40% H_2/Ar, middle of furnace	710	1100
2100 °C, Vendor C, 40% H_2/Ar, bottom of furnace	715	1080

Table 8.2 Oxidation and corrosion potentials of PAFC separator plates made with different materials.

Material	$E_{oxidation}$ (mV)	$E_{corrosion}$ (mV)
Vendor X, moulded PPS/graphite	445	820
Vendor X, moulded heat-treated graphite	545	1030
Vendor Y, moulded heat-treated graphite	640	1015
Glassy carbon heat-treated by Vendor X	835	1160
50/50 graphite/phenolic heat-treated by Vendor X	725	1115
Pyrolytic graphite	875	1240

experimentally and is called out explicitly here, to illustrate its use as a possible acceleration factor.

Equation (8.2) is in the form of a Tafel equation with Tafel slope, $RT/\alpha_a F$ ($\times 2.303$). The anodic transfer coefficient, α_a, for carbon corrosion in hot phosphoric acid[3] is essentially 1 giving a Tafel slope of 95 mV per decade. The area, A, exposed to electrolyte is explicitly shown in Equation (8.2) and will become important in section 8.8.4. Because water is the product of both the PAFC and PEMFC reactions, stabilisation of carbon by reducing water concentration is not practical, or even plausible. In addition to the product water, the cathode potential in the range of 0.6–1.2 V versus SHE makes this portion of the cell quite susceptible to carbon corrosion. Experience has taught only two ways to retard the corrosion reaction: stabilisation of the carbon by graphitisation and separation of reactants by treating the carbon with Teflon® to make it hydrophobic. And with that, the Teflon®-carbon bond is only temporary, the bond delaminates slowly in hot phosphoric acid.[4]

A cross-section of the repeating unit of a PAFC cell is shown as Figure 8.1, with the carbon components noted. To obtain useful voltages, the unit cell is repeated some hundreds of times. An important element in the cell is the separator plate, which needs to be impermeable to the flow of phosphoric acid.[5]

Individual cells in a stack are separated by bi-polar plates. This term is derived from the bi-polar environment in which these plates operate; oxidant on one side and fuel on the other. The high conductivity of the plates could lead to high currents for electrochemical corrosion reactions and component failure

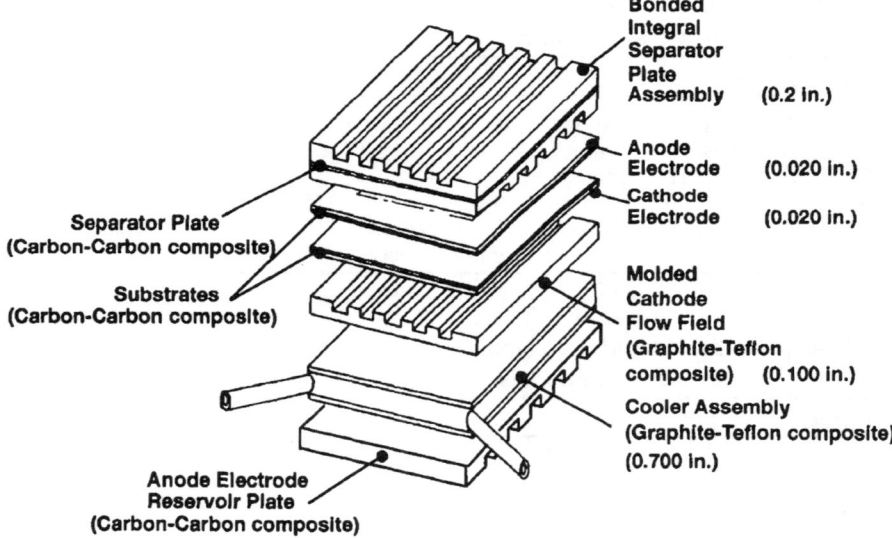

Figure 8.1 The repeat unit of a PAFC-stack showing coolers. Stacks can have between five and ten cells per cooler and up to 40 coolers.

under certain conditions. Therefore, the stability of these plates and the conditions in which they are stable must be determined.

Bi-polar plates serve multiple functions: provide reactant flow across the planform, separate reactants, afford electronic contact between cells, and maintain an electrolyte barrier between cells. The last function is present only in liquid electrolyte cells, not in PEMFCs. Electrolyte is actively pumped between cells electrochemically because the ionic current is not completely carried by protons, the active species in PAFCs. About 11% of the current is carried by anions, $H_2PO_4^-$, which corresponds to mass transport of acid and is termed 'shunt current.' Shunt currents lead to electrolyte flooding on the anode side of each substack and dry-out on the cathode side. Flooding and dry-out lead to performance degradation; dry-out also leads to cross-over and cell failure. UTC Power has determined the maximum amount of electrolyte transfer that can be tolerated before cell failure.

UTC Power's PAFC bi-polar plate is a laminate of the flow fields bonded together with a carbon separator plate as shown in Figure 8.2.

As such, the bi-polar plate is termed the integral separator plate (ISP). Therefore, the multiple functions of the ISP are broken into separate components. Gas supply is performed by flowfields. Providing uniform gas supply throughout the stack platform can be difficult to achieve due to manufacturing tolerances. However, with proper selection of materials it is not a first-order life-limiting problem. The primary function of the separator plate – stopping electrolyte migration – has proven to be one of the most demanding requirements of any component. Carbon corrosion in separator plates can cause filling of pre-existing pores with electrolyte and creation of new pores filled with

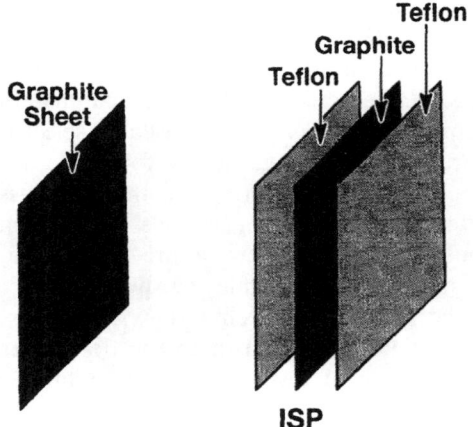

Figure 8.2   Lamination of a graphite separator plate with two graphite–Teflon[®] composite flow fields to form a PAFC ISP.

electrolyte, both will conduct electrolyte since these pores are necessarily interconnected. Opening of new pores increases the surface area of contact between electrolyte and separator which increases the pre-exponential term A in Equation (8.2) and accelerates further degradation. Therefore, any accelerated-degradation model must take this into account.

Carbon corrosion causing filling of pores leading to electrolyte migration will cause fuel-cell failure above a pre-established amount. This is the failure mode that will be used to illustrate how accelerated testing supported achieving the design lifetime of PAFCs.

8.3   Experimental Design

A thorough understanding of fundamentals of the degradation mechanisms is required to plan and execute successful accelerated testing. Early in the product development process, specific rates for each component may not have been quantified, so first-order estimates of their magnitudes are often used to begin testing. There are different approaches to measuring the rate of any degradation mechanism, but the fall into two broad categories of out-of-cell testing and in-cell testing. Some out-of-cell tests are relatively simple and can be quite fast, thus lending themselves to screening. Other tests are more complicated, require more expensive equipment, and take longer. Finally, cell testing requires design and construction of special tooling and test stands is the most complex and requires the most time to complete. Qualification of new materials requires this progression of tests. A separator plate from an alternate vendor constitutes a new material since the processing and resulting micro-structure likely differs from the present separator. As with all planned experimentation, a little screening up front will help ensure that the next phase is well-directed.

With carbon corrosion, Equation (8.1) gives us the chemistry and thermo-dynamics while Equation (8.2) gives us the kinetics. In PAFC, simple testing done early with heated acid baths determined weight loss measured over time at fixed concentrations. These screening tests guided material selection. Once subscale/proof of concept testing had been used to select materials, full-scale short-stack hardware was used to gather performance and life data. Short-stack experience is invaluable in pointing out the weaknesses of cell or stack design and execution. Refinement of models used to analytically predict boundary conditions are made as new failure modes are identified. Rather than simple acid bath tests, subscale hardware was used to potentio-statically control the voltage of 5 cm×5 cm specimens. A broad range of voltages and time at potential help map the domain to ensure that con-ditions used to do subsequent testing are representative of the worst-case conditions.

Commercial decisions about warranty require technical and production readiness that is well-documented and represents full-scale hardware. The final stage of PAFC accelerated testing required fabrication of new subscale tooling that subjected samples to the in stack conditions on both sides. This 'cell and one half' configuration exposes the separator plate to actual conditions with all the physics and electrochemistry that is important in degrading the component. The reader will see in the following sections that attention to detail in all the above paid dividends. Enough time has passed so that full-scale PAFC cell stack assemblies have now met or exceeded the lifetimes (> 50 000 h) predicted by accelerated testing accomplished in approximately 1000 min.

8.4   Testing

8.4.1   First Generation Testing: Pot Tests and Corrosion Potential Measurement

In the late 1960s UTC Power PAFC material evaluation testing of electrically conductive samples began with submerging multiple samples in 85%, 97% or 105% H_3PO_4 under a N_2 blanket over acid. Sample temperature varied from 149 °C (300 F) to 204 °C (400 °F). Samples were periodically removed to mea-sure their weight loss/gain. In the 1970s at UTC power, the potential of the sample was controlled at selected potentials between 700 and 900 mV. This test provides a history of weight loss versus long time at various temperatures and potentials for the material. The visual observations made on samples after testing were qualitatively valuable. Subscale cell and short-stack testing as well as work being done in other laboratories supplemented the weight loss testing to move forward the development effort. Table 8.3 reports typical results of this testing. The soak testing only modestly accelerated the time required to acquire data. The temperatures and voltages were close to the operating conditions. The relationship between weight loss and product life was not able to be quantified with this approach.

Table 8.3 Weight loss and observations of graphite samples soaked in hot H_3PO_4.

Sample ID	Potential (V)	H_3PO_4 (%)	Temperature (°C), and comment(s)	Time (h), % wt loss	Time (h), % wt loss	Time (h), % wt loss	Time (h), % wt loss
Graphite DA-8	0.75	105	400	500, 1.50	1000, 0.6	3000, 0.7	5000, 3.6
Graphite	0.0	85	300; No change in appearance	162, 0.87	1056, 0.81	–	–
Graphite no. 108	0.0	95	300; 20 wt% solid phenolic	241, 0.23	1001, 0.46	–	–
Graphite no. 109		95	300; 15 wt% solid phenolic	216, 0.12	1001, 0.00	–	–
Graphite no. 110		95	300; 10 wt% solid phenolic	216, 0.00	720, 1.64	1001, 3.64	–
Graphite no. 360	0.0	85	300; badly blist, black residue	168, 0.0	–	–	–

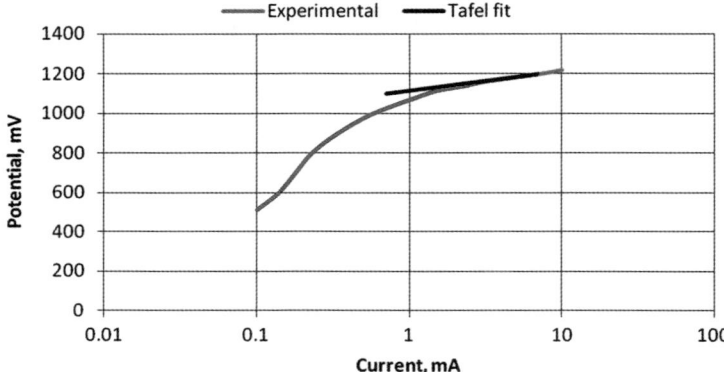

Figure 8.3   Typical corrosion potential sweep for a UTC Power separator, the corrosion potential is ∼1100 mV.

By the late 1970s at UTC power, work on pressurised systems for electric
utility application had pushed carbon corrosion work forward. Quantifying
corrosion potential replaced oxidation potential as the preferred voltammetric
sweep characterisation. Oxidation potential is the potential where corrosion
reaction is initiated; below that potential the material is stable and corrosion
potential is where the reaction rate increases exponentially. A typical sweep and
data fit are shown in Figure 8.3. Operation at or near the corrosion potential is
clearly not acceptable for products that will be durable.

Table 8.1 shows oxidation and corrosion potentials on separator plate
samples with differing heat treatments. There is no clear advantage to the
higher heat treat temperatures, likely because the starting materials are already
graphitic. Table 8.2 shows a much larger range of oxidation and corrosion
potentials on separator plate samples made with different materials. Difficulty
in finding low-cost materials that had high enough oxidation potentials to
protect against corrosion in all areas of the cell forced efforts to develop
techniques to quickly obtain quantitative data that would be useful in predicting product lifetimes.

Pot tests and corrosion potential tests address stability of materials and
components giving the ε (volts) and E_C values of Equation (8.2) for these
materials. However, nearly all materials are metastable so the next step in
accelerated testing of materials for PAFCs is to address the rates of corrosion,
i.e., the $i_{cc,0}$, n and α_a values in Equation (8.2).

8.4.2   Second Generation Testing: Potential Control and
         Monitor Current

The increasing availability of potentiostats facilitated anodic polarisation
studies of separator plate and other cell components. The ability to run
unattended helped map the time dependence of separator plate corrosion

throughout the potential range of interest (0.9–1.2 V versus reversible hydrogen electrode (RHE) in the same solution). Focused effort on ambient pressure drove the choice of 99% H_3PO_4 and 200 °C. Typical data collected for 1000 min is shown in Figure 8.4. Note that at potentials below 1 V the corrosion current decreases an order of magnitude from $2 \, \mu A \, mg^{-1} \, C^{-1}$ to $\sim 0.1 \, \mu A \, mg^{-1} \, C^{-1}$ over 1000 min. The corrosion rates at higher potentials also initially decrease, but then increase and exceed their initial values, with the 1100 mV potential sample reaching almost $20 \, \mu A \, mg^{-1} \, C^{-1}$ at 1000 min. As might be expected, the higher potentials give the highest corrosion currents. The corrosion rate with potentials between 1100 mV and 1200 mV appear to be asymptotic after 200 min.

Understanding this behaviour is essential to the design of accelerated tests for PAFC-separator plates. The conclusion after many years of studying carbon corrosion at UTC Power, is that this observed increase in current versus time for potentials > 1000 mV is due to etching of the carbon surface to increase the area, A, in Equation (8.2). The carbon surface area in contact with electrolyte can be measured and is proportional to double-layer capacitance. This has been done for carbon substrates and electrolyte reservoir plates. When a plot of $\mu A \, cm^{-2}$, current per unit area in contact with electrolyte, versus time is plotted, the current reaches a steady state. This steady state $\mu A \, cm^{-2}$ follows Equation (8.2) and yields values for $i_{cc,0}$ and n. The observed value for $\alpha_a \, F/RT$ is equivalent to a Tafel slope 95 mV per decade, RT/F, as expected and is consistent with other carbon materials.[3]

The case for separator plates is more complicated than for carbon substrates and electrolyte reservoir plates. Figure 8.5 presents the surface area normalised corrosion rate.

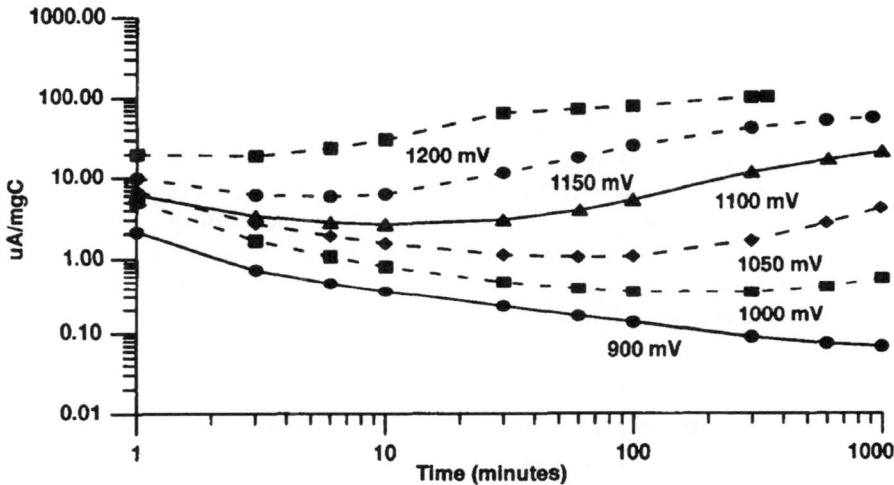

Figure 8.4 Corrosion current versus time for a UTC power separator plate in 204 °C in 99 wt% phosphoric acid.

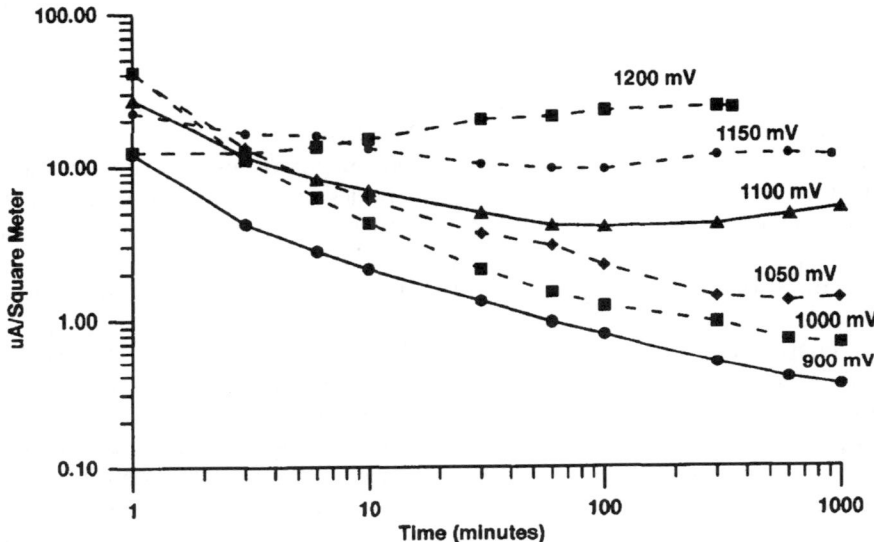

Figure 8.5 Corrosion current per unit area vs. time for a UTC Power separator plate in 99 wt% phosphoric acid at 204 °C.

On this basis, the corrosion rate has levelled off at the highest potentials, but is still decreasing at 1000 min at 900 mV. In the case of separator plates, while carbon corrosion is still opening up porosity and increasing the sample surface area, this area is connected to the sample surface through very thin, tortuous pores. There is significant resistance and iR drop (i is the current) within a pore so that at the higher potentials with higher current, the E_{app} is not constant within the pore. Therefore, while the current/area reaches steady state, the fit to Equation (8.2) is not as good as with other components.

The 1000-min testing results in very low weight losses and project to only fractions of a per cent in 100 or 1000 h. Another important difference in the case of separator plates, is the separator plate has some closed porosity at beginning of life, even small weight losses can lead to failure as through-plane holes defeat the separator function Figure 8.6 shows the weight loss for several conditions, including separator plates in predicted service. The impact of the water partial pressure, as expected based on Equation (8.2), is also shown in Figure 8.6. The lowest weight-loss line of data presented in this figure takes us to the final phase of PAFC accelerated testing.

8.4.3 Third Generation Testing: Inter-cell Effects

As noted before, the bi-polar plate operates with cathode potential on one side and anode potential on the other side. Since the plate is an electronic conductor, this potential difference cannot be electronic within the plate. In reality, the potential changes within the electrolyte. The effective E_{app} in Equation (8.2) is not a constant through the plane of the separator; it is large on the side facing

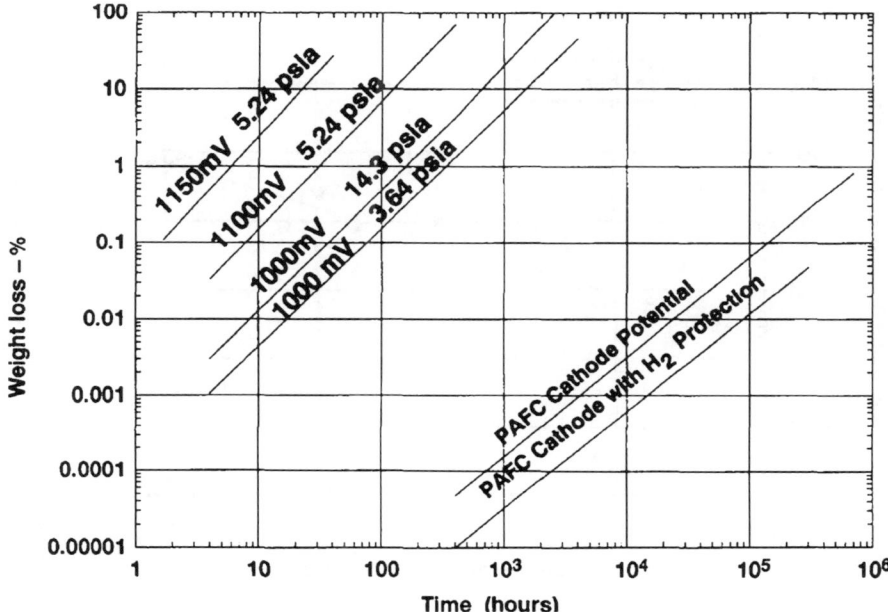

Figure 8.6 Projected weight loss vs. time for PAFC separators based on Equation (8.2) in 99% phosphoric acid at 200 °C.

the cathode and low on the side facing the anode. Therefore, inter-cell effects play a large role in the accelerated testing.

The importance of the closed porosity (10–25%) in the separator plates might not be immediately obvious, even more so that this porosity might actually work to the advantage of developers. As the carbon corrosion consumes the walls of the closed pores, they are no longer a barrier to acid, which can further intrude into the separator plate from the cathode side of the cell. As the corroding front encounters the hydrogen side of the separator, the local potential changes and the reaction rate diminishes. The lowest line in Figure 8.6 shows how the stack degradation rate is less than would be expected based on the cathode potential alone. This reality dictated that a new set of hardware be constructed to ensure that stack conditions be represented in subscale hardware. 'Cell and one half' hardware introduced a second anode to sandwich the cathode between two anodes. This configuration can be used to accelerate the corrosion process by raising the cathode potential, but maintain the beneficial effect of the hydrogen on the other side of the separator plate.

The importance of filling through pores with electrolyte is that once these pores fill and are subjected to the inter-/multi-cell potential, they will allow inter-cell transport of electrolyte. Eventually this inter-cell acid flux will empty one cell (allowing cross-over) and overfill another cell (impeding diffusion). In parallel with the carbon corrosion experiments, a significant additional characterisation effort was under way on full-scale components with field experience. Acid flux

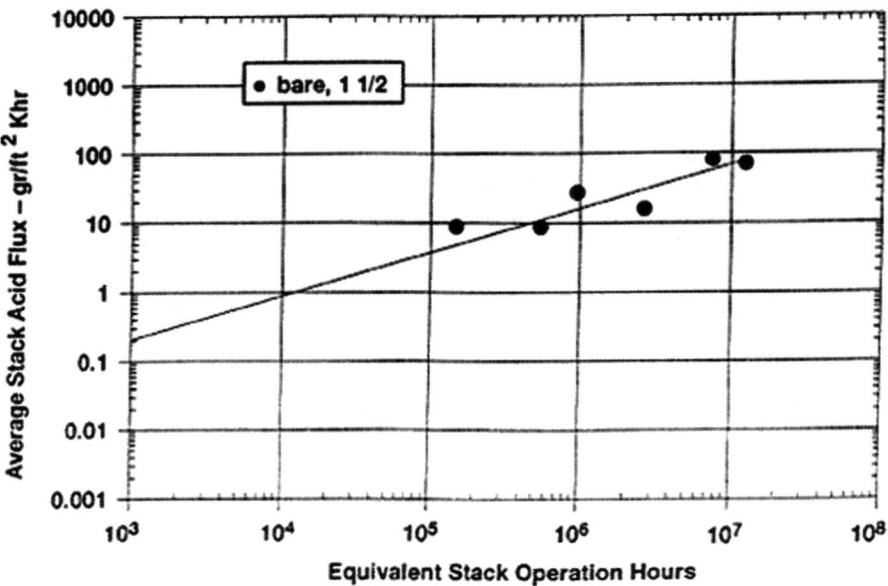

Figure 8.7   Acid migration rate through a 'bare' UTC Power separator plate. 'Bare'
refers to the fact the plate was not laminated into an ISP. The plate was
aged in the 'cell and half' and acid migration rate estimated from ionic
resistance measurements through the plate in the laboratory.

data was gathered on separator plates operated in 200 kW PAFC power plants.
Figure 8.7 introduces 'equivalent stack operation hours' as the time scale; this is
the culmination of the previous figures where weight loss or current/area were
presented as a function of time at potential. Laboratory corrosion rates of Figure
8.5 on a surface area basis and actual PAFC cell stack aged data were correlated
and combined with acid flux measured on laboratory and stack-aged samples to
make this transformation. Fleet durability data on the 200 kW power plants is
shown as Figure 8.8. The number of units that have exceeded the 40 000-h cell
stack assembly design life speaks to the utility of this accelerated testing.

8.5   Conclusions

The risks of a potentially life-limiting failure mode of PAFC cell stacks, inter-
cell acid transfer, were successfully mitigated by use of accelerated testing.
Potential controlled carbon corrosion testing of carbon–graphite separator
plates was used to accelerate the aging of samples and reduced that test time to
1000 min. Acid transfer rates measured on corroded samples coupled with
samples from field units produced a correlation predicting product design
lifetimes would be met. Subsequent field data on production units has proven
that the predictions were correct.

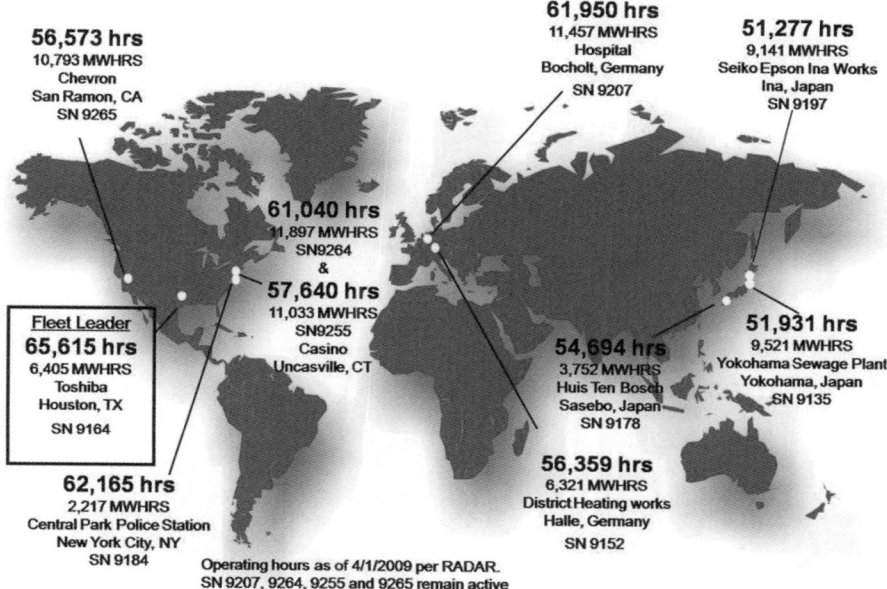

56,573 hrs
10,793 MWHRS
Chevron
San Ramon, CA
SN 9265

61,950 hrs
11,457 MWHRS
Hospital
Bocholt, Germany
SN 9207

51,277 hrs
9,141 MWHRS
Seiko Epson Ina Works
Ina, Japan
SN 9197

61,040 hrs
11,897 MWHRS
SN9264
&
57,640 hrs
11,033 MWHRS
SN9255
Casino
Uncasville, CT

Fleet Leader
65,615 hrs
6,405 MWHRS
Toshiba
Houston, TX
SN 9164

54,694 hrs
3,752 MWHRS
Huis Ten Bosch
Sasebo, Japan
SN 9178

51,931 hrs
9,521 MWHRS
Yokohama Sewage Plant
Yokohama, Japan
SN 9135

62,165 hrs
2,217 MWHRS
Central Park Police Station
New York City, NY
SN 9184

56,359 hrs
6,321 MWHRS
District Heating works
Halle, Germany
SN 9152

Operating hours as of 4/1/2009 per RADAR.
SN 9207, 9264, 9255 and 9265 remain active

Figure 8.8 PAFC PC25C® Cell Stack Durability as operating hours and MW hours as of 1 April 2009.

Acknowledgements

The authors wish to express thanks to the Materials Research Society for permission to use Figures 8.1 to 8.7 and to Carl Rohrback Jr for Figure 8.8.

References

1. A. Appleby, Corrosion in Low and high temperature fuel cells – an overview *Natl. Assoc. Corrosion Eng.*, 1987, **43**(7), 398–408.
2. M. Pourbaix, *Atlas of Electrochemical Equilibria in Aqueous Solutions*, National Association of Corrosion Engineers, Houston, TX, 1970, p. 453.
3. K. Kinoshita, *Carbon: Electrochemical and Physiochemical Properties*, John Wiley, New York, 1988, p. 328.
4. D. Wheeler, F. Luczak, R. Fredley and N. Cipollini, Carbon and fluorinated carbon materials for fuel cells, in *Materials Research Society Symposium Proceedings, Materials for Electrochemical Energy Storage and Conversion II – Batteries, Capacitors and Fuel Cells*, Materials Research Society, Warrendale, PA, 1997, vol. 496, pp. 39–147.
5. R. P. Roche, US Patent No. 5 268 239 (7 December 1993).

Part 5
Hydrogen Generation and Reversible Fuel Cells

Introduction

Fuel cells were discovered as the reverse effect of electrolysis (splitting of water by electrical current) by Schönbein and Grove in 1838 and 1839, respectively. That the functions of fuel cells and electrolysis can be integrated into a single device is scientifically obvious, because electrochemical reactions always have a forward and a backward reaction. However, whether or not this would yield practical devices is another issue. Such a scheme would allow the use of a compact unit to both produce hydrogen for electricity storage purposes and re-electrify it by the same unit. High-temperature fuel cells used as electrolysis units, for instance, offer extremely high and attractive conversion efficiencies close to 100%, if the steam necessary for the process can be obtained from waste heat sources. Another way of producing hydrogen with fuel cells is by modifying the operation of internal reforming (high temperature) fuel cells. By influencing the fuel utilisation within these fuel cells, hydrogen can be produced as a by-product, thus making the fuel cell attractive as a tri-generation unit producing both electricity and heat, but also hydrogen as a third product.

A third way of producing hydrogen by fuel cells – yet in an indirect way from carbon monoxide – is by using a direct carbon fuel cell. Thermodynamics allows for the intriguing concept of a fuel cell that produces hydrogen instead of consuming it and converting heat into power instead of producing waste heat. In the next section we will first describe the option of electrolysis while in the second article we cover the option of hydrogen production with fuel cells. Direct carbon fuel cells have been covered in Chapter 6.

CHAPTER 9

Electrolysis Using Fuel Cell Technology

A. BRISSE,[a] J. SCHEFOLD,[a] C. STOOTS[b] AND J. O'BRIEN[b]

[a] European Institute for Energy Research, Emmy-Noether-Strasse 11, 76131 Karlsruhe, Germany; [b] Idaho National Laboratory, P.O. 1825, MS 3870, Idaho Falls, ID 83415, United States

9.1   Introduction

Hydrogen is used as raw material in many industrial and chemical processes. The first commercial technology of hydrogen production was the alkaline electrolysis of water, developed in the late 1920s. In the 1960s, industrial production of hydrogen shifted gradually towards a fossil fuel based feedstock. At present, around 600 billion Nm^3 of hydrogen are produced worldwide per year.[1] About 96% of this production is based on fossil fuels, mainly natural gas and coal. The use of these fuels results in significant greenhouse gas emissions and contributes to global warming. Moreover, there is a general agreement that global hydrocarbon production will peak in the near future, a factor which should favour the development of hydrogen as an energy carrier.

Hydrogen can be produced *via* the classical electrolysis of water at temperatures between 70 and 90 °C or, alternatively, by using fuel cell technology.[2] There are three types of water electrolysers based on fuel cell technology: (1) the proton-exchange membrane or PEM electrolyser (referring to its solid polymeric electrolyte membrane); (2) the solid-oxide ion (O^{2-}) conductor or SOC

RSC Energy and Environment Series No. 2
Innovations in Fuel Cell Technologies
Edited by Robert Steinberger-Wilckens and Werner Lehnert
© Royal Society of Chemistry 2010
Published by the Royal Society of Chemistry, www.rsc.org

electrolyser (referring to its solid ceramic electrolyte); and (3) the electrolyser using solid-oxide proton conductors.

In a fuel cell, electrical energy is generated by the exothermic oxidation of hydrogen. In the reverse electrolysis operation of such a cell, heat is required to reduce steam using electrical energy, Equation (9.1):

$$H_2O \xrightarrow{\text{Electricity+heat}} H_2 + \tfrac{1}{2}O_2 \qquad (9.1)$$

Operational temperatures of fuel cells vary widely from around 80–120 °C (proton-exchange membrane fuel cells, PEMFC) to about 700–1000 °C (solid oxide fuel cells or SOFCs). The free energy required for the reaction (ΔG) decreases with increasing temperature while the free enthalpy (ΔH) remains almost constant (Figure 9.1). This thermodynamic relation explains the particular interest for performing electrolysis at higher temperature. The relation is, in principle, unfavourable for fuel cell operation at higher temperature, as it implies an increased heat release, and therefore lower chemical-to-electrical energy-conversion efficiency, already for an ideal (loss-free) cell. For the electrolysis mode at such a cell, on the other hand, the relation means an increased heat uptake with increasing temperature. In other words, ideal fuel cell efficiency, always below 100%, drops with increasing temperature, while the (electrical-to-chemical) energy-conversion efficiency of an ideal electrolysis cell, always above 100%, increases with increasing temperature. (Note that the term 'energy conversion efficiency' for the HTE refers to the electrical-to-chemical energy conversion throughout this chapter.) Because real SOFCs achieve competitive (chemical-to-electrical) energy-conversion efficiencies despite the less favourable thermodynamic conditions, one can expect that solid oxide electrolyser cells (SOECs) achieve much higher efficiencies, provided that the

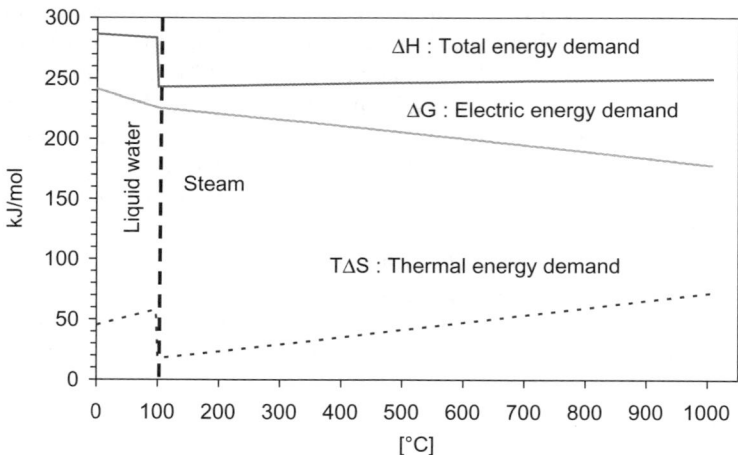

Figure 9.1 Electrical, heat and total energy requirements for water electrolysis as a function of temperature.

cell over-voltages are of comparable magnitude, *i.e.* that the cells can be operated reversibly. Operation cell voltages of about 1.1-1.3 V are presently achieved with state-of-the-art SOECs, as compared to cell voltages above 1.6 V for low temperature PEM electrolysers (*cf.* section 9.3.2). Because current efficiencies approach 100% in both cases, the difference in cell voltage is almost linear in the energy consumption for the reaction.

The first part of this chapter (section 9.2) refers to the technologically more mature PEM electrolysers. In the second part (section 9.3), high-temperature electrolysis (HTE) using SOECs is discussed. Cell and stack performance is analysed as well as durability, system demonstration, and development issues including the question of to what extent SOFC technology can fulfil the requirements of SOEC operation, before an SOEC technology may eventually develop.

9.2 Low-temperature Electrolysis

9.2.1 Introduction

The first PEM electrolysers were developed in 1966 (General Electric, USA)[3] for undersea life support systems. The British Royal Navy adopted this technology in the early 1980s for the submarine fleet. The main applications today are the production of pure hydrogen for analytical instrumentation, welding, metallurgy of high purity metals and alloys, and the manufacture of pure substances for the electronics industry and analytical chemistry. These PEM systems present several advantages over traditional alkaline electrolysis technology, such as higher energy-conversion efficiency, a more compact design,[4] a faster start and stop capacity, and improved response to fluctuations in the electrical power input (hence the application flexibility relevant to capturing intermittent renewable electricity supplies, such as wind and solar power).[5]

PEM electrolysers use PEM fuel cell technology. So far, the development of PEM electrolysers benefits from the progress made with PEM fuel cells. The heart of the electrolyser is a solid polymer membrane as electrolyte. This membrane is a perfluorinated ion exchange polymer membrane (commonly Nafion® from DuPont de Nemours), containing sulfonated-acid groups (SO_3H). These groups are converted into their conjugated base (SO_3^-) under an electrical field according to Equation (9.2).

$$H_2O + SO_3H \rightarrow SO_3^- + H_3O^+ \qquad (9.2)$$

The membrane is connected with the anode and the cathode, which consist of highly porous graphite squeleton impregnated with electrocatalytic particles and then covered with a solution of the electrolyte ionomer. Protons are conducted from the anode to the cathode through the electrolyte where hydrogen is formed. The membrane also serves as diaphragm to separate the produced H_2

at the cathode (Equation (9.3)) and O_2 at the anode (Equation (9.4)):

$$\text{Cathode}: \quad 2H^+ + 2e^- \rightarrow H_2 \qquad\qquad (9.3)$$

$$\text{Anode}: \quad H_2O \rightarrow \tfrac{1}{2}O_2 + 2H^+ + 2e^- \qquad\qquad (9.4)$$

Different electrocatalysts are used for the anode (*e.g.* RuO_2, IrO_2) and cathode (*e.g.* Pt). Noble metal oxides as electrocatalysts are well established in many industrial electrochemical processes in the form of dimensionally stable anodes as first developed by Beer.[6] Ruthenium is recognised as the most active oxide for the oxygen evolution reaction;[7] however, it suffers from instability and therefore it is stabilised with another oxide such as IrO_2. Tantalum is a known addition to the electrodes, and Ir–Ta oxides have been suggested as efficient electrocatalysts for oxygen evolution in acidic electrolytes due to their high activity and corrosion stability.[8] The electrical contact and the mechanical support in the PEM are established with porous backings like metallic meshes (*e.g.* titan).

9.2.2 Commercial Systems

Small-to-mid-scale PEM electrolysers are commercially available from a number of companies, the main ones being Giner Electrochemical Systems,[9] Teledyne Energy Systems,[10] Norsk Hydro[11] with StatoilHydro Hydrogen Technologies formed by the merger of Statoil with the oil and gas division of Norsk Hydro (Norway). Hydrogenics Corporation[12] (Canada) offers PEM electrolysers with capacities between 1.1 and $30\,Nm^3\,h^{-1}$ of hydrogen. Hamilton Sundstrand[13] (USA) and Stuart Energy System formed a strategic alliance in 2002 to develop PEM hydrogen generation products. Proton Energy Systems[14] offers PEM electrolysers at capacities of 0.5 and $6\,Nm^3\,h^{-1}$ of hydrogen at pressures between 15 bar and 30 bar. In Japan, Shinko Pantec[15] is developing PEM electrolysers from 0.5 to $100\,N\,m^3\,h^{-1}$. Wellman Defence[16] in the UK produces oxygen with PEM electrolysers.

9.2.3 Performance

Operation voltages of PEM electrolysers vary typically between 1.65 and 1.75 V at $1\,A\,cm^{-2}$ and $80\,^{\circ}C$.[17–19] Results were published with cell voltages below $1.6\,V$,[20] for instance, within the World Energy NETwork, WE-NET.[21] In that programme a cell voltage of 1.56 V was achieved at $1\,A\,cm^{-2}$ and $80\,^{\circ}C$, using a $52\,\mu m$ thin membrane of an area of $2500\,cm^2$.

The power consumption of PEM electrolysers is about 3.9–$4.1\,kWh\,N^{-1}\,m^{-3}$ of hydrogen.[22] Hydrogen purity is above 99.98% and operating pressures are up to 30 bar. The loading of noble metals is around 0.3–$1.0\,mg\,cm^{-2}$ on the cathode and 1.5–$2.0\,mg\,cm^{-2}$ on the anode.[23,24] Long-term operation of $60\,000\,h$ has been reported for a commercial system.[14]

9.2.4 Development Issues

Major development fields of PEM electrolysers are a higher cell surface, a further increase in cell efficiency (>95%), and cost reduction. An important issue with respect to cost reduction is a maintained or improved catalytic activity of the electrodes under reduced catalyst loading.[21] Present PEM electrolysers have a typical hydrogen production capacity of up to $10\,N\,m^3\,h^{-1}$. The high capital costs of PEM units stem from the membranes, the noble-metal electrocatalysts, the construction materials (commonly a Ti structure), and the water purification. Despite the high capital costs, the dominating cost share of hydrogen produced by PEM electrolysis comes from the electricity cost ($\sim 70\%$). This means that compared to alkaline electrolysis, lower power consumption compensates the higher initial investment to some extent.

An important development path consists in the increase of the operating temperature to values above 100 °C (typically 120 °C) which requires advanced or alternative membranes. Higher operation temperatures would mean an increase in the PEM electrolyser performance by accelerating the kinetics of the electrochemical reactions[25,26] and reducing the membrane resistance. In addition, operation at higher temperature would facilitate the removal of heat from loss processes.

9.3 High-temperature Steam Electrolysis

9.3.1 Introduction

The feasibility of high-temperature steam electrolysis with tubular SOECs was demonstrated in the 1980s within the project HOTELLY (High Operating Temperature ElectrolYsis), led by Dornier and Lurgi.[27–30] The company Westinghouse demonstrated the HTE at about the same time period using similar technology.[31] The Japan Atomic Energy Research Institute studied the possibility of using heat generated by a gas turbine power plant to produce hydrogen using a HTE system.[32] At the end of that decade, HTE development was abandoned as consequence of both a low oil price and a high cost of the electrolyser stacks, which left insufficient short-to-mid-term perspectives for an industrial application.

In recent years, renewed interest in HTE has arisen from the search for alternative methods of energy conversion owing to increasing costs and availability limitations of fossil fuels. These SOEC activities benefit from the vast progress of planar SOFC technology during the last two decades. Planar cells offer advantages in terms of power density and energy-conversion efficiency. Moreover, a certain cost reduction was achieved as a consequence of beginning industrial production of cells.

The primary motivation for HTE is the potential of a reduced demand for electrical energy as compared to electrolysis at low temperature. The reduced energy demand stems from the above-mentioned favourable thermodynamic and kinetic conditions of the electrochemical reaction. This may allow electrical-to-chemical energy conversion efficiencies even exceeding 100%, as already recognised in early work[27,31] (see below). The free energy of the reaction, ΔG, decreases from $\sim 1.23 \, eV \, (237 \, kJ \, mol^{-1})$ at ambient temperature to $\sim 0.95 \, eV$ at $900 \, °C$ $(183 \, kJ \, mol^{-1})$ while the free enthalpy term remains essentially unchanged $(\Delta H \sim 1.3 \, eV$ or $249 \, kJ \, mol^{-1}$ at $900 \, °C)^{33}$ (*cf.* Figure 9.1). Part of the energy required for an ideal (loss-free) HTE can thus be provided by heat. Increasing ohmic and/or reaction losses in a real HTE system increase the demand for electrical energy and decrease the demand for an external heat supply until, finally, the reaction becomes exothermic as in low temperature electrolysis. Hence, three modes of operation are distinguishable in HTE: thermal-neutral, endothermic and exothermic. The HTE operates at thermal equilibrium (the thermal-neutral mode) when the electrical energy input equals the enthalpy of the reaction. In that case, the entropy necessary for the water splitting equals the heat generated by the loss reactions, the process is isothermal and adiabatic, and the energy conversion efficiency is 100%. In the exothermic mode, on the other hand, the electric energy input exceeds the ΔH term, which corresponds to an efficiency below 100%. Finally, in the endothermic mode the electric energy input remains below the enthalpy term. Therefore, heat must be supplied to maintain the temperature. This mode means energy conversion efficiencies above 100%.

The availability of an external heat source, as well as the temperature of that source, influence the design of a HTE system. Without any heat supply, the electrical energy required for (water) electrolysis amounts to at least the free reaction enthalpy at ambient temperature ($\Delta H \sim 1.42 \, eV$ or $287 \, kJ \, mol^{-1}$). With an external heat source for steam evaporation, the minimum requirement of electrical energy drops to the free enthalpy for steam electrolysis of $\Delta H \sim 1.3 \, eV$ or $249 \, kJ \, mol^{-1}$ (*cf.* Figure 9.1). In both cases, the design goal is to approach the thermal-neutral mode; that is, to limit the thermal losses to a value required to compensate for the endothermic reaction. This leaves state-of-the-art cells a rather wide margin for cell over-voltages (above 0.44 V for water electrolysis and at least 0.3 V for steam electrolysis). This margin may be used for a lowering of the temperature or an increase in the current density, as compared to the SOFC operation of the cell. With an external supply of high-temperature heat, on the other hand, the design goal of the HTE system is to reduce the over-voltages as far as possible to allow for a significant uptake of heat, and, thus, to reach endothermic operation with a corresponding electrical-to-chemical energy conversion efficiency above 100%. The achievement of low over-voltages in this operation mode will require higher cell temperatures (800 °C or above).

The HTE cells with electrolytes with O^{2-} conduction operate at temperatures above 650 °C. As in fuel cell technology, electrolytes with proton conduction might be used at intermediate temperatures (about 500–600 °C). However, so far, no practical demonstration seems to be reported. In such a proton ceramic electrolyser cell (PCEC) unconverted water would leave the cell in the exhaust of the oxygen side while it escapes at the hydrogen side in the SOEC (Figure 9.2). The PCEC compared to the SOEC lower operation

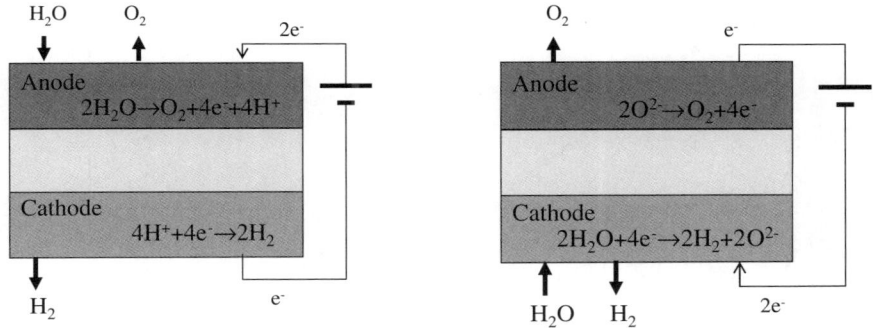

Figure 9.2 High temperature electrolysis cell using (a) a proton electrolyte conductor (PCEC) and, (b) an O^{2-} ionic electrolyte conductor (SOEC).

temperature means a higher free energy of the reaction, ΔG, as well as, most likely, increased reaction over-potentials. Both factors render an endothermal operation (with external heat supply) less likely.

It has been demonstrated by several research projects since 2004 that the planar SOFCs developed in recent years can be operated as SOECs.[34] Cells of both commercial and research type and including the common designs were tested (electrolyte-, hydrogen electrode-, and metal substrate-supported). As under fuel cell operation, the non-electrolyte supported cells showed the highest performance owing to the low resistance of the thin electrolyte layer. A high current density of $-3.6 \, \text{A cm}^{-2}$ at a cell voltage of 1.48 V and 950 °C cell temperature was reached with such a cell by DTU-Risoe (Denmark).[35]

9.3.2 Solid Oxide Electrolyser Cells

Virtually all development work on SOECs done within the last few years refers to SOFCs operated in the electrolysis mode, *i.e.* the fuel cells are used without modification.[36–39] As with SOFC technology in general, much emphasis in SOEC is put on planar cell technology. Compared to the tubular cells operated at temperatures above 900 °C, state-of-the-art electrode supported cells with classical yttria-stabilised zirconia (YSZ) as a dense thin-film electrolyte provide maintained or even improved performance at reduced temperature in both modes (800 °C or lower).

The standard cell design, commonly included in the above-mentioned research projects, is a Ni/YSZ cermet as hydrogen/steam electrode on a Ni/YSZ supporting substrate and a composite of strontium-doped lanthanum manganite (LSM) or –ferrite (LCSF) as oxygen electrode. The performance of cells with LSM electrodes is described in the following.

Voltage–current density (U–j) data of a research cell in Figure 9.3 show similar performance in the SOFC and the SOEC mode, *i.e.* similar curve slopes or area specific resistances (ASRs). This demonstrates the reversibility of the operation. The voltage offset at zero current results from a change in the

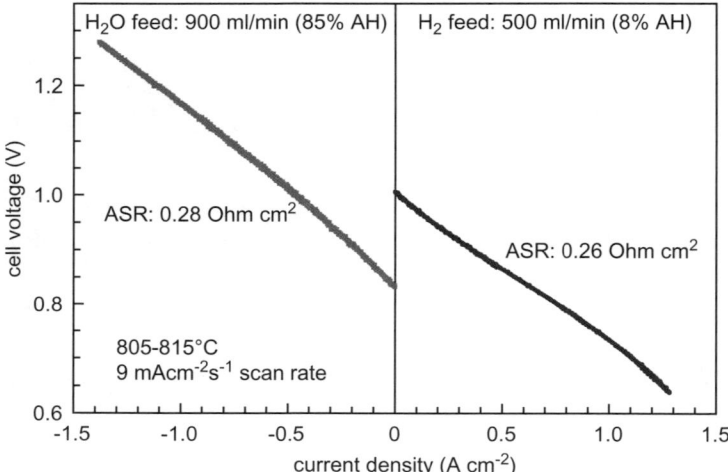

Figure 9.3   Initial voltage–current behaviour of an anode-supported SOFC (research cell) reversibly operated in the SOFC and SOEC modes (circular cell with 45 cm² area). Measurements are taken in an unsealed configuration. Forward and reverse scans are shown.

absolute humidity (AH) in the feed gas to the hydrogen/steam electrode. The cell voltage in the SOEC remains below 1.3 V over the covered current density range. Hence, endothermic operation is feasible even under large current densities (here $|j|$ up to 1.4 A cm^{-2}). The SOEC cell performance fulfils the requirements for practical operation, such as high current density, a relatively large cell area (45 cm²), and a moderate cell temperature ($\sim 810\,°C$).

The U–j behaviour depends on the humidity of the feed gas to the hydrogen/steam electrode, as shown for a commercial cell in Figure 9.4. Higher humidities reduce the cell voltage, in agreement with the Nernst equation. Under practical operation the AH value of the (H$_2$) feed gas should be as high as possible; it is usually limited to about 90% in order to keep a reducing environment to protect the cells in the absence of a current flow. As in the previous figure, endothermic operation is feasible over a wide current density range with cell voltages of around 1.0 V. Fast electrode kinetics is assured by the higher operation temperature of 890 °C.

A levelling-off of the cell voltage occurs when the steam conversion rate becomes too high (Figure 9.4). The maximum possible steam conversion depends somewhat on the cell provider and is linked with the porosity of the H$_2$ electrode.[34,37] Steam conversion rates under constant current operation are usually set to 40–70%. Higher steam conversion rates mean a reduced amount of water to be recirculated in an electrolyser system, but they also mean increased cell voltages.

Though precautions can be taken to prevent operation under steam starvation, one may ask about the cell behaviour beyond the range of voltage rise as a

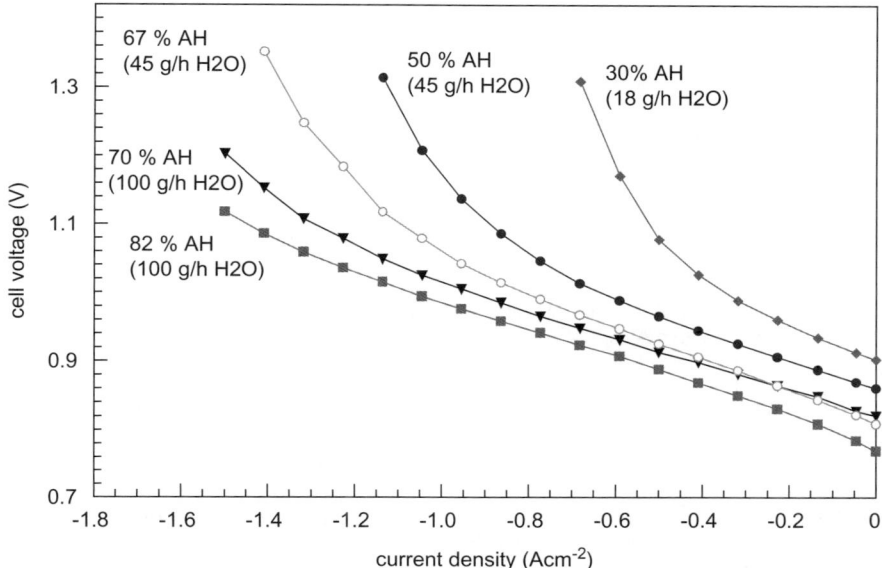

Figure 9.4 Voltage–current behaviour of a commercial anode-supported SOFC from HTceramix in the electrolysis mode as a function of the absolute humidity AH at the hydrogen/steam inlet ($T = 890\,°C$). Other measurement conditions are as in Figure 9.3.[43]

consequence of steam starvation as shown in Figure 9.4. Higher cell voltages might cause cell destruction *via* electrolyte reduction and they might lead to increased electronic conduction in the electrolyte.[40,41] To discuss this point for cells with a YSZ electrolyte, operation was extended to current densities, j, corresponding to steam conversion rates above 100% and, alternatively, the steam supply was interrupted under constant current conditions.[41] In both cases, cell voltages saturated at $\sim 1.9\,V$ at $810\,°C$, without indications for electrolyte decomposition during experimental runs lasting several days. The mechanism limiting the cell voltage was therefore attributed to electronic conduction in the YSZ electrolyte. Sufficiently ionic electrolyte conduction was extrapolated for normal SOEC operation. Nonetheless, the results indicate that the ionic transfer number in the SOEC mode is lower than in the SOFC mode. Owing to the high activation energy of electronic conduction in YSZ $(3.9\,eV)$,[41] electronic conduction may no longer be negligible for temperatures approaching $1000\,°C$.

An example for operation beyond the range of voltage rise due to steam starvation is shown in Figure 9.5. The decrease of the slope of the s-shaped $U–j$ curve at highest current density magnitudes is a consequence of increasing electronic conduction. Such intrinsic limitation of the cell voltage provides a cell protection against insufficient steam supply up to the limit given by the drastic increase of joule heating in the electrolyte.

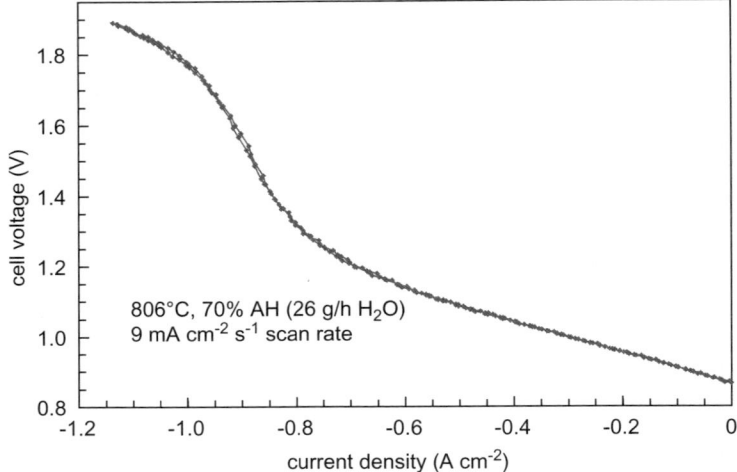

Figure 9.5 Voltage–current behaviour of commercial anode-supported SOFC from HT ceramix in the electrolysis mode (circular cell with $45 \, cm^2$ area). The forward and reverse scans shown virtually coincide. Other measurement conditions are as in Figure 9.3 (compare Mogensen and Jacobsen[40]).

Because the voltage drop across the electrolyte is larger in the electrolyser mode than in the fuel cell mode, electrolytes with a smaller voltage operation window than YSZ may not be usable in the electrolysis application.

So far, a critical issue with respect to the use of SOFC in the SOEC mode is long-term stability. While (voltage) degradation rates in the constant current operation mode below 1% per 1000 h are demonstrated for SOFCs (and stacks) running tens of thousands of hours,[42] degradation in the SOEC mode is higher. Moreover, available results are limited to experiments rarely exceeding 1000 h. Degradation rates in the 1% range are occasionally reported under small current densities ($|j| < 0.3 \, A \, cm^{-2}$).[34] Higher $|j|$ values, which approach the design target for cells in commercial systems of around $-1 \, A \, cm^{-2}$, cause faster degradation. Sudden failure of the SOECs, on the other hand, is rarely observed.

In the Figure 9.6 operation data over about 400 h at $-0.5 \, A \, cm^{-2}$ are shown for an electrode supported cell from HTceramix. The degradation rate is about 16% per 1000 h, a high value not sufficient for practical operation.

A number of degradation features are under discussion, most of them known from the SOFC operation. The problem to tackle consists in the identification of the specific degradation processes responsible for the faster degradation in the SOEC mode. Discussed major degradation features are (1) delamination of the oxygen electrode, (2) oxygen evolution in closed pores in that electrode leading to a loss of electrode material, (3) deactivation of the electrodes by changes in the microstructure (*e.g.* Ni coarsening or oxidation) or by poisoning, and (4) electrolyte ageing. Poisoning may occur by impurities in the steam

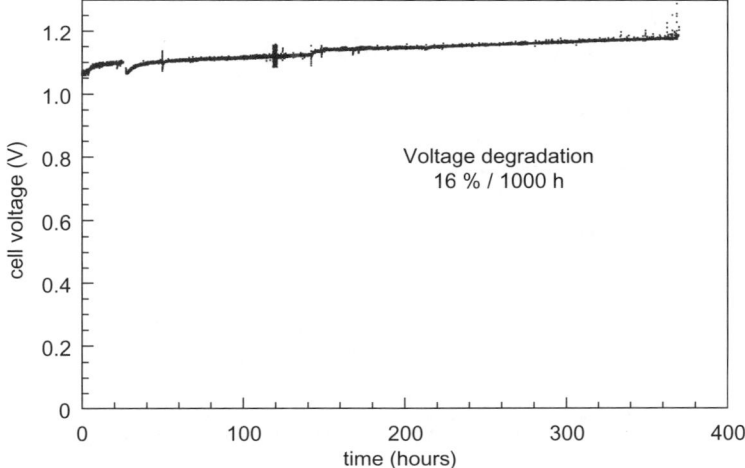

Figure 9.6 Long-term SOEC operation at $j = -0.5\,\mathrm{A\,cm^{-2}}$ of a commercial anode-supported SOFC from HT ceramix (810 °C, 70% AH, circular cell with $45\,\mathrm{cm^2}$ area).[43] Measurement conditions as in Figure 9.3.

supply or by material evaporation, *e.g.* from the seals, the cell housing, or the tubing. In view of the widely varying degradation rates in reports, one can expect that the influence of a specific degradation feature largely depends on the cell type and operation conditions.

Delamination of the oxygen electrode was frequently reported, but it seems to be less critical in cells using composite electrodes of LSM/YSZ type or the mixed ionic/electronic conductor LSCF. Degradation of the hydrogen electrode composed of a nickel/YSZ cermet is treated in several works, *e.g.* using impedance spectroscopy.[43] That electrode seems to be particularly sensitive to silica poisoning coming from the glass sealing.[44] The higher operating voltage in the SOEC mode means a more oxidising environment for the oxygen electrode which might affect the electrode stability. It also means a larger gradient in the oxygen partial pressure across the cell which could affect the known ageing of the YSZ electrolyte. Poisoning of the oxygen electrode with Cr vapour, well known from SOFC operation, might be less critical owing to the inverted flow direction of the oxygen.

Impedance spectroscopy (IS) is essentially the only *in situ* tool so far used for the monitoring of cell degradation during long-term experiments. IS data are later complemented by the results from post-mortem analysis. *In situ* tool here means that a small AC current signal of varying frequency is superimposed to the steady state DC current. Operation therefore remains in the steady state, unlike, for example, *U–j* scanning. As an example, the time evolution of the impedance of a HTceramix cell is shown in Figure 9.7. Voltage degradation amounts to about 10% during the time window shown. Impedance at the high-frequency limit remains almost independent of time, *i.e.* the ohmic

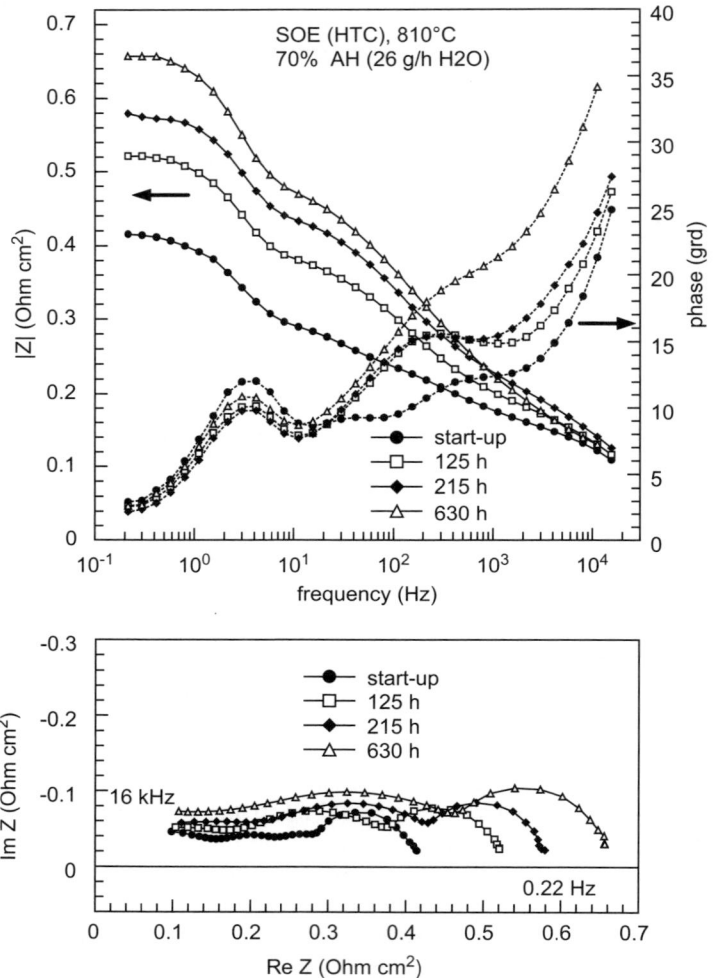

Figure 9.7   Impedance of a HTceramix cell as function of the operation time under a
DC current density of $-0.5\,A\,cm^{-2}$ in the magnitude/phase (left) and the
complex plane (right) presentations.[43] Other experimental conditions as in
Figure 9.3.

contributions such as electrolyte and contact resistances hardly vary with time.
Also, the impedance term at around 3 Hz, which comes from steam conversion
within the hydrogen/steam electrode, only shows minor time dependence. Cell
degradation is mainly reflected by an increase in the impedance contribution at
intermediate frequencies, where interfacial charge transfer processes commonly
appear. Impedance at low frequency is ohmic, it represents the ASR value at
$j=-0.5\,A\,cm^{-2}$ which rises from $0.42\,\Omega\,cm^2$ after start-up to $0.66\,\Omega\,cm^2$ after
630 h operation.

The large variety of existing SOFCs and also the ongoing SOFC development mean that much experimental work remains to be done to obtain a more complete understanding of the behaviour in the SOEC mode. Nonetheless, the use of SOFCs as electrolyser cells may only present an early development step, justified by cell availability (without the need for specific SOEC development) and notably by the promising results encountered with most cell types.

9.3.3 High-temperature Electrolyser Stacks

Based on the SOEC technology discussed previously, high-temperature electrolysis stacks have been developed. Only a few results and experiments are found and reported in the literature. Many of them have been carried out at Idaho National Laboratory (INL) with stacks developed by Ceramatec Inc.[45] In 2004, INL operated an instrumented 10-cell stack between 800 and 900 °C under different conditions, varying steam inlet fraction, gas flow rates and current density. Experimental results have been favourably compared with results from a three-dimensional computational fluid dynamics model. Detailed modelling results show that heat transfer and gas flow within the stack are first-order parameters that determine the optimal stack operating conditions. Then a durability test of 1000 h was performed on a 25-cell stack (cell active area of 64 cm^2).[46] An initial rapid degradation followed by a subsequent slowing of the degradation rate was observed. This two-step degradation led to a rate close to 20% per 1000 h. Internal resistances are moreover significantly higher within the stack compared to single cells since the per-cell ASR at 800–850 °C ranged between 1.3 Ω cm^{-2} at the beginning of the test to 2.5 Ω cm^{-2} at the end whereas ASR values below 0.5 Ω cm^{-2} are classically obtained on single cells at similar temperatures (*cf.* section 9.3.2). It was observed that this difference could be due to the joining of cells and interconnects with a degradation of the intermediate conductive layer made of lanthanum strontium chromite.[46]

The European Institute for Energy Research has tested a stack during 5000 h.[47] This stack was a R-design unit from the company Htceramix that integrated five rectangular second generation Indec (H.C. Starck)[48] anode-supported cells (active area per cell of 50 cm^2) and 1 mm thick coated F17TNb interconnect plates with a $MnCo_2O_4$ protective layer. The stack operated during 2000 h with 50% absolute humidity in the inlet gas after which the water content was increased to 67%. Cell degradation rates vary largely from cell to cell during the first 2000 h with an average stack-degradation rate of approximately 14% per 1000 h. That rate thereafter drops to 6% per 1000 h, a value closer to those typically found in the cell testing at the applied current density (*cf.* section 9.3.2). In principle, the differences in the cell voltages at 50% AH can arise from a non-homogeneous steam distribution within the stack, as well as from non-uniformity of the cell properties defining the diffusion over-potential and the maximum possible steam conversion. However, the fact that the cell behaviour is rather uniform at 70% absolute humidity indicates that the steam distribution is at the origin of the observed differences.

9.3.4 System Development

Most HTE testing to date has concentrated upon quantifying material and cell performance and has not attempted to address balance-of-plant (BoP) issues such as thermal management (feed-stock heating, heat recuperation, and high-temperature gas handling), hydrogen recycle, multiple-stack hot zone design, multiple-stack electrical configurations, materials of construction, and others.

INL has developed a 15 kW HTE test facility, termed the Integrated Laboratory Scale (ILS) HTE test facility, in an attempt to study some of these issues and assess the readiness of HTE technology for pilot-scale testing. Photographs of the skid with the components identified are presented in Figure 9.8. The ILS support system actually consists of three parallel systems that supply feedstock, sweep gas streams, and electrical power semi-independently of each other to each of the three modules. All three modules were located within a single hot zone enclosure. Heat recuperation and hydrogen product recycle are also incorporated into the facility. A single hydrogen recycle system was shared amongst the three modules.

Initially, a single 5 kW solid oxide module developed by Ceramatec[45] was used for the first operational test. No heat recuperation or hydrogen recycle was initially incorporated. First operation of this facility was in 2007 and resulted in over 400 h of operation with an average hydrogen production rate of approximately $0.9 \, N \, m^3 \, h^{-1}$.[49] Full testing of an expanded (incorporating heat recuperation and hydrogen recycle) INL ILS experimental facility with three modules (720 cells total, each of $64 \, cm^2$ active area) commenced in September 2008, and continued for 1080 h. The test average H_2 production rate was approximately $1.2 \, N \, m^3 \, h^{-1}$ based upon current, with a peak measured value of over $5.7 \, N \, m^3 \, h^{-1}$. This corresponded to a peak electrolyser power of over 18 kW.[50] Details of the design and initial operation of this facility are documented in papers by Housley *et al.* and Stoots *et al.*[51–54]

Testing the ILS facility provided useful BoP data and revealed some interesting problems. Firstly, the importance of avoiding unwanted condensation cannot be understated. Such condensate accumulating at some point and suddenly flowing can result in system transients, or even accelerated degradation in cell performance. In the case of INL ILS testing, although the hydrogen recycle system was found to be effective and eliminated the use of any external hydrogen source for maintaining reducing conditions in the cells during steady-state operation, testing revealed problems with condensation forming in the hydrogen recycle system. The effect of this was similar to subjecting the cells to multiple reduction/oxidation cycles and resulted in accelerated degradation of cell performance. Mid-test correction of this condensation problem allowed partial recovery cell performance.

Secondly, internal temperature measurements in the whole stacks must be carefully taken. The cell-to-cell potential is only 1–2 V, the cell-to-ground potential (and hence the bias voltage for grounded thermocouples) for a 10-cells stack could be 13 V or more. For relatively short stacks, this is usually not a problem for a data acquisition system. However, ILS testing involved

ID	Component
1	Hot zone enclosure lid
2	Power supply and instrument racks
3	Electrical distribution cabinet
4	Data acquisition and control monitors
5	Deionized water system
6	Steam generator
7	Steam and H_2 superheaters
8	Air compressor
9	Patch panel
10	Product finned cooler
11	Steam condenser
12	Mass flow controllers
13	H_2 vent
14	Air and O_2 vent

Figure 9.8 Right side view of the INL ILS facility, with major components labelled.

multiple 60-cell stacks, resulting in possible bias voltages of 80 V or more. In early testing, intra-stack thermocouples lost insulation from ground and passed this high bias voltage, causing problems with measurement and control signals. This problem was accounted for in the final three-modules test data acquisition and control system.

Each module employed two heat recuperators, one for the hydrogen/ steam inlet and outlet streams and a second for the air sweep inlet and outlet streams. Heat transfer analyses indicated that combining all four streams into a single heat exchanger would have resulted in more efficient heat recuperation. However, to prevent leakage of hydrogen into the sweep gas, separate heat exchangers were employed. Even this simple, non-optimised attempt at heat recuperation resulted in a reduction in total electric heater power by 50%.

Parametric studies of various HTE plant layouts and operating conditions have been conducted using system-level engineering process modelling. All of the BoP components that would be present in an actual HTE plant (pumps, compressors, heat exchangers, turbines), and the electrolyser, are modelled within the process analysis software UniSim.

System modelling has provided interesting insight into HTE efficiency trends. One such result has to do with the use of a sweep gas. Use of a sweep gas reduces the cell Nernst potential, and improves the efficiency of the electrolyser itself. However, contrary to solid oxide fuel cell operation where a sweep gas is required for cooling the cells, HTE can be operated without a sweep gas and produce pure oxygen. This pure oxygen could be saved as a valuable commodity or be supplied to some other process such as an oxygen-blown gasifier. Furthermore, systems modelling has shown that not using a sweep gas avoids the additional mechanical work required to compress the sweep gas, thereby improving the overall process efficiency.[55] There are also, however, significant disadvantages to not using a sweep gas to dilute the evolved oxygen. Pure oxygen at high temperatures must be handled with care and it may be simpler and more cost effective to avoid the issues related to high temperature pure oxygen by incorporating dilution with a sweep gas.

9.3.5 Summary

Hydrogen can be produced by the electrolysis of water with reversibly operated fuel cells, namely PEMFCs at low temperature (below around 120 °C) and SOFCs at high temperature (800 °C range). The known main characteristics of each of these technologies also hold under the reverse electrolysis operation. For instance, the PEM electrolysers require expensive precious metal catalysts and polymer membranes, but long-term stability is demonstrated at these commercially available electrolysers. Electrolysis at high temperature with solid oxide electrolysis cells, on the other hand, allows the use of less expensive ceramic cell materials while maintaining favourable reaction kinetics to the extent that reversible SOFC/SOEC operation becomes feasible at the same cell, *i.e.* without modifications in the electrodes.

A major advantage of electrolysis at high temperature is the lower free energy required for the reaction, which should translate to a largely reduced electricity consumption. This is confirmed by the use of recently developed SOFCs with improved performance, notably from the planar cell technology, as electrolysis cells. Such SOECs can be operated with and without external heat supply; already without heat supply the energy conversion efficiency is above the one of low temperature cells. With external heat supply, the electrical-to-chemical energy conversion efficiency of the cell can exceed 100%. Operation voltages of around 1.0 to 1.3 V at 800 °C are achieved with state-of-the-art cells, as shown in this contribution. This has to be compared to voltage values of at least 1.6 V at PEM cells.

Examples for a successful stack operation in the SOEC mode during a few thousands of hours are given as well as first demonstrations of integrated systems. Many SOFCs seem to operate well in the SOEC mode. Therefore, the development of high temperature electrolysis may rely for some time on SOFC technology, before a specific SOEC technology evolves as a new branch of solid oxide cell technology.

At present the main bottleneck for industrial application of SOEC technology is cell degradation which is generally faster than in fuel cell mode. Therefore, research efforts are required on this issue dealing with the identification of the degradation features and their suppression.

The high energy-conversion efficiency also renders SOEC technology attractive in the context of the storage of renewable energy (*e.g.* wind or solar). The SOEC compared to SOFC operation commonly lower heat flow (to or from the cell depending on the operation mode) may facilitate a cell operation with cyclic electricity supply, which is often generated by renewable energy sources.

References

1. International Partnership for Hydrogen Economy (IPHE), Implementation Liaison Committee, Scoping papers, 2005, www.iphe.net/docs/Scoping_Papers/Combined_Scoping_Papers.pdf [Accessed 5 March 2010].
2. P. A. Lessing, in *Materials for the Hydrogen Economy*, ed. H. Jones Russel and J. Thomas, CRC Press Taylor & Francis Group, Boca Raton, 2007, p. 37.
3. V. N. Fateev, S. A. Grigoriev, R. Blach and S. V. Ostrovsky, in *Proceedings of the 1st European Hydrogen Energy Conference,* , September 2–5 2003, Grenoble, France, Association Française de l'Hydrogene, available on CD.
4. R. Oberlin and M. Fischer, in *Proceedings of the 6th World Hydrogen Energy Conference, Miami Beach, Florida*, ed. T. Veziroglu, N. Getoff and P. Weinzierl, Pergamon Press, Oxford, 1986, p. 333.
5. A. F. G. Smith and M. Newborough, Report to the carbon trust and ITM-Power PLC, 2004, http://www.h2fc.com/ [Accessed 5 March 2010].
6. H. Beer, UK patent 1.147.442, 1969.
7. S. Trasatti, *Electrochim. Acta*, 1984, **29**, 1503.
8. Y. Roginskaya, O. Morozova, E. Loubnin, A. Popov, Y. Ulitina, V. Zhurov, S. Ivanon and S. Trasatti, *J. Chem. Soc. Faraday Trans.*, 1993, **89**, 1707.
9. Giner Electrochemical Systems, http://www.ginerinc.com/ [Accessed 5 March 2010].

10. Teledyne Energy Systems, http://www.teledyneenergysystems.com/ [Accessed 5 March 2010].
11. Hydrogen Technologies, http://www.electrolysers.com/ [Accessed 5 March 2010].
12. Hydrogenics Corp., http://www.hydrogenics.com/ [Accessed 5 March 2010].
13. Hamiltonsundstrand, http://www.hamiltonsundstrand.com/ [Accessed 5 March 2010].
14. Proton Energy Systems, http://www.protonenergy.com/ [Accessed 5 March 2010].
15. Shinko Pantec, http://www.pantec.co.jp/ [Accessed 5 March 2010].
16. Wellman Defence, http://www.wellman-defence.com/ [Accessed 5 March 2010].
17. E. Rasten, PhD thesis, NTNU Trondheim, Norway, 2001.
18. J. Sedlak, R. Lawrence and J. Enos, *Int. J. Hydrogen Energy*, 1981, **6**, 159.
19. H. Takenak, E. Torikai, Y. Kawami and N. Wakabayashi, *Int. J. Hydrogen Energy*, 1982, **7**, 397.
20. A. T. Marshall, S. Sunde, M. Tsypkin and R. Tunold, *Int. J. Hydrogen Energy*, 2007, **32**, 2320.
21. WE-NET, http://www.enaa.or.jp/WE-NET/index.html [Accessed 5 March 2010].
22. S. A. Grigoriev, V. I. Porembsky and V. N. Fateev, *Int. J. Hydrogen Energy*, 2006, **31**, 171.
23. M. Yamaguchi, M. Horiguchi and T. Nakanori, in *Proceedings of the 13th World Hydrogen Energy Conference*, June 12–15 2000, Beijing, China, ed. Z. Q. Mao, IAHE, Miami, FL, p. 274.
24. A. Hashimoto, K. Hashizaki and K. Shimizu, in *Proceedings of the 14th World Hydrogen Energy Conference*, June 9–13 2002, Montreal, Canada, available on CD.
25. J. Zhang, in *PEM Fuel Cell Electrocatalysts and Catalysts Layers Fundamentals and Applications*, ed. J. Zhang, Springer, London, 2008.
26. WELTEMP project (Water ELectrolysis at Elevated Temperatures), European Framework Program 7, http://www.weltemp.eu/ [Accessed 5 March 2010].
27. W. Dönitz and E. Erdle, *Int. J. Hydrogen Energy*, 1985, **10**, 291.
28. E. Erdle, W. Dönitz, R. Schamm and A. Koch, *Int. J. Hydrogen Energy*, 1992, **17**, 817.
29. G. Dietrich and W. Scafer, *Int. J. Hydrogen Energy*, 1984, **9**, 747.
30. K. H. Quandt and R. Streicher, *Int. J. Hydrogen Energy*, 1986, **11**, 309.
31. N. J. Maskalick, *Int. J. Hydrogen Energy*, 1986, **11**, 563.
32. S. Kubo, S. Kasahara, H. Okuda, A. Terada, N. Tanaka, Y. Inaba, H. Ohashi, Y. Inagaki, K. Onuki and R. Hino, *Nucl. Eng. Des.*, 2004, **233**, 355.
33. J. Larminie and A. Dicks, *Fuel Cell Systems Explained*, John Wiley, Chichester, UK, 2nd edn, 2003, p. 428.
34. European Institute for Energy Research (EIFER) Hi2H2 project (Highly Efficient, High Temperature, Hydrogen Production by Water Electrolysis),

2004–2007, within the European Framework Program 6, http://www.hi2h2.com/ [Accessed 5 March 2010].

35. M. Mogensen, S. H. Jensen, A. Hauch, I. Chorkendoff and T. Jacobsen, in *Proceedings of the 7th European Solid Oxide Fuel Cell Forum*, July 3–6 2006, Lucerne Switzerland, ed. U. Bossel, P0301, available on CD.
36. Y. Shin, W. Park, J. Chang and J. Park, *Int. J. Hydrogen Energy*, 2007, **32**, 1486.
37. A. Brisse, J. Schefold and M. Zahid, *Int. J. Hydrogen Energy*, 2008, **33**, 5375.
38. A. Hauch, S. H. Jensen, S. Ramousse and M. Mogensen, *J. Electrochem. Soc.*, 2006, **153**, A1741.
39. J. S. Herring, J. E. O'Brien, C. M. Stoots, L. Hawkes, J. J. Hartvigsen and M. Shahnam, *Int. J. Hydrogen Energy*, 2007, **32**, 440.
40. M. Mogensen and T. Jacobsen, *ECS Trans.*, 1315, **25**, 2009.
41. J. Schefold, A. Brisse and M. Zahid, *J. Electrochem. Soc.*, 2009, **156**, 897.
42. European Institute for Energy Research (EIFER), Real SOFC project (Realising Reliable, Durable Energy Efficient and Cost Effective SOFC Systems), 2004–2008, within the European Framework Program 6, http://www.real-sofc.org/news_extended/news_04-2008 [Accessed 5 March 2010].
43. J. Schefold, M. J. Garcia, A. Brisse, D. Perednis, M. Zahid, in *Proceedings of the 8th European Fuel Cell Forum*, June 30– July 5 2008, Lucerne, Switzerland, ed. U. Bossel, PA1101, available on CD.
44. A. Hauch, J. S. Højgaard, J. B. Bilde-Sørensen and M. Mogesen, *J. Electrochem. Soc.*, 2007, **154**, A619.
45. Ceramatec Inc., http://www.ceramatec.com/ [Accessed 5 March 2010].
46. J. O'Brien, C. Stoots, J. Herring, G. Hawkes and J. Hartvigsen, INL Technical Report INL/ICON-06-11716, November 2006, http://www.inl.gov/technicalpublications/Documents/3562840.pdf [Accessed 10 June 2010].
47. A. Brisse, A. Hauch, M. Mogensen, G. Schiller, U. Vogt and M. Zahid, in *Proceedings of the World Hydrogen Energy Conference*, June 15–19 2008, Brisbane, Australia, ed. Cuppar Associates Inc., http://www.proceedings.com [Accessed 10 June 2010].
48. INDEC, http://www.hcstarck.com/ [Accessed 5 March 2010].
49. K. G. Condie, C. M. Stoots, J. E. O'Brien and J. S. Herring, INL Technical Report INL/EXT-07-13626, December, 2007, http://www.inl.gov/technicalpublications/Documents/3874581.pdf [Accessed 5 March 2010].
50. C. M. Stoots, J. E. O'Brien, K. G. Condie and J. J. Hartvigsen *Int. J. Hydrogen Energy*, 2010, **35**, 4861.
51. G. Housley, K. Condie, J. E. O'Brien and C. M. Stoots in *Proceedings of the ANS Embedded Topical: International Topical Meeting on the Safety and Technology of Nuclear Hydrogen Production, Control, and Management, Boston, Massachussetts*, June 24–28 2007, Boston Massachusetts, ed. Curran Associates, Inc., http://www.proceedings.com/ [Accessed 10 June 2010].

52. C. M. Stoots and J. E. O'Brien, in *Proceedings of the 2008 International Congress on Advances in Nuclear Power Plants*, June 8–12 2008, Anaheim, California, http://www.icapp.ans.org/ [Accessed 10 June 2010].
53. C. M. Stoots, J. E. O'Brien, K. Condie, L. Moore-McAteer, G. K. Housley, J. J. Hartvigsen and J. S. Herring, *Nucl. Technol.*, 2009.
54. G. K. Housley, J. E. O'Brien and G. L. Hawkes, in *Proceedings of the 2008 ASME International Congress and Exposition*, October 31 – November 6 2008, Boston, Massachusetts, IMECE 2008–68817, available on CD.
55. J. E. O'Brien, M. G. McKellar, E. A. Harvego and C. M. Stoots, in *Proceedings of the International Conference on Hydrogen Production*, May 3–6 2009, Oshawa, Canada, ICH2P-09, http://www.inl.gov/ [Accessed 10 June 2009].

CHAPTER 10
Hydrogen Production by Internal Reforming Fuel Cells

KAS HEMMES

TU Delft fac. TPM section T&DO (Technology Dynamics & Sustainable Development), Jaffalaan 5, NL-2628 BX Delft, Netherlands

10.1 Introduction

Internal reforming fuel cell concepts have been developed based on the molten carbonate fuel cell (MCFC) as well as on the solid oxide fuel cell (SOFC). In these concepts effective use is made of the high temperature of the fuel cell and the surplus heat produced by it. The high temperature allows the reforming of hydrocarbons to hydrogen by a reaction with steam. This is an endothermic reaction consuming part of the (waste) heat produced in the fuel cell, thereby increasing overall efficiency. For the proper operation of a fuel cell stack it is necessary to ensure a minimum partial pressure of hydrogen in each of the cells inside the stack and at the outlet side of each of these cells. According to the Nernst equation the equilibrium potential decreases if the partial pressure of the reactants decreases (pH_2 and pO_2) and the partial pressure of the products increase (pH_2O). For a hydrogen/oxygen fuel cell the Nernst equation yields the following equilibrium potential difference between anode and cathode, also called the open cell voltage (OCV):

$$V_{eq} = E_0(T) + \frac{RT}{2F} \ln \frac{(pH_2)_a (pO_2)_c^{1/2}}{(pH_2O)} \tag{10.1}$$

Because of the logarithmic nature of the equation the OCV theoretically would go to infinite values because the argument of the logarithm goes to the infinite

RSC Energy and Environment Series No. 2
Innovations in Fuel Cell Technologies
Edited by Robert Steinberger-Wilckens and Werner Lehnert
© Royal Society of Chemistry 2010
Published by the Royal Society of Chemistry, www.rsc.org

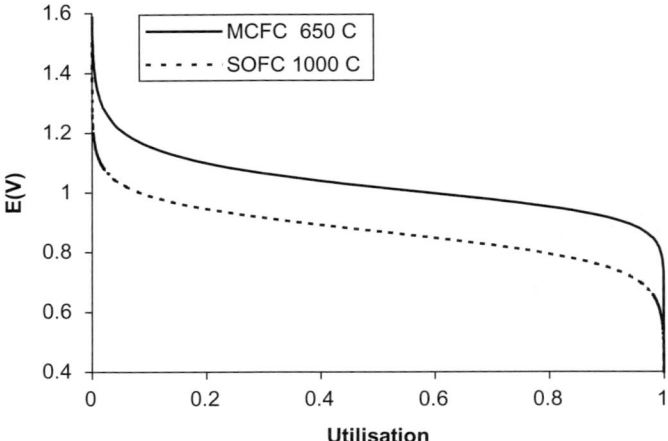

Figure 10.1   Calculated reversible cell voltage between anode and cathode of an
MCFC and SOFC as a function of hydrogen utilisation in the hypothetical case of using pure hydrogen as input fuel gas.[1]

values, when no water (reaction product) is present. On the other hand when no
hydrogen (or oxygen) would be available the argument of the logarithm would
go to zero and the logarithm itself to minus infinite and so would the OCV.
These theoretical values for the OCV as a function of utilisation (percentage of
the inlet hydrogen that is electrochemically converted) are depicted in Figure
10.1. based on pure hydrogen as the fuel.[1] Normally the fuel cell would not be
operated with 100% hydrogen but instead some water will always be contained
in the gas. In the figure this would be equivalent to a non-zero utilisation and
the steep decrease in the potential would be avoided. The further decrease as a
function of hydrogen utilisation is relatively modest, and depends on the
temperature. It is only at very high utilisation of the hydrogen that a rapid
decrease in potential occurs again. Therefore in a fuel cell the fuel can be utilised up to about 98%. However, to safeguard that some hydrogen will always
be available at the end of each cell in the stack, often total fuel utilisation is kept
below 95%.

   Therefore in normal operation some hydrogen (a few per cent) is still present
in the off-gas of a fuel cell. This will of course decrease the overall efficiency
because some hydrogen is not converted into a power. In order to increase
efficiency of the fuel cell, this amount is kept as low as possible. The hydrogen
still present in the off-gas is often converted into process heat by catalytic
combustion; for example to pre-heat the input gas streams to the fuel cell.

   However, when we treat hydrogen as a useful product, we can optimise the
operation of the internal reforming fuel cell system to produce more hydrogen.
This can be done either by decreasing the electric power output and/or by
increasing fuel input. These are the two essential 'control knobs' of fuel cell
operation.[1] It leads to many operation modes for the production of combined

heat and power *plus* hydrogen. The internal reforming fuel cell system thus becomes a tri-generation energy system. The fact that the internal steam reforming reaction takes away heat produced by the fuel cell and converts it into a useful product, namely hydrogen, allows for operation of the fuel cell at much higher current and power densities than can be normally achieved. At higher current and power density, heat losses are increased and the fuel cell efficiency would drop below acceptable levels in conventional operation with high fuel utilisation. However, in the hydrogen production concept discussed here, part of the increased losses in the form of heat is converted into a useful output by the use of this waste heat to produce more hydrogen. So the total system efficiency in terms of electric power plus hydrogen remains high even at higher power density as we explain in more detail in the sections below, but first an overview of the international developments on this topic will be presented.

10.2 International Developments Reported in Literature

One of the earlier papers on hydrogen and power co-production with internal reforming fuel cells is by Vollmar *et al.* of the Siemens Corporation.[2] They proposed converting the hydrogen produced by an internal reforming SOFC, in the manner described above, in a polymer electrolyte fuel cell. Obviously this can only be done after cooling the off-gas stream and separating the hydrogen from the off-gas. Or, alternatively, CO may be completely removed from the off-gas to avoid poisoning the Pt catalysts in the polymer fuel cells. A highly efficient system is obtained in this way for three reasons. Firstly, the Nernst loss in the SOFC is minimised as explained in the section above (see Equation (10.1)). Secondly, almost full utilisation of the hydrogen can be achieved in the polymer fuel cell better than in the SOFC. Thirdly, although Nernst loss due to a high utilisation occurs in a polymer fuel cell as well, it is much lower due to its lower temperature and the linear dependence on temperature of the Nernst loss.[1,3]

Leal and Brouwer also investigated the co-production of hydrogen in a series of papers.[4–6] In their most recent paper they address the important issue of heat distribution within the SOFC at different fuel utilisations. They find that maximum internal SOFC temperature differences range from 150 K for the $u_f = 60\%$ case to 100 K for the $u_f = 85\%$ case. They also confirmed the synergistic relation between high hydrogen production and high electrochemical efficiency. They also convincingly report that by the internal use of heat the parasitic losses in the air flow (through the cathode) for cooling are reduced. A different route in hydrogen and power co-production using a SOFC is followed by Granovskii *et al.*[7] They include membranes to externally reform the natural gas directly yielding hydrogen as a product which can partly be used to fuel the SOFC. Secondly, the remaining hydrogen in the anode off-gas is combusted in a membrane reactor using an oxygen-conducting membrane. By proper heat integration using heat exchangers and by integration with compressors and expanders as in gas turbines the mass and energy balances can be made fit to the basic laws of physics. The whole system was analysed for its energy efficiency and compared to two conventional separate methods for the production of hydrogen and electricity.

Also fuel cell developers like Fuel Cell Energy have recognised the possibility and the advantages of the co-production of hydrogen and power by internal reforming fuel cells.[8] Although the system is not included in their standard product portfolio Fuel Cell Energy is describing a Hydrogen Energy Station Vision in one of their brochures.[9] They conclude:

Co-production of electricity and hydrogen offers a potentially lower cost option for the hydrogen infrastructure. High temperature fuel cell power plants can produce electricity, hydrogen and heat at an overall efficiency of 80–85% and offer ultra-low emissions. The electricity produced on-site can power the local grid, while hydrogen can be used for fuel cell vehicles or industrial customers. The fuel cell power plant cost is paid for by the on-site power sold to the grid, and the by-product hydrogen is available practically for "free" except for the cost of its separation. The ability to generate multiple products and the flexibility to vary the hydrogen and electricity production to match with the on-site demand helps to improve the overall economics of on-site hydrogen generation.They have also developed an interesting way to separate the hydrogen from the reformed fuel in an electrochemical hydrogen separator (EHS). Using the reversibility of the electrochemical reactions hydrogen can be transformed into protons that are selectively transported through the membrane of, for example, a polymer electrolyte fuel cell. Next, they are converted back in to hydrogen molecules again at the opposite electrode. This process requires some DC electric power which of course is readily available from the internal reforming fuel cell to which the separation unit is coupled. If the EHS is mechanically well-constructed and the membranes strong enough the electric driving force can be used to pump the hydrogen from a relatively low partial pressure in the anode off-gas mixture to high pressures at the other side of the membrane.

10.3   Co-production of Hydrogen and Power

In the following the co-production of hydrogen and power with an internal reforming fuel cell system is further explained using an IR-SOFC system design as shown in Figure 10.2.

System calculations by Hemmes *et al.* on a tri-generation system show that the electric power output can be doubled compared to standard operation under the co-production of about an equivalent amount of hydrogen in terms of (chemical) energy per second.[10] Therefore total output in terms of hydrogen and electric power can, in principle, be quadrupled, albeit in that case the fuel

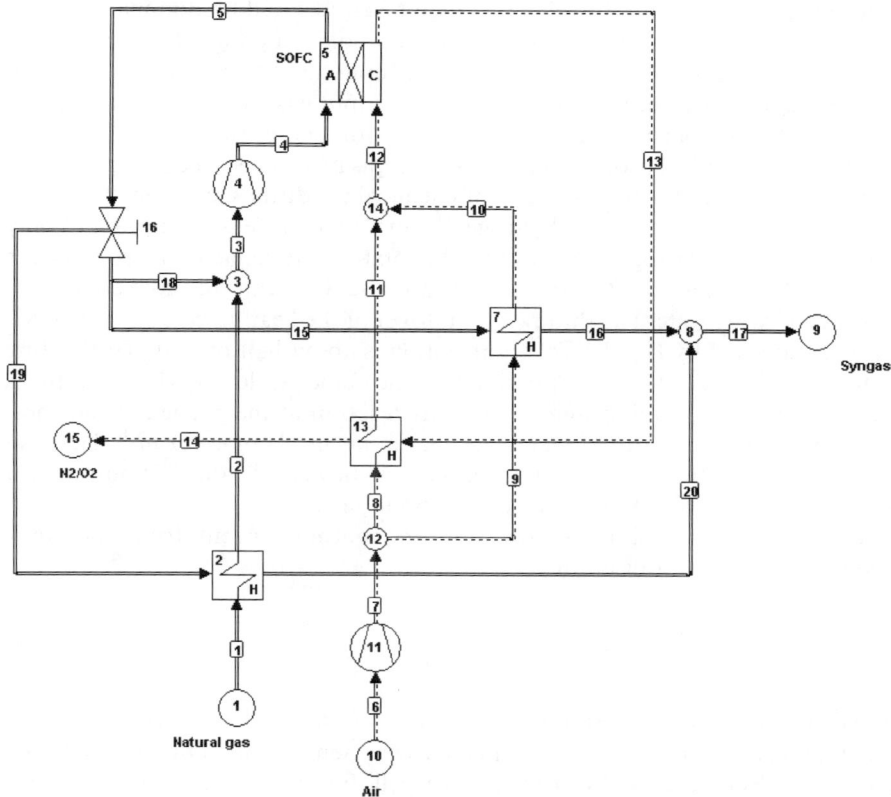

Figure 10.2 Cycle tempo flow sheet diagram of an internal reforming SOFC system for the co-production of hydrogen and power.

input also has to be increased of course! In addition, high temperature heat is still produced for combined heat and power applications. The system was modelled using the flow-sheet program 'Cycle Tempo' developed at TU Delft.

The flow sheet system layout is shown in Figure 10.2. It is designed to be as simple as possible and the design is not optimised in any way (neither for efficiency nor economically). The composition of the natural gas is chosen such as to represent the gas composition from the largest Dutch source 'Slochteren', which was chosen as the standard gas composition in the Netherlands. It is a low calorific gas containing a relatively high concentration of nitrogen (14%). Recycle loops are applied to the anode as well as to the cathode gas streams. The recycle ratio of the anode outlet gases is regulated by valve #16. It is the only optimisation performed within this system design. We found that at a ratio of 0.4 (*i.e.* 40% of the anode off-gas is recycled) the highest total efficiency is obtained over a wide range of fuel utilisation from 60% to 95%. In practice a better temperature distribution inside the fuel cell stack is obtained by recycling, because recycling provides the necessary heat for the endothermic reforming reaction that often due to a high reaction speed predominantly takes place at the inlet side of the fuel cell stack. Secondly, recycling also provides the necessary steam for the reforming reaction. Input gas streams are preheated in heat exchangers by output gas streams. The output gas stream through pipe 17 contains the hydrogen that is still mixed with other gases such as CO, CO_2 and H_2O. In the chemical process industry such a gas mixture is called syngas. It is a useful gas mixture for the synthesis of chemical products such as methanol. In the analysis the separation of hydrogen from this mixture is excluded and only the amounts of hydrogen and CO in the mixture are depicted as such in the results. We included CO production because CO can be converted into hydrogen by the water gas shift reaction. Most of the heating value in the anode off-gas is carried by H_2, the CO contribution is about half of that. The heating values of H_2 and CO are approximately the same while the H_2/CO ratio is determined by the equilibrium of the shift reaction at the particular temperature and pressure. Further details can be found in Hemmes *et al.*[10] Here we present some of the results depicted as electric power, H_2 and CO output as a function of fuel utilisation for a certain input gas flow.

The combination of input gas flow and output current from the stack determines the fuel utilisation:[1]

$$u_f \equiv \frac{i}{i_{in}} \qquad (10.2)$$

in which u_f is the fuel utilisation, i the current density and i_{in} the (hypothetical) maximum current density that would have been achieved if all the fuel had been converted electrochemically inside the fuel cell. It is called the 'equivalent input current density', and can simply be calculated from the number of moles of fuel per second entering the fuel cell (m_{in}) multiplied by the number of electrons (n) per molecule that is produced electrochemically at the anode and Faraday's constant (F), divided by the active area of the fuel cell:

$$i_{in} = \frac{nF.m_{in}}{A} \qquad (10.3)$$

where F is Faraday's number. The fuel cell model in the Cycle Tempo flow sheet program correctly takes into account the so-called Nernst loss. This loss accounts for the lower hydrogen partial pressure at the end of the fuel cell where it results in a lower Nernst potential and consequently lowers the local driving force for the reactions.[1] According to Standaert *et al.* the overall Nernst loss can simply be approximated by a factor proportional to the fuel utilisation.[3] Therefore the cell voltage can be approximated by:

$$V_{cell} \approx OCV - \tfrac{1}{2}\alpha u_f - i.r \qquad (10.4)$$

In this equation α is the slope of the Nernst potential as a function of utilisation (in volts).[1] All ohmic, kinetic and diffusion losses are simply modelled as an ohmic resistance, r, for convenience. In spite of its simplicity, Equation (10.4) is shown to be quite accurate.[11] It reflects two sources of losses in (high temperature) fuel cells: the reversible Nernst loss (the second term in Equation (10.4)) and the irreversible quasi-ohmic losses (third term). Although utilisation and thus Nernst loss are also proportional to current density (see Equation (10.2)) the utilisation can be varied independently by adjusting the input gas flow expressed here as the equivalent input current density. The electric power output, P, can be simply calculated as:

$$P = iV_{cell}.A \tag{10.5}$$

in which A is the active area of the fuel cell. (Note that here we refer to the active area of the fuel cell as the geometric macroscopic two-dimensional size of the prorous electrodes and not to the internal microscopic reaction surface of the porous electrodes, which is much larger.)

Because the fuel cell essentially has two independent 'control knobs', many operating conditions can be achieved with different hydrogen production, electric power production and heat production at different efficiency rates. To structure the results we impose three main modes of operation:

- High-efficiency mode
- Constant-current mode
- High-power mode.

However, it should be emphasised that more operating conditions and modes are possible.

10.3.1 Mode 1: High-efficiency Mode

In the high-efficiency mode, the input flow rate of the fuel gas is kept constant at the level of conventional operation. The fuel utilisation, u_f, is then decreased by reducing the current density. In this case the input flow is kept constant at a value of about '2 MW equivalent' (in the example shown here) in order to match the arbitrarily chosen size of the fuel cell in conventional operation. u_f is set at values between 60% and 95%. In practice this can be done by adjusting the electric load to obtain a current density as determined by Equation (10.2). Although sometimes lower utilisation can be obtained in the Cycle Tempo simulations, only results between 60% and 95% are presented here, because for all modes feasible operating conditions could be obtained within these limits. If the utilisation becomes too low, the fuel cell can not provide sufficient waste heat for the endothermic reforming reaction. So in practice such a low utilisation (high hydrogen production rate) cannot be reached.

Figure 10.3 depicts a graph of efficiency versus fuel utilisation for the high-efficiency mode. Since the gas efficiency (defined as the sum of the power output of H_2 and CO relative to the power input flow) is increasing more than the electric power efficiency is decreased, total efficiency increases at lower

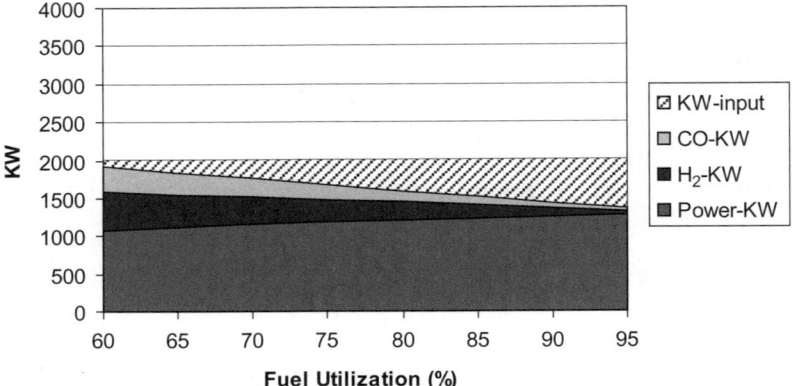

Figure 10.3   Electric power output and H_2/CO fuel output versus fuel utilisation for
the high-efficiency mode.

utilisation and low power output. Yet the total power output – defined as the
sum of hydrogen and electric power – is larger than in conventional operation.
With respect to electric power output only, it is somewhat lower but still
comparable with conventional operation. Although this mode of operation is
the most efficient mode, it is not the most economic as will be explained later.
The calculations nicely show that it is possible to trade power for hydrogen, but
it is not a one-to-one trade-off. In other words, the sum of electricity and
hydrogen power is not a constant. Heat loss across the system boundaries is
greatly reduced, leading to a maximum efficiency of 94% at an utilisation of
60%. Note that this is the total efficiency for the production of hydrogen and
power *not including heat* as is usually done in the definition of total efficiency of
combined heat and power (CHP) systems! On the other hand, in these simulations we applied the first law of thermodynamics (energy conservation) so the
efficiency of the tri-generation of hydrogen, power and heat is 100% by definition. No convective and radiation heat losses to the environment are taken
into account since the exact engineering circumstances of a 'real' system remain
unknown at this stage of consideration. Also, the heat is leaving the system
through both the anode off-gas as well as through the cathode off-gas at different temperatures. Whether or not this heat can be usefully applied and to
what extent is also not included in this analysis. Therefore we focus on
hydrogen (and CO) and electric power output only. (see Figure 10.4)

10.3.2   Mode 2: Constant-current Mode

In the constant-current mode the current density is kept constant. Fuel utilisation is now decreased by increasing the input flow (see Equation (10.2)). The
current density is fixed at $1500\,A\,m^{-2}$ representing conventional operation at

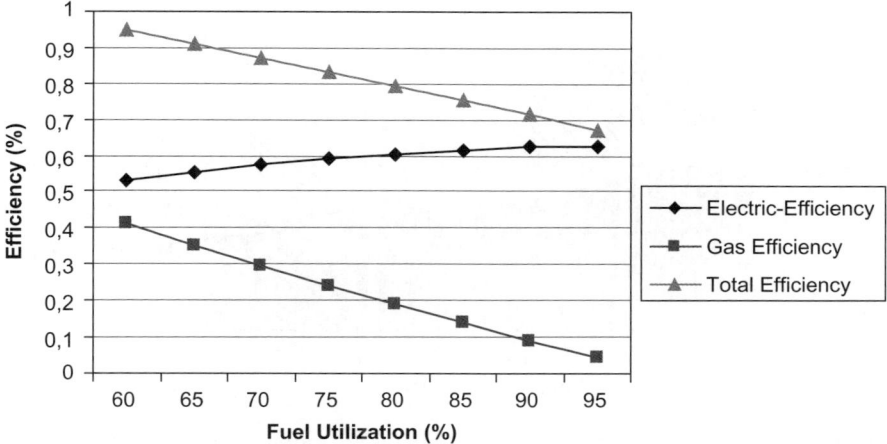

Figure 10.4 Efficiency versus fuel utilisation for high efficiency mode.

reasonable power density. When setting the fuel utilisation (u_f) Cycle Tempo is allowed to change the input flow to meet both the fixed numbers of i and u_f. Because the results of this operating mode are between the high-efficiency and high-power modes, only the results of the latter two are presented here.

10.3.3 Mode 3: High-power Mode or Constant (Low) Voltage Mode

A particularly interesting operational mode from an economic point of view is the high-power mode. Maintaining a low cell voltage (by applying a large electric load) results in high current densities. In high-power mode the cell voltage is fixed at the low value of 0.5 V. The power density is around its maximum value since the cell voltage is around half the OCV (see Figure 10.8). As in the constant current mode, the fuel utilisation, u_f, is decreased by increasing the input gas flow. The results for the high-power mode are depicted in Figures 10.5 and 10.6. As can be seen from the results in these figures, very high power output values are obtained and the high power mode must be seen as an extreme operation mode.

At $u_f = 60\%$, the current density, $i = 4400 \, \text{A m}^{-2}$, *i.e.* almost three times the standard operating current density. Due to the low cell voltage, the fuel cell stack might suffer from large heat dissipation. Yet this heat is advantageously used for the reforming reaction and a much larger amount of natural gas can be reformed than in the previous modes. It may be surprising, but a large electric output is still achieved. This is because at the low cell voltage of 0.5 V the system operates near the maximum in the power versus current density curve (*cf.* Figure 10.8), yet at the expense of a lower electric efficiency. At low u_f this is partly compensated by the higher H_2/CO production, therefore still a total

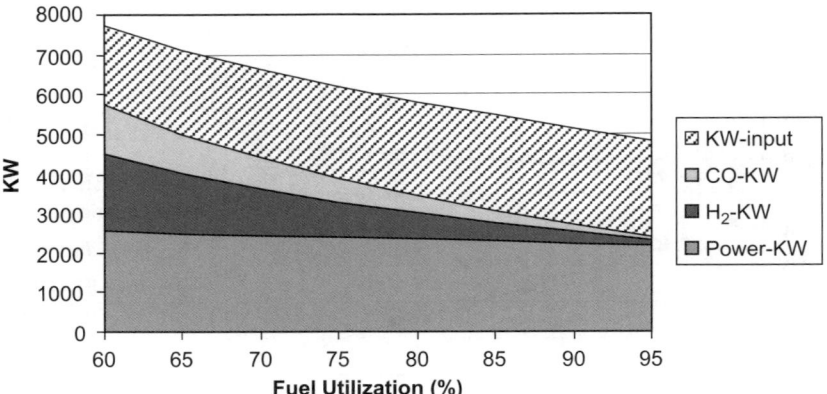

Figure 10.5   Electric power output and H_2/CO fuel output versus fuel utilisation in high-power mode.

efficiency (in terms of $H_2/CO + $ power) of almost 75% can be achieved. It can be concluded that with the same fuel cell around twice as much electric power output can be obtained: *i.e.* 1260 kW_e in high efficiency mode versus up to 2500 kW_e in high power mode (see Figure 10.5). While simultaneously and in addition to this high electric power output another 3100 kW in the form of H_2/CO is produced, yielding a total useful output of 5600 kW, *i.e.* 4.5 times the power output in conventional operation with the same fuel cell (*i.e.* 1260 kW_e). This can be explained through Figure 10.8. As shown in the figure when the I–V curve is linearly decreasing, the power output curve is a parabola with its maximum at a cell voltage of half the OCV. In relation to this, the following two points should be noted:

- A linear I–V curve is usually a good approximation for MCFCs up to maximum power although sometimes activation losses can be noticeable at low current densities. To a lesser extent this also holds for SOFCs.
- In the flow sheet simulations the fuel gas composition that enters the fuel cell is not constant but depends on the parameters of the recycle loop that change in different operation modes and u_f settings and therefore the OCV also slightly changes.

This power maximum is normally not achieved because it would yield too high losses and too low conversion efficiency. However, by effectively using the waste heat for hydrogen production we can choose to operate at maximum power density without sacrificing too much in overall efficiency. This clearly shows the advantages of the co-production mode of operation for internal reforming fuel cells. Another way of looking at it is by noting that although at lower utilisation there is a shift towards more H_2 production the electric power output still increases as well! This may seem contradictory, but can be explained

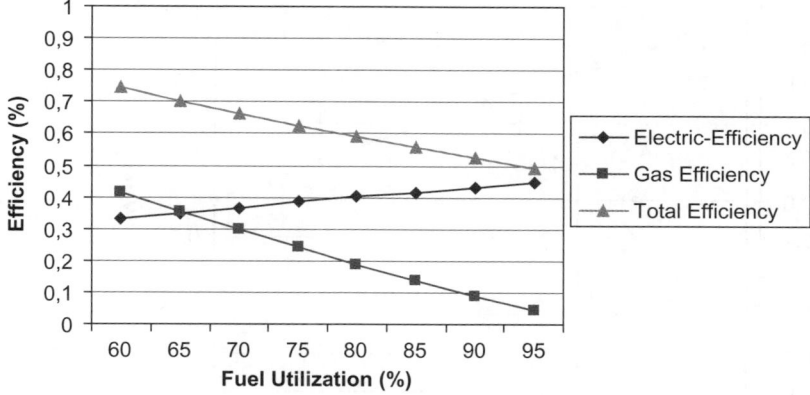

Figure 10.6 Efficiency versus fuel utilisation for high-power mode.

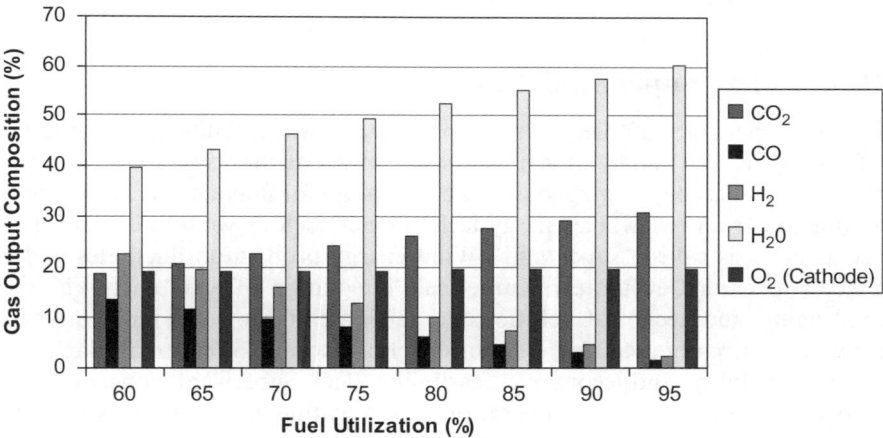

Figure 10.7 Gas output composition versus fuel utilisation for high-power mode.

since in this high power mode utilisation is decreased by increasing the fuel (natural gas) input flow, hence more joules per second flow into the system allowing an increase in both H_2 and power output. Moreover, due to the higher partial pressure of hydrogen at the outlet at lower utilisation, Nernst loss is significantly reduced and the cell voltage and thus fuel cell efficiency improved (see Equation (10.4)). Figure 10.7 shows the gas output at the anode and cathode respectively as a function of fuel utilisation. At the anode outlet the gas mixture contains H_2, H_2O, CO and CO_2 as shown Figure 10.7, and at the cathode outlet we find O_2 and N_2 (not shown in Figure 10.7). Obviously as fuel utilisation is reduced from 95% to 60% both H_2 and CO output increase.

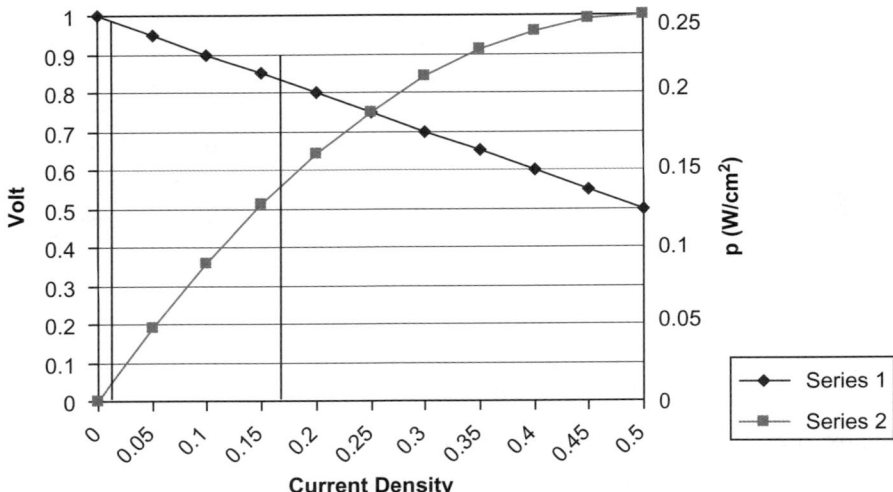

Figure 10.8   Simplified cell voltage and power-density curves for a high-temperature fuel cell.

10.4   The 'Superwind Concept'

The characteristics of fuel cells, in particular the flexibility of an internal reforming fuel cell in a hydrogen and power co-production system as described above, can be used advantageously to compensate for fluctuations in electricity production from renewable energy technologies such as wind and solar. This new concept is called 'Superwind'. Wind energy production fluctuates and is difficult to predict accurately. Large-scale integration of wind energy is thus challenging and technical solutions to facilitate it are very much needed. However, currently available solutions such as storage of electricity lack efficiency, flexibility and economic feasibility. The Superwind concept is an innovative solution for the integration of fluctuating energy sources into the grid. It consists in the integration of a wind turbine with an internal-reforming fuel cell (MCFC or SOFC) as shown in Figure 10.9. When compensating a peak in wind energy by decreasing the power output from the fuel cell, while keeping the input fuel stream (natural gas, biogas or a mixture thereof) constant the flow sheet calculations as presented above show an increase in hydrogen production that is three to four times than the wind peak that it is compensating in terms of energy per unit time. It seems as if the peak in wind energy has been converted with 300% to 400% efficiency into hydrogen (and CO). Of course, the first law of thermodynamics has not been violated. Instead, what happened is that heat produced inside the fuel cell is effectively converted into chemical energy by the internal reforming reaction, while also less waste heat is produced due to lower polarisation losses and lower Nernst loss. So the system produces less heat and of course the sum of all energy inputs remains equal to the sum of all energy outputs. We can see this peculiar phenomenon

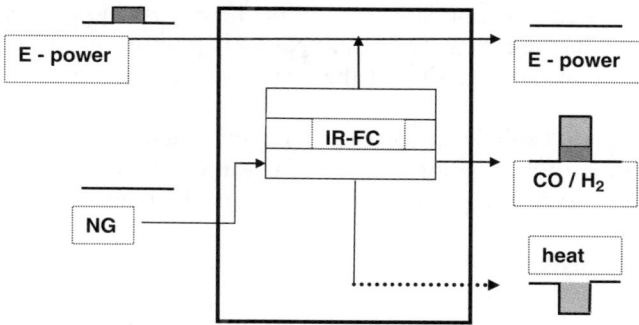

Figure 10.9   Schematic representation of the Superwind concept.

already from Figure 10.3. When decreasing utilisation (going from right to left in Figure 10.3) the hydrogen (and CO) output increase more than the power output decreases. This is also reflected in the steeper increase in gas efficiency relative to the decrease in electric power efficiency in Figure 10.3 (again going from right to left in the figure, reflecting decreasing utilisation).

The Superwind concept is an example of a more general class of energy systems called multi-source multi-product systems.[12] Its main advantage lies in the high degree of flexibility and its high-efficiency. Peaks in wind energy production can be compensated by a decrease in electricity production from the fuel cell. Contrary to other electricity production units the fuel cell does not stand idle when compensating for peaks in wind energy, but instead it can switch to hydrogen production. This results in an even higher instead of a lower total system efficiency. Up to more than 90% efficiency for the production of electricity and hydrogen can be obtained as explained above and shown in Figure 10.4. It is important to emphasise that in the Superwind concept there is <u>no</u> storage of electricity. The fluctuations in wind energy production are compensated for by using the flexibility in power production of the fuel cell.

Moreover, the production of hydrogen allows the fuel cell to be operated at almost double power density compared to standard operation. So, in principle, with the same fuel cell a double electric power output can be obtained in co-production with an equal amount of hydrogen in energy terms per unit time. Therefore the useful power output of the same fuel cell (not counting heat) can almost be quadrupled. Consequently the capital expenses per kW_e of the fuel cell are about halved. Capital expenses per kW useful products (power + hydrogen) can thus be decreased by about a factor of 4. Nevertheless it must be emphasised here that these are theoretical possibilities. The fuel cell stack must be re-designed to ensure proper internal heat transfer and heat management to prevent too large thermal loads, temperature fluctuations and material degradation when operating the fuel cell in dynamic mode.

The technical and economic feasibility of the Superwind concept has been investigated.[13] Summarising the advantages and applications of the Superwind system we see that the flexibility of the system can be used to:

- Minimise the imbalance between the predicted and actual electricity production, thus avoiding penalties for wind turbine owners.
- Produce a large amount of electricity when the demand – and therefore price – is high.
- Adjust the hydrogen production to an ever-growing hydrogen market (for transport).
- Double the electrical output of the fuel cell compared to standard operation thus reducing the fuel cell cost per kW by 50%. The resulting larger waste heat production is effectively used for the production of hydrogen from biogas or natural gas.
- Supply any remaining waste heat to the bio-gas production process to speed up that process.
- Deliver heat on demand to a nearby residential area or industry.

Superwind ensures that:

- Wind energy will become a reliable source of renewable energy.
- There is no conversion and storage of wind energy so there are no associate losses. Similarly, solar energy can be introduced without the drawbacks of this fluctuating power source.
- Energy is saved due to the high system efficiency and control of power output without loss of efficiency.
- Biomass can be integrated much more efficiently into the energy and transport system by the simultaneous production of hydrogen and electric power.
- A large market is created for high-temperature fuel cells which will lead to price reductions for fuel cells and subsequent larger markets leading to a higher overall conversion efficiency for the production of electricity (70–80%) and lower transport costs because of the distributed generation in medium-sized units.
- A clean transport sector can develop based on hydrogen as an energy carrier.

10.5 Hydrogen Production from Carbon Using a Direct Carbon Fuel Cell

A very interesting but more exotic form of hydrogen production using fuel cells is a concept in which carbon is electrochemically converted into carbon monoxide and subsequently reacted with steam to form carbon dioxide and hydrogen in the well-known shift reaction. The electrochemical reaction takes place in a direct carbon fuel cell (DCFC). This type of fuel cell has attracted the attention of a growing number of researchers in recent years (see Chapter 6).

However, contrary to the focus on carbon dioxide production, chosen by most researchers, here we do not see carbon monoxide as an unwanted by-product but have tried to optimise carbon monoxide formation for its favourable thermodynamics (even more favourable than for carbon dioxide production as will be explained next); and secondly, for the possibility of converting it into hydrogen in the well-known shift reaction with steam. Assuming yttrium-stabilised zirconium oxide is the electrolyte of a direct carbon fuel cell we have the following (electrochemical) reactions:

$$\begin{array}{ll} \frac{1}{2}O_2 + 2e^- \rightarrow O^{2-} & \text{(cathode reaction)} \\ C + O^{2-} \rightarrow CO + 2e^- & \text{(anode reaction)} \\ H_2O + CO \rightarrow CO_2 + H_2 & \text{(shift reaction)} \\ C + H_2O + \frac{1}{2}O_2 \rightarrow CO_2 + H_2 & \text{(overall reaction)} \end{array} \tag{10.6}$$

because of the overall positive entropy production in the electrochemical conversion from carbon to carbon monoxide we find the fuel cell efficiency defined as $\Delta G/\Delta H$ to be larger than 1. This does not mean that the first law of thermodynamics is violated, it merely means that in the definition of fuel cell efficiency we do not take into account the option that heat has to be supplied to the reaction and that it is an energy input of the process next to the chemical energy input (enthalpy). This possibility was proposed by us in 1998 and is further explained below.[14]

For reactions in which more gaseous species are formed than consumed like:

$$C_{(s)} + \tfrac{1}{2}O_{2(g)} \rightarrow CO_{(g)} \tag{10.7}$$

the entropy increases and $\Delta S > 0$. This has large consequences for the efficiencies of fuel cells in which Equation (10.7) is the overall cell reaction as will be shown below. The intrinsic maximum energy conversion efficiency (η) of an electrochemical cell is defined as the ratio between the free energy (ΔG) and enthalpy (ΔH) of the cell reaction:

$$\eta = \frac{\Delta G}{\Delta H} \tag{10.8}$$

ΔG is the maximum energy, which can be converted to electrical work, while ΔH is the total chemical energy stored in the fuel. Since $\Delta G = \Delta H - T\Delta S$, η can alternatively be written as:

$$\eta = \frac{\Delta H - T\Delta S}{\Delta H} = 1 - \frac{T\Delta S}{\Delta H} \tag{10.9}$$

Then for $\Delta S > 0$ and $\Delta H < 0$, as is the case for Equation (10.7) efficiencies higher than 100% are obtained with Equation (10.9). Although this is well known, the fact that in processes with $\Delta S > 0$, heat can be converted into power with an efficiency higher than the Carnot efficiency seems to be overlooked. So fuel cells do not only circumvent Carnot processes in the conversion

Table 10.1 ΔH and $T\Delta S$ values for the formation of CO gas taken from literature,[15] and the resulting efficiencies calculated with Equation (10.9)

T (K)	ΔH (kJ mol^{-1})	$T\Delta S$ (kJ mol^{-1})	η
600	–110.185	54.309	1.49
800	–110.949	71.450	1.64
1000	–112.021	88.240	1.79
1200	–113.245	104.551	1.92

of chemical energy into power. Also the conversion of heat into power is not limited by the Carnot efficiency. This is not a violation of the second law of thermodynamics because a process with negative entropy change is coupled to a process with positive entropy change. In the limiting case of a reversible process the overall entropy change is zero. The heat converted into power equals $T\Delta S$ ($\Delta S > 0$), just as in an ordinary fuel cell the reversible heat production equals $-T\Delta S$ ($\Delta S < 0$). The ΔH and $T\Delta S$ values for the formation of CO gas as well as the efficiencies obtained from these values with Equation (10.9) are listed in Table 10.1.

Clearly, a considerable fraction of the power, which increases with temperature, produced with a DCFC stems from the $T\Delta S$ term. This is the great advantage of direct carbon fuel cells compared to conventional fuel cells and heat engines. In conventional fuel cells gaseous reactants are used and $\Delta S < 0$ and consequently efficiencies lower than 100% are obtained.

So in a multi-source multi-product process chemical energy as well as high temperature heat are coverted into hydrogen, power and (waste) heat.[16]

Whereas for the direct carbon fuel cell aimed at predominantly carbon dioxide production the temperature has to be kept as low as possible, here in the case of electrochemical gasification higher temperatures are preferred. Firstly because carbon monoxide formation is promoted above carbon dioxide formation at higher temperatures in accordance with the Boudouard equilibrium. Secondly, the amount of heat that can be converted into power increases proportionately because of the term TdS in the fuel cell efficiency equation (also see Table 10.1). Thirdly, polarisation losses in general decrease with increasing temperatures in particular reaction kinetics are improved at higher temperature, which in general is the weak point of a DCFC.

In a paper presented at the ASME Fuel Cell Conference in 2004 we calculated the open cell voltage of a direct carbon fuel cell as a function of temperature, taking into account the ratio between the equilibrium partial pressures of carbon monoxide and carbon dioxide, according to the Boudouard equilibrium.[17,18] The OCV was found to increase more rapidly than just linear as expected in first instance (similar to the linear decrease of Gibbs free energy and OCV for a hydrogen oxygen fuel cell as a function of temperature). So at higher temperatures the fuel cell efficiency increases significantly and the amount of high temperature heat that has to be supplied to the direct carbon fuel cell is increased accordingly (Figure 10.10).

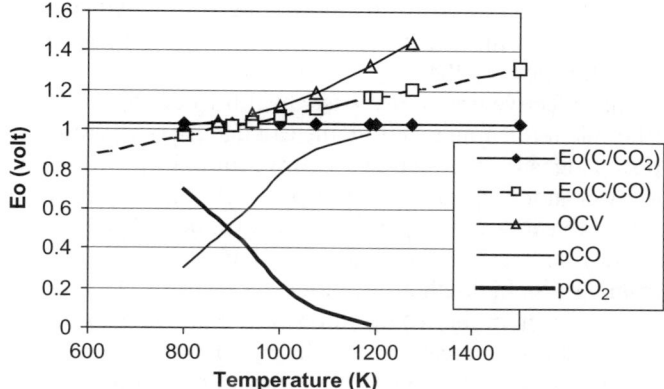

Figure 10.10 Standard potentials for the oxidation of carbon to CO and CO_2 and the open cell voltage of a carbonate DCFC assuming $p_{CO} + p_{CO_2} = 1$ atm. at the anode and $p_{O_2} = p_{CO_2} = 1$ atm. at the cathode.

This concept of electrochemical gasification has been calculated in an early paper by us in which we proposed concentrated solar power to be the source of high temperature heat needed in this process.[14] However, the concept is still in the conceptual design phase. Nevertheless, the development of direct carbon fuel cells for carbon dioxide production also makes the realisation of high temperature, monoxide producing direct carbon fuel cells more feasible.

The use of concentrated solar power can also provide the heat for the production of clean and reactive carbon for the direct carbon fuel cell from methane (natural gas) by thermal decomposition. Flow sheet calculations on the integration of carbon concentrated solar power with a Direct carbon fuel cell have been performed.[19]

System studies still need to be performed to determine overall performance of these systems. In the design a proper heat management must be applied in which the production of steam for the shift reaction can benefit from waste heat from the total system. Although the concept is quite exotic and still far from being achieved, thermodynamics shows that it is possible to conceptualise a fuel cell that produces hydrogen instead of consuming it and convert heat into power instead of producing waste heat!

10.6 Conclusions

By flow sheet calculations on an internal reforming SOFC system it has been shown that a very flexible co-production system can be designed that can operate in conventional mode producing mainly electric power and heat, as well as in co-production mode in which it can produce hydrogen, next to electric power and relatively little heat. The heat losses can be used directly and very effectively inside the fuel cell for the endothermic reforming reaction to achieve very high overall efficiency of up to 95% in terms of hydrogen and electric power produced (not counting waste heat as is usually done in conventional combined heat and power efficiency definitions).

Because of the effective use of any waste heat for the production of hydrogen in the endothermic reforming reaction one can choose to operate the fuel cell at very high power density. Compared to conventional operation the same fuel cell in high-power mode can deliver twice as much electric power, while on top of that a comparable amount of power in the form of hydrogen is co-produced. System efficiency in terms of hydrogen and electric power output is then decreased from 95% in high-efficiency mode, to a nevertheless respectable value of 75% in high-power mode. (Note, however, that in high-power mode the input of fuel gas is also increased accordingly.) In the latter case 25% of the input is dissipated as heat that can still be used in conventional combined heat and power applications, however.

Although flow sheet calculations on a SOFC system showed that the concept is feasible in terms of energy and mass balance of the total system, the dynamic operation and high power operation of the fuel cell possibly require adjustments in the design of the fuel cell stack to ensure the integrity and endurance of the stack. A demonstration project is thus necessary to test the technical performances of state-of-the-art fuel cell technology in practice. The flexibility of the fuel cell allows for different operation strategies to optimise profits under often strongly fluctuating market conditions for electricity and hydrogen. Based on data of past electricity prices on the Amsterdam power exchange market, APX, different operation strategies were simulated and analysed on economic performance.[13]

The flexibility of this tri-generation system allows for the operation of the systems to meet varying demands for hydrogen and/or to optimise operation for high economic efficiency taking into account fluctuating market prices for hydrogen and electric power. This tri-generation system promises to be a viable stepping-stone in the transition towards a more sustainable energy system. It can accommodate fluctuating renewable energy sources like wind energy and solar by complementary production thus minimising grid congestion problems. Also the transport sector can profit from this system since it can provide flexible amounts of hydrogen for fuel cell vehicles as well as power for electric battery vehicles adopted to fluctuations in demand in the short term and an expected increase in average demand over time, without the need for building new installations when demand increases or suffering low economic profits in the early phases when demand is low.

References

1. K. Hemmes, *Fuel cells*, in *Modern Aspects of Electrochemistry*, ed. R. E. White, B. E. Conway and C. G. Vayenas, Kluwer Academic–Plenum Publishers, New York, 2004, vol. 37.

2. H. E. Vollmar, C. U. Maier, C. Nolscher, T. Merklein and M. Poppinger, *J. Power Sources*, 2000, **86**(1–2), 90–97.

3. F. Standaert, K. Hemmes and N. Woudstra, *J. Power Sources*, 1996, **63**(2), 221–234.

4. E. M. Leal and J. Brouwer, in *Proceedings of the 3rd International Conference on Fuel Cell Science, Engineering, and Technology, Am. Soc. of Mech. Eng.*, New York, 2005, pp. 449–458, ISBN 0-7918-3764-5.

5. E. M. Leal and J. Brouwer, in *Proceedings of the ASME Advanced Energy Systems Division, American Soc. of Mechanical Engineers*, New York, 2005, vol. 45, 481–490 ISBN 0-7918-4211-8.

6. E. M. Leal and J. Brouwer, *J. Fuel Cell Sci. Technol.*, 2006, **3**(2), 137–143.

7. M. Granovskii, I. Dincer and M. A. Rosen, *Chem. Eng. J.*, 2006, **120**(3), 193–202.

8. www.fuelcellenergy.com

9. www.fuelcellenergy.com/knowledge-library.php

10. K. Hemmes, A. Patil and N. Woudstra, *J. Fuel Cell Sci. Technol.*, 2008, **5**(4), 041010–1-041010-6.

11. S. F. Au, N. Woudstra, K. Hemmes and I. Uchida, *Energy Conversion Manage.*, 2003, **44**(14), 2297–2307.

12. K. Hemmes, J. L. Zachariah-Wolff, M. Geidl and G. Andersson, *Int. J. Hydrogen Energy*, 2007, **32**(10–11), 1332–1338.

13. A. L. Vernay, G. Steenvoorden and K. Hemmes. Superwind: a Feasibility Study. Integrating Wind Energy with Internal Reforming Fuel Cells for Flexible Coproduction of Electricity and Hydrogen. Final report for Senternovem, November 2008. Project number: NEOH 02010 Senternovem, 2008, TU Delft, Delft.

14. W. H. A. Peelen, K. Hemmes and J. H. W. De Wit, Carbon – *High. Temp. Mater.* 1998, P-US2:(4), 471–482.

15. I. Bahrin, *Thermochemical Data of Pure Substrates*, Wiley-VCH, Weinheim, 1995.

16. K. Hemmes, J. L. Zachariah, M. Geidl and G. Andersson, *Int. J. of Hydrogen Energy*, 2007, **32**(10–11), 1332–1338.

17. K. Hemmes and M. Cassir, in *Proceedings of the Second International Conference on Fuel Cell Science, Engineering and Technology, 1 June 2004, Rochester, NY*, Am. Soc. of Mechanical Engineers (ASME), New York, 2004, p. 395–400, ISBN 0-7918-4165-0.

18. K. Hemmes and M. Cassir, *J. Fuel Cell Sci Technol.*, (accepted for publication 2010).

19. G. Cinti and K. Hemmes, in *Third ASME European Fuel Cell Technology & Applications, The 'Piero Lunghi' Conference (EFC09)*, Dec 2009, Rome, Italy, Am. Soc. of Mechanical Engineers, New York, 2009.

Part 6
Outlook

Introduction

As with all technical progress and developments, the introduction of fuel cells to the commercial market follows completely different rules to engineering practice. 'Real' consumers expect 'real' products which offer added value, are easy to handle, last a long time (enough time) and offer value for money. This is often not understood by technology developers who prefer to firmly believe in their technology and its benefits, thereby forgetting that market customers need to be persuaded and convinced to spend their money. Often design over-rides technology and the example of hand-held phones gives a good impression of how mediocre and over-expensive technology can overcome any cost considerations, if it 'hits the right keys' with the consumers. Technology marketers have to start thinking from a consumer's perspective if fuel cells are to be successfully turned into products that sell by themselves. This also includes a re-assessment of the issues of what targets are most important for achieving first market success. In the context of politics this, for instance, affects the cost targets set for market entry and the conditions arranged for market incentives and subsidies programmes. It has to be acknowledged that markets today are biased towards the incumbent technologies and that a variety of hidden subsidies exist that are not recognised; for instance, the external costs of energy services that are carried by society, not by technology users. Fuel cell market access therefore also needs be seen in the context of energy market regulation as a whole.

CHAPTER 11

Products, Not Technology: Some Thoughts on Market Introduction Processes

ROBERT STEINBERGER-WILCKENS

Institute of Energy Research, Project Fuel Cells, Leo-Brandt-Str., 52425 Jülich, Germany

11.1 Introduction

Fuel cells today are on the verge of market entry. Tremendous technological progress has been made in the past few years and the general focus in fuel cell development has been on testing fuel cell components and systems, proving their technology readiness level and adjusting the performance to the requirements of first (niche) markets. Nevertheless, in a consumer market, technology is rarely bought as such, but in the form of consumer products or appliances. It is therefore worth looking into the ways new technologies enter the market and what factors influence success and failure of such market entry. It will be shown here that the fixation on technological progress is acceptable in the early phases of technology development, but that general market introduction needs more than this, especially when the main benefit of a new technology concerns environmental aspects.

RSC Energy and Environment Series No. 2
Innovations in Fuel Cell Technologies
Edited by Robert Steinberger-Wilckens and Werner Lehnert
© Royal Society of Chemistry 2010
Published by the Royal Society of Chemistry, www.rsc.org

11.2   Background

Fuel cells have the potential for achieving a considerable increase in energy efficiency in many areas of energy conversion. Electrochemical electric power production is not thermodynamically limited by the Carnot efficiency and could theoretically aim at 100% efficient devices.

On the other hand, steam and gas turbines, internal combustion engines and heating boilers have been providing power and heat to households, the industry and economy for the last 150 years more or less reliably. The pending market entry of fuel cells is thus a struggle of a new and somewhat unproven technology to replace a well established, cost effective adversary of long standing. It goes without saying, therefore, that fuel cells will experience a high pressure at market entry due to their competition with technology that has for decades been optimised towards cost goals whilst at the same time attempting to generate returns on the high development costs.

The market entry problem is not inherent to fuel cells but affects them more severely than other technologies because of their – at least partly – disruptive character (see below). Costs at market entry are high due to (initially) expensive materials and non-standard manufacturing processes. Also, fuel cells in many areas currently only offer marginal extra performance for the end user – apart from the obvious environmental benefits – since they merely 'replace' conventional equipment of the same function (producing heat and electricity) without offering immediate access to new fields of application or extra user performance (*i.e.* higher consumer value). High efficiency and environmental benefits alone are traditionally not regarded as consumer purchase incentives, although some awareness of environmental issues definitely exists and this trend is increasing.[1] Moreover, the environmental benefit is of an abstract nature, in many cases not accessible to direct observation by the customer, and therefore is a matter of abstraction and ethics.

The question is, whether fuel cell technology can rely on marketing its main advantages – efficiency and environmental benefits – or whether other ways of entering consumer markets are more promising. Therefore this chapter will be inspecting alternative ways of viewing fuel cell systems and seek for ways of promoting market entry using traditional marketing tools.

11.3   Technology Phasing-in versus Disruptive Development

One much-cited problem of fuel cells is the disruptive character inherent to their fuel supply. In particular, mobile (vehicle and portable) applications rely heavily on hydrogen or methanol supply. Whereas the latter can today be at least purchased at chemists, in harbours or at building supply stores, the former has no public infrastructure that were worth mentioning. This immediately implies the chicken-and-egg problem that has been faced by many new technologies, especially in the automotive markets, but also in telecommunications.

This is a problem less prominent with stationary applications, or vice versa: these have avoided the fuel question by concentrating on natural gas driven appliances.

It is very instructive to have a look at the past and at the development history of a number of well-known technologies.[2] From a vast choice of examples the following have been chosen for a brief qualitative inspection of the way a technology disruption was either avoided or overcome:

- The transition from coal, biomass, wind and solar energy to petrol(eum) from the middle of the nineteenth century
- Introduction of natural gas vehicles in the 1990s
- Introduction of unleaded petrol in the 1970s in Europe
- Introduction of mobile telephones in the 1990s.

These examples highlight some of the driving forces behind technological change – be it market-driven or regulation-driven – and give some indication of the possible market development options for fuel cells.

The late nineteenth century saw the large-scale introduction of mineral oil derivatives as an energy carrier for stationary and, later, mobile applications. This process spread over more than half a century until the necessary infrastructures had been established, oil wells developed and exploitation technology been advanced towards cost effectiveness. World oil production rose from 70 thousand tons in 1860 to 21 million tons in 1900, a factor of 300 within 40 years.[3] The 'oil revolution' succeeded due to the far superior energy storage density, the comparatively easy handling and the parallel development of automobiles. Speaking in strictly technical terms, oil products were predestined for mobile applications. In stationary applications they hardly display any technical advantages compared to solid or gaseous fuels and subsequently never played more than a marginal role in electricity generation, apart from the cases where there was no access to the natural gas grid. Although electric vehicles had a leading edge in technical development up to 1912, petrol engines could take the lead due to the longer range the liquid energy carrier could offer.[4] In the end it was the convenience and performance of oil products that beat the other energy competitors, not market price: the initially higher cost was offset by easy use, high controllability, higher efficiency (as compared to coal firing), and suitability for mobile and portable applications. The technical benefits of liquid fuels for vehicles bring oil products close to being in an 'enabling' position for many sectors of the mobility markets as we experience them today. Widespread individual mobility will probably not be possible once oil has left us. The difficulties in establishing alternative fuels – of which difficulties many are based on technical deficiencies under today's given transport regime – further prove the symbiotic relationship of today's road, air and marine transport systems and oil-based fuels. The 'added value' of higher convenience for the individual has only recently been offset by the implied social and political costs as these become manifest to the individual in the way of taxation, levies and high oil prices.

In contrast to the rapid development of petrol-driven (and diesel-driven) vehicles, the introduction of natural gas as a vehicle fuel has met a rather mixed response. Replacing higher (liquid) hydrocarbons by methane clearly has an environmentally benign effect due to the more favourable ratio of hydrogen to carbon atoms. Therefore a larger part of the heating value is attributed to burning the hydrogen (resulting in water as a reaction product) than the carbon. In addition, gas engines often emit less noise, fewer pollutants like SO_2, NO_x, or carbon monoxide, and no particles. This has led governments in many countries to offer subsidies in the way of direct grants for vehicle purchase or retro-fitting, or reduced energy taxes on gaseous fuels. In other countries, for instance Italy, the price difference (including all taxes) between natural gas and liquid fuels is such that gas vehicles have been economically attractive for several decades already. Nevertheless, even subsidies and attractive costs cannot distract from the necessity of establishing a completely new fuel infrastructure. Natural gas as a vehicle fuel can be seen as a 'moderately disruptive' technology since this fuel is generally not distributed and processed by the same companies as liquid fuels and the dispensing and storage technologies (at the filling station and on board) are very different. On the other hand, the vehicle, traffic flows and infrastructures remain intact which makes the introduction of natural gas vehicle fuels an interesting precedent for the introduction of hydrogen fuel.

In Germany, the process was seriously hampered by the refusal of oil and gas companies to build up a natural gas refuelling infrastructure. The main argument being that there were no customer vehicles, which was a self-fulfilling prophecy since there were no filling stations.[2] Only when the company ARAL announced in 2004 that it would equip 1000 out of the 15 000 German petrol stations with natural gas pumps, was a sufficient level of filling stations reached to lead natural gas vehicles to a moderate break-through.[5] In Canada, on the other hand, the introduction of gas vehicles was a failure.[6]

The comparison of the introduction of petrol and natural gas fuels shows that a disruptive technology may or may not succeed, depending on a number of factors:

- Added customer convenience and value
- Potential for recovering the cost of new infrastructure from market prices
- Recognition and acceptance of new market opportunities by market players.

In essence, if customers are prepared to pay a considerable surcharge for the new opportunities and higher performance gained, companies will invest in new infrastructures. Still, this also requires that market players actually perceive this development and are prepared to invest. With natural gas pumps, for instance, it was not the traditional oil companies that installed the first filling posts but natural gas distributors, thus taking opportunity of developing a new market. Rather, with the exception of ARAL, the oil companies were very reluctant to become involved. This is a typical effect of disruptive developments since the incumbent technology companies have invested in the traditional infra-structures and would have to re-invest in new and abandon 'old' assets in order to partake in the new development.

Two further facts obstructed the market introduction for natural gas: firstly, the customer 'added value' was limited, apart from the subsidies, since gaseous and liquid fuels essentially gave the same (or similar) performance and the main advantage, the environmental issue, is a somewhat abstract concept. Secondly, the cost benefit was considerable, but was based on fuel tax reductions that were subject to regular review, at least in Germany. A customer could therefore not be sure that the price advantage would persist throughout the lifetime of the vehicle. This situation underlines the importance of clearly spelling out the benefits of a new technology to the consumer and securing stable regulatory and political framework conditions.

The third example concerns unleaded petrol which was introduced in Germany in 1972 in order to reduce the then dangerously high lead pollution levels. This move occurred against the combined resistance of the whole automobile industry, petrol companies and surrounding European Union (EU) countries. The issue addressed was similarly abstract as in the case of natural gas fuels but differed insofar as the effect of lead pollution was well known to consumers and began to result in immediate health problems in the 1960s.[7] Regulations were passed that provided for a phasing out of leaded petrol over the period stretching to 1988, all other EU countries following suit. Convinced by the ensuing dramatic reduction of lead content in blood samples in the EU most countries worldwide have now switched to unleaded petrol. Today's exhaust gas treating catalysts would also be impossible with leaded fuels.

In contrast to the case of natural gas, the introduction process was driven by government regulation and met high public acceptance. Although the price of unleaded petrol was initially slightly higher, a growing number of drivers switched over long before regulations completely banned leaded fuels, thus indicating that they were prepared to pay a certain (limited) extra cost for environmental benefits without immediate individual advantages.

This case shows that a market situation can be over-ridden by government regulation, thus avoiding hesitant market take-up of a societally desirable technology. The need for investment lay less in the retrofitting of petrol pumps – the existing pumps and tanks could be further employed after cleaning – but in adapted engines and refinery processes. The effect was therefore less 'disruptive' for the infrastructure. Nevertheless, the transition did mean that certain vehicles with inadequate engine designs had to be scrapped, thus causing a considerable disruption and interference with owners. But in this specific case government action had a wide societal support due to the considerable impact on infant health, in particular. Resistance with car companies and owners quickly wore off. It should be pointed out, though, that the disruptive element on the consumer side leads to a higher destruction of societal wealth since the consumer has no alternative to scrapping his car, whereas the infrastructure operator may find various ways of transforming and refurbishing his installations.

Mobile phones may serve as the fourth, and final, example of an infrastructure disruption. They are still by far one of the most expensive ways of communicating. Even as late as 2010, costs per minute are up to ten times higher than with telecommunication by cable.[8] Nevertheless, in Germany in 2001 the market share of mobile telecommunications for the first time surpassed that of cable telecom.[9] Mobiles have gained market dominance for a variety of reasons. In the first place, they (supposedly) secure availability of anyone in any place for instant communication. This has become an element of value in a society of short termed and 'fast' economic activity but also of growing (individually perceived) insecurity, allowing parents, for instance, to instantly locate and contact their children, or persons in danger to contact help. Last but not least, possession of a mobile has developed into a status symbol, by now not merely by displaying the fact of owning one, since everyone 'has one', but by showing off the additional functions and ring-tones.

The mobile phone offers convenience and security to the individual. It matches (and sustains) some of today's lifestyles that rely heavily on instant communication. Nevertheless, a network of repeating and transmission stations had to be built and maintained and the purchase of frequency bands especially for digital services necessitated unprecedented investments in network infrastructure. It can be hypothesised therefore that the 'added values' experienced by the consumer (in flexibility, status and connection to digital life) are considered so significant that the high costs are accepted. In return, market prospects were valued so highly by the grid operators that investments rocketed, although much potential in digital services is still dormant today. Maybe this gives a good example of how expectancy engineering by governments – leading operators to a firm belief in the value of the frequency bands on sale – can initiate the growth of the necessary infrastructure (hint: hydrogen filling stations), regardless of whether this is then an economic success or the first wave of operators go bankrupt. Once the investment is made and the transition can begin, the history of the infrastructure is of little importance.

For the case of fuel cells, these examples (of many more) can prove the following:

- Massive investment in infrastructure is possible, if stakeholders believe or are led to believe in the economic potential of the developing structures.
- The high cost of new technology is not an issue if consumers recognise considerable (added) value in the novel appliances.
- State regulation can be a key to rapid infrastructure development and transition if public and market understanding of its necessity is underdeveloped.
- Successful disruptive transitions can occur at an infrastructure or product level; in the first case the service operators invest into new or refurbished equipment and will refinance this from future income; in the second, a certain amount of societal wealth will be destroyed or rendered useless.
- New players may profit more from the new development than traditional suppliers.

Disruption itself is not the problem, it is the way it is sold to the customers and markets.

11.4 Battling Incumbent Technology

The introduction of new technologies, though, will not leave the incumbent technology unaffected. (Note that the term 'incumbent', indicates the existing, conventional or traditional technology as the dominant form.) The bulk of road vehicles sold today display an average tank-to-wheel energy conversion efficiency in real driving cycles of 15–18%; the average power station electrical efficiency in Germany is as low as 37%.[10,11] Fuel cells could improve these figures to something above 30% and 60%, respectively. Roughly speaking, this is a factor of 2 and justifies speaking of an 'efficiency leap' or change in 'technology paradigm'. The difference is motivation enough for car makers to dramatically improve efficiency and emission levels of their products. As the competitor fuel cell rears its head over the horizon, automobile and power industries strive to exploit the margins for advanced developments in internal combustion engines and power station technology, thus narrowing the advantages fuel cells can offer. This 'sailing ship effect', first identified by Rosenberg and Howell, has been seen in a number of industries, most notably as the competition between sailing and steam ships during the nineteenth century, and can be seen as a positive indicator that the incumbent technology views the new technology as a serious threat.[12,13] A tremendous development push in sailing ship technology was perceived that improved the competitiveness of sailing in an attempt to outperform the engine-driven vessels. In vain, as history teaches us.

Fuel cells will thus face a serious problem of continuously increasing technical performance of the conventional alternatives, as these are improved in parallel. Examples include:

- EU vehicle emission regulations EURO5, in force from 2008, with very low emission levels, to be followed by even stricter EURO6
- Increases of best-of-their-kind power plant electrical efficiencies to 40% (300 MW scale gas turbine), 52% (hard coal fired), and 60% (combined cycle)
- Extension of a single battery charge to several hours of laptop operation, or a week or more of mobile phone use.

Many fuel cell developments that seemed promising initially will fail to show superiority over conventional technology under the benchmarks listed. Although some need to be debated in more detail (efficiency levels for power stations and engines, for instance, are generally given at the optimum point of operation, not for annual cycles of operation, including transients, part load and stand-by/idling *etc.*), the high quality of today's incumbent technology in many of the named aspects reflects a shift of paradigm towards more environmentally sustainable systems which, again, is (was) a field that fuel cells are determined to occupy.

Nevertheless, this development only partially addresses the relevant issues. Stepping back and taking an end-of-pipe point of view, we have to consider that a fuel cell vehicle running on hydrogen produced from renewable energy sources has no operational emissions and a very low environmental footprint from vehicle production and provision of hydrogen. In an attempt to 'not compare apples with pears', we will now have to find a comparable vehicle with similar performance, which is impossible. The battery electric vehicle running on renewable electricity will be more energy efficient but will have a shorter range. The petrol or diesel vehicle will emit CO_2 (amongst other pollutants) and noise; so will the hybrid vehicle, although at a somewhat lower level. The vehicle running on bio-diesel or bio-ethanol will claim a closed loop of CO_2 but eventually have an effective footprint of greenhouse gases due to considerable emissions and energy consumption in the production process. No other vehicle technology can provide the same level of emission reduction at the same environmentally low impact and a comparable price. Even if a synthetic fuel-driven diesel engine were considered that receives its fuel from a centrally operating synthesis plant with carbon capture and storage, thus not emitting any CO_2 to the atmosphere (though still to the environment, albeit an underground storage), fossil resources would still be depleted and total energy consumption would be higher.

The importance of taking the point of view of comparing equal with equal can be visualised by monetarising the differences between the examples listed above in the following mind experiment: Imagine a vehicle with a standard performance in speed, range, transport capacity, *etc.* that has no emissions (as a total system, thus taking global effects into account). List all the components necessary to achieve the goal and attribute a price tag. One of the first findings will be that a petrol or diesel car will not achieve the goal of freedom of emissions at all and thus has to be excluded from the comparison for reasons of failure to perform. Therefore the exercise of turning external cost (of emissions) into internal costs (by charging for the avoidance of all emissions or calculating the cost of avoidance) is extremely helpful. Only when consumers adopt this point of view by demanding the appropriate properties from products, will truly environmentally friendly technologies be able to develop sufficient market shares. The same will apply when state regulation calls for a complete monetarisation of external costs.

11.5 Paradigm Shifts and Succession of Generations

Technology development always takes place by progression, transition and quantum leaps. In what way these may be motivated was discussed above. The more 'disruptive' a development is – regardless of whether the disruption refers to behavioural changes (no cost), changes in equipment (consumer investment) or infrastructure (operator investment) – the more it will call for support from a 'paradigm shift'. With this term I would like to indicate a change in thinking, attitude, approach *etc.* in any element of the parallelogram of forces between suppliers, market and consumers. In any case it will lead to a disruption or quantum change in whatever circumstances. Various examples for paradigm shifts in the past are:

- The change from long-term planning of production to just-in-time production and delivery, leading to increasing congestion on motorways which are now turned into mobile warehouses
- Increased public perception and valuing of environmental performance of commodities, leading to partially functioning (unsubsidised) markets in ecological products
- Governmental and societal pressure to include full cost of services into market pricing, leading to more competitive markets for renewable electricity, environmentally benign products *etc.*

A change in consumer behaviour will be necessary in introducing fuel cells to the market. This paradigm shift will need to motivate consumers and customers to value environmental aspects higher than a number of traditional considerations made today when buying a product: short-term low personal expenses (maybe at the cost of long-term high public expenditure, *i.e.* structural subsidies), low sensitivity to the environmental footprint of products (again at high long-term societal cost), preoccupation with appearance and design of products (maybe at the cost of long lifetime, usefulness *etc.*), little consideration for energy consumption of appliances (with ensuing long-term high expenses for energy services) *etc.* Without a shift in these consumer considerations, any environmentally superior product will lack acceptance and acquire disrespect for its qualities, resulting in a refusal to recognise the full market value and a continuation of the market bias in favour of incumbent technology. A paradigm shift towards improved environmental performance of, for instance, cars will inevitably tip the scales towards the fuel cell vehicle.

One very simple way of achieving paradigm shifts, though not very well controllable, is that of succession of generations. It has been well observed and discussed that young people have a higher concern for environmental issues than persons well into their vocational lives. This is an expression of more consideration for 'justice' and the well-being of the society as a whole as opposed to the individualised quest for personal advantage. This tendency is not prominent in every single generation of youngsters but can be taken as statistically proven. (The quotation by Francois Guisot, 1787–1874, 'Not to be a republican at twenty is proof of want of heart; to be one at thirty is proof of want of head' comes to mind.) But even stripped of its conservative political connotation the above is true: next generations will grow up into technical surroundings unknown to their elders at the same age. They will more easily adapt to these (for them) prerequisites as well as adopt them for their everyday chores. This effect is one reason for the triumphant success of mobile phones since a new and fresh perception of their potential quickly led to new habits and languages (SMS-speak), a culture of documentation (using the phone's camera), or the need/desire for access to digital services at any time and any place. All of which was not visible at the inception of the basic technology.

Some hope can be placed in young generations viewing fuel cell appliances as products in their own right – not 'just' replacements of conventional gadgets – with a number of uses unique to this technology. What is more, a new and fresh appraisal of fuel cell technology may result in uses and advantages not dreamed of today by technology developers. A similar evolution can be traced in modern digital equipment – including mobile phones, the internet, the personal computers *etc.* – whose potential uses and powers were only foreseen by a visionary few. This new approach will even lead to regarding traditional technology as inferior and non-performing, or 'not cool' – in any case undesirable – which brings us back to the 'paradigm shift' postulated in the opening sentences of this section.

11.6   Transforming Technology into Products

Let us next have a look at the way technology evolves. Three main phases can be identified, building on the concept of 'technology readiness levels':[14]

1. Technology development from scientific principle to laboratory demonstration (TRL 1 to 4)
2. First demonstration and niche market products are brought to the public (TRL 5 to 9)
3. The technology breaks through to the (mass) market and wins a reasonable share in market sales or at least firmly establishes itself in niche areas.

The main point to make here is that in many areas, especially those concerning technology, the two first phases are dominated by an 'engineering approach' where a new development is driven by the belief in the technological potential and arguments for supremacy over competing technology are based on technical (and maybe environmental) performance alone. The third phase, though, will be dominated by considerations that have nothing much to do with the technology as such but rather with how it appeals to specific markets through aspects like usefulness, durability, cost, added value *etc.* In this phase the general customer has to be convinced by the desirability of the fuel cell product. This could easily be done if fuel cells were cheap, but as indicated above, this is decidedly not yet the case today. Therefore other arguments have to be used to sell fuel cells. Admittedly, car customers still tend to discuss such topics as the pro's and con's of diesel versus petrol vehicles, *i.e.* very 'technical' issues, but then, automobiles are a market segment with a high affinity of customers to technology.

The two most obvious advantages of fuel cells – low emissions and high efficiency – are difficult to market since both are relatively 'abstract' and hardly accessible to direct experience by the customer. Environmental and efficiency aspects may also play a role but this is limited to a market share of 10–15%,[2] a situation that may change under the worldwide threat of climate change, although such a market impact is not yet proven. Consumers are becoming increasingly aware of energy consumption, but this awareness is predominately linked to energy cost, less to the actual consumption of energy. Therefore fuel cost plays more of a role than fuel amount and a fuel cell vehicle running on hydrogen will be difficult to compare to a conventional vehicle due to a lack of consumer experience with consumption and fuel pricing and no 'gut feelings' or intuitive appraisals available for immediate comparison of the alternatives. Similar is true for today's natural gas vehicles where the common knowledge of consumption figures still needs to build and any individual driver will need to collect fuel consumption and cost from other drivers of natural gas vehicles in order to directly compare figures and form some kind of 'standard'.

Stripping fuel cells of their two main advantages leaves an expensive product with doubtful usefulness to the customer. This hands a rather tough job to the sales department. Since the situation is plainly as it is and will have to be accepted by companies developing the market, the question is what can be done to change it?

Leaving aside the technical perception of fuel cells as a technologically superior product for the moment and taking the side of the consumer viewing an appliance on the sales shelf, helps in visualising the approach. The first question a buyer will have towards a new product is 'What can I use THIS for?!?' The obvious answer of 'Well, produce electricity …' (applying to any fuel cell application, to have the most general approach) will not really help the buyer since this is neither novel, nor revolutionary, nor 'sexy' in any way. Rather (intentionally mixing arguments from all fuel cell application areas for clarity), one could answer along lines of:

1. Drive a noiseless car
2. Drive a car without exhaust gases
3. Drive a car with high acceleration
4. Drive a car with no gears
5. Recharge your computer/mobile phone/battery gadget independently from sockets
6. Use your laptop on long distance flights/boat trips/in forest cabins, *etc.*
7. Use electrical power tools on building sites, independently of mains power supplies
8. Use electrical power tools on road building sites
9. Freely use electrical appliances in parks, gardens, forests, woodland areas, *etc.*
10. Supply lighting in parks, gardens, forests, woodland area, sports arenas, *etc.* with no mains connection
11. Supply electricity/lighting to 'dachas', forest cabins, garden sheds, *etc.* without the need to invest in expensive grid connections with low usage hours
12. Use your fuel cell 'heating boiler' as an uninterruptible power supply.

None of these arguments builds on fuel cells as such, but concentrates on aspects that would not be possible with other technologies (apart from battery electric vehicles with reference to points 1 through 4; as far as the discussion here is concerned, these can be considered in the same way as fuel cell vehicles). Nevertheless, they may serve well in convincing potential customers. All of them aim at either superior technical performance (points 1–4) or improvements to everyday handling of other products, *i.e.* battery operated devices, or the reliability of electricity supply. Points 5–11 especially offer solutions to limitations many travellers, building workers, sports enthusiasts or gardeners experience today. Often enough, light or electricity is needed in places that are far enough away from any mains socket to either prevent any electrical appliances form being used – reverting to two-stroke petrol engine chain saws or engine-generators being employed – or making the installation of expensive mains connections necessary that may well only be used a few hundred hours a year. Fuel cells would open up new possibilities, for instance in the use of electrical appliances on boats, during long-distance travel or out in the countryside, regardless of access to mains electricity. Point 12 again refers to a completely new aspect, which is using the domestic electricity production device (fuel cell or photovoltaics) as a self-reliant electricity supply in case of grid failure, a prospect that many consumers may find extremely attractive although the risk of grid failure is still low in Europe compared to the US.

Exploiting the arguments developed above in marketing campaigns can be quite rightly left to the sales department. Fuel cells as a whole can, on the other hand, benefit tremendously from a kind of 'backwards engineering'. By visualising the fuel cell as a product delivering the above attributes, specific consumer requirements and expectations can be targeted. Eventually, the product will contain a fuel cell, but this may not even be visible, nor may it be prominent in product marketing. A famous example of this is displayed in Figure 11.1. The advertisement run by Smart Fuel Cells in 2008 shows a young lady happily using a hair dryer on a camper van. A fuel cell is nowhere to be seen and the headline runs 'energy everywhere' in the sense of items 9 and 10 above. Regular campers will realise that using a hair dryer on the camper van DC power supply will be almost impossible and many vans rely on sockets being supplied on-site of camping grounds and at parking lots to run the AC equipment. Although the DC/AC converter could also be supplied by the vehicle DC grid, it would most probably serve to run the battery low within no time whereas fuel cells sustain operation by being regularly refuelled.

The imagination of fuel cell developers is now required not only to improve the properties of fuel cell devices towards more reliable operation, but also to 'invent' uses the consumer will gratify by buying the product. Who would have imagined 20 years ago that mobile phones would serve as miniature TV sets, music download centres, cameras, e-mail & SMS terminals, data transmitters, *etc.* The first mobiles were bulky and cumbersome to handle but served well in applications on building sites, remote communication *etc.* (as long as there was access to the transmission system). Today's phone designs are essential in developing and sustaining the market for communication and even creating

Figure 11.1 Advertisement run by Smart Fuel Cells (SFC). Reproduction courtesy of SFC.

new trends and desires. This is all very clear in retrospect; the question is, whether it is possible to foresee such developments and develop and design appliances to future purposes. Of course, this cannot be answered readily and there are just as many examples of visionary developments (the early internet visionaries and Apple computer's Steve Jobbs may serve as two examples) as there are of total failures (naming the Beta 2000 video format as just one). Nevertheless, a new and valuable view of fuel cell market entry would be supplied if developers took the aspect of consumer perception more seriously.

11.7 Added Value, Special Markets and Allowable Cost

Consumer perception of product value has a decisive influence on the price a buyer will be prepared to pay for a commodity. In the internet world today this is a problem in that many services were introduced free of charge (*e.g.* free news portals, google search, Wiki's, *etc.*) in order to 'bait' customers and then turn these offers into a pay-for-use service, once they have been accepted and integrated into everyday life. Obviously, the acceptance of charges by the users will be a direct indication of the monetary value they allow in return for a useful commodity. It is therefore interesting to observe the resistance met by many providers in introducing paid services and the generally low esteem internet users thus have of these offerings.

The willingness of users to pay a certain (additional) sum for useful services can be taken as a direct measure of the so-called 'added value'. This is the additional value a customer can derive from a commodity in comparison to competing, incumbent or conventional items or services. The added value can then be expressed as a monetary figure in the form of a higher allowable market price as compared to the competitors. One main example of this mechanism is a higher price for a quality product on the basis that the consumer assumes or knows from experience that the item will last longer or serve better. The added value, basically 'desirability', is not directly related to the material value, and not even necessarily to the technical performance of the commodity. Consequently, given that the added value is estimated high enough by the buyer, even high-priced goods can be placed in the market as long as they offer a consumer value that allows for the extra cost.[2] Product features that offer performance unobtainable with conventional equipment will spur the market introduction of a new product considerably, winning it the label 'enabling technology'.

On the other hand, the usual approach to the market entry of fuel cells is to call for dramatic cost reductions, thus directly and undisputedly accepting the target cost of conventional products. As was explained above, this is a completely wrong approach in that it refuses to acknowledge the many aspects of the deviant character of fuel cell technology. Cost targets have been set by the US American SECA programme as well as by the European Commission.[15,16] These targets are removed from today's fuel cell equipment prices by a factor of 10 or more, indicating there is a tremendous need for cost reductions. Viewed from the angle of organising market introduction of new technologies, it could be stated that by narrowing down discussions to pure cost comparisons whilst ignoring the broader connotations of environmental footprint *etc.*, creates a further bias towards incumbent technologies with their inherently long history of cost optimisation. The added value a competitive product gains is effectively 'zero', a situation that postulates total consumer disregard of any advantages the novel product may have, even assuming the consumer would find the novel product 'useless'. This assumption is in itself a bias towards incumbent technology, as the discussion of paradigm change and the possible shifts in customer perception have shown.

Much development effort is effectively invested into market competitiveness in the sense of costs breaking even with competing technologies.[2] Due to the simple point that this means the same price can be charged for the new technology as for the incumbent competitors, it is impossible to regain any development costs unless the new technology is inherently cheaper, which fuel cells are clearly not. Therefore cost competitiveness will return no investment in the short term. On the other hand it can be found that aspects such as design, use in a recreational context, lifestyle issues *etc.* might return very much more income as compared to 'pure' engineering aspects.

Figure 11.2 shows the decrease of a hypothetical fuel cell product's price over time as compared to the cost of the competing conventional technology (for instance as a comparison between a fuel cell-driven car *versus* the petrol internal combustion engine vehicle). Fuel cell cost drops from first prototypes

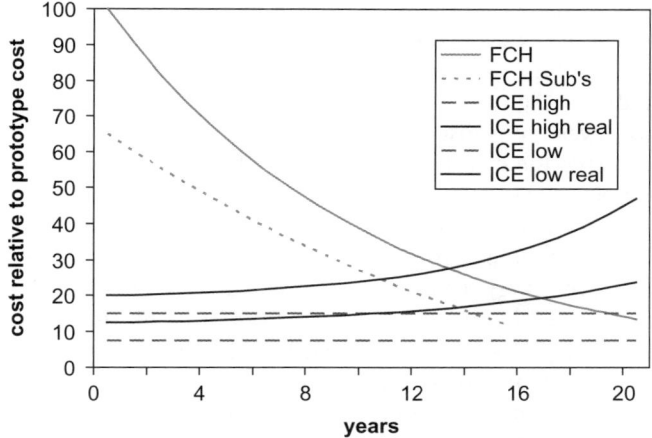

Figure 11.2   Generic cost development curve and target market entry cost for a fuel
cell product (upper curves, grey) and the competing conventional technology (lower curves, black). The dashed lines indicate lower and upper
range of conventional technology price, assuming no change of price
over the range of 20 years shown here. The drawn, upwards curving lines
indicate the respective total cost of ownership, assuming an increase in
cost with years, also including surcharges, for instance from carbon
taxing. The fuel cell curves show the total cost, beginning with prototype
cost and the same curve with market introduction subsidies deducted
(dotted line).

down to a level where it can compete with conventional technology by
increasing numbers being produced (economy of scale) and by technological
and manufacturing improvements, for instance by using cheaper materials and
manufacturing processes. Several issues have to be considered in such a comparison, though. Firstly, the cost of the incumbent technology, contrary to
statements from most sources, is neither constant, nor is it a fixed figure.
Remaining in the picture of vehicles, it is rather obvious that passenger cars
today might have wildly varying prices, depending on their quality, level of
luxury, size *etc.* A similar situation exists for electricity generating equipment,
which may sell at a cost between 400 and several thousand euros per kW_{el}.
Therefore a bandwidth of conventional technology cost has to be assumed.
This is also in accordance with the statement that new technology should first
target high-price niche markets. Furthermore, the cost of conventional technology will change over time, mostly with an upwards tendency. Finally, in cost
comparisons the total cost of ownership has to be taken into account. It was
said earlier that in decisions on vehicle choices a customer will most probably
not only consider the vehicle price itself, but include petrol consumption, taxes
and insurance in his or her reasoning. Given that fuel cell vehicles are more
energy efficient, this would allow for a certain extra cost of purchase being
partly compensated for by savings in fuel (if the hydrogen fuel were sufficiently

cheap, of course). The cost of ownership should include the real societal cost, also considering the external costs of using the vehicle, as discussed earlier. The upper dashed line Fig. 11.2 depicts a programme for market entry for fuel cells which will reduce the cost of equipment. In the early stages this will hardly play a role, since even subsidised units will remain far above market prices. Nevertheless, once the cost has fallen reasonably within the range of conventional technology, the subsidies help to compete earlier, although they will have to be phased out as soon as competitiveness has been achieved, as shown.

Figure 11.2 depicts the market entry as the point in time where fuel cell and conventional technology cost curves meet; this is strongly influenced by assumptions on the incumbent technology cost. It is thus important to clearly state what the target costs are supposed to be and how they are derived, otherwise a biased comparison is drawn which can easily lead to a wrong perception by the public, which again can influence political decisions.

The historical comparison with the market entry of technologies and commodities actually shows that cost and economic competitiveness alone did, and do not, determine market development for new products. The 'added value' offered to consumers and end users is critically important and may in some cases, as was shown above, compensate for otherwise completely uneconomic performance (as for instance with mobile phones). This fact lays special importance on defining exactly what the 'added value' of fuel cells is and exploiting and maintaining this characteristic in market entry strategies.

With many of the successful 'new technologies' of the past the cost issue was not the driving force in market introduction. These technologies, when first introduced, were definitely not cheaper than existing technology, and did not in all cases offer any obvious extra consumer performance (a thermopane double-glass window is still 'just a window pane', unleaded petrol drives a car the same way leaded does *etc.*). What they did offer was some 'added value' that was estimated so highly by the consumer (or not, in some of the cases) that the extra cost for the new product was accepted. Only later, after the 'new product' has been firmly established in the market, does the formation of a market price begin which may eventually lead to a competitive price. As an example, today's insulated glass windows are the standard product with high volume production and low prices, whereas single-pane glass is just as expensive due to the low turnover and extra handling effort as a 'non-standard' product. It is therefore paramount to obtain some understanding of what the 'additional cost' level might be in determining the 'allowable cost' or surcharge acceptable in first market entry.

The allowable cost issue is a vital problem to solve in a mass market entry. With fuel cells, at the time of writing, there is no discussion of a near-term mass market since we are still speaking of the first small series of equipment being available, if not prototypes. The above-mentioned company Smart Fuel Cells is one of the few examples to the contrary, having sold 15 000 units of Direct Methanol Fuel Cell 250 W power supplies between 2007 and 2009.[17] This picture will not change in the short term. On the other hand a market urgently needs to be developed to (1) actually sell equipment and enter the 'real world' of commercial applications (including all aspects of O&M, handling, durability, quality, customer satisfaction, product guarantees *etc.*); (2) earn revenue from selling equipment; (3) gather all the experience coming from 'real world' applications; and thus (4) bring fuel cell technology forward by a decisive step. This last step determines whether or not fuel cells can achieve any market potential at all – in whatever niche market – on their own balance of consumer value. Only when this first market entry has been accomplished and a firm market basis established is there any substance to build on for a mass market expansion, *i.e.* markets relying on the quality of the product and a customer acceptance of the cost–benefit relationship offered. The expected negative photovoltaics sales development after expiry of feed-in tariff programmes in various countries may serve as an example of regulated market entry that may – if not managed properly – not result in a self-sustained market participation of the new technology.

In achieving this first step of market entry, the maximum available cost margin needs to be claimed. The ranking of added value could be assumed as shown in Table 11.1. Most of this ranking is more or less obvious. Meeting the need for independence, improved handling (weight, noise, smell *etc.*), electricity supply in off-grid situations and secure power supply will achieve a maximum of allowable costs. On the other hand application in central power stations will require cost effectiveness which will only be reached in the long term, unless added value is generated by the necessity for an environmental portfolio management for power producers.

Fuel cells offer potential environmental benefits due to lower emissions as compared to conventional equipment – or even zero emissions in some cases – higher efficiency, higher rate of electricity to heat produced and modular and lightweight portable electricity supply. These characteristics need to be developed and optimised beside, and maybe even prior to, the final struggle for cost competitiveness since they are the major constituent of the 'added value' fuel cells offer. More importantly, they are also the means by which this currently high-priced technology may be established in contradiction to purely economical considerations.

In several segments a market for fuel cell applications is more or less evident. In these cases the market introduction problem is shifted to that of supplying functional equipment at realistic prices (mirroring the added cost the consumer will accept). Areas concerned are:

- Off-grid electricity supply (radio and telecommunication relay transmitters, traffic signalling, lighting)
- Recreational vehicles (boats, camping vehicles *etc.*)
- Uninterruptible power supply
- Battery replacers
- Range extenders for battery electric vehicles.

The next possible market could be vehicles where the character of the novel propulsion could also be modelled into a 'fun' (= 'lifestyle') quality. The often

Table 11.1 Ratings of various aspects of fuel cell 'added value'.

Added value	Sector addressed	Application	Consumer ranking
Independence	Lifestyle	Portable applications of methanol FC	High
Recreation	Lifestyle/usefulness	Low-noise/low pollutant emission remote energy supply (camping, boats *etc.*)	High
Driving fun	Lifestyle/usefulness	Electric vehicle performance, vehicle appearance	High
Remote power	Usefulness	Mobile electricity supply for outdoor workmen, forestry workers *etc.*, electricity supply in remote areas	High
High quality power	Usefulness	Secure power supply	High
Freedom from particle emissions	Usefulness/ethics	Urban traffic electric vehicles, moored marine vessels	Medium to high
Independence/ sustainability	Lifestyle/ethics	Small-scale residential CHP	Medium to high
Power stations	Ethics/economics	Low-emission electricity production	Medium to low

formulated view from the car industry that a fuel cell vehicle needs to equal a conventional vehicle in appearance is unconvincing because this actually aims solely at a 'replacing' function which mobilises least 'added value' with the consumer.

This calls for vehicles which are decidedly different to conventional designs. They could offer novel and exciting 'looks', thus providing an 'added benefit' for customers valuing appearance, especially if this is different from standard consumer expectations, allowing them to distinguish themselves from the 'broad mass'. They could offer new possibilities of usage and extend the ways a product is integrated into everyday life. One example of such an approach is the General Motors 'Autonomy' concept vehicle (Figure 11.3) presented at the Detroit Motor Show in 2002. On the one hand it attempts to offer a modern design that will mirror the character of a novel product in its outer appearance, thus appealing to customers with a technical interest, but also those valuing design aspects. What is more, the design exploits some dedicated properties of fuel cell vehicles. Fuel cell vehicles will always be electric vehicles which means that many functions like steering, braking *etc.* will be performed

Figure 11.3 General Motors 'Autonomy' concept vehicle which separates the function of the driving system (chassis) from the vehicle hull through a fully electrical interface. Reproduction courtesy of GM.

electrically, *i.e.* 'by wire'. The concept of 'drive by wire' is not unique to electric vehicles – it has also been discussed in the context of other vehicles – but is especially easy to realise and logical in an electric vehicle. Once the interface between driver and vehicle has been integrated into a set of wires/plugs, this could be extended into a kind of 'bus' system where the driver can plug into the control interface from different places on the chassis of the vehicle. As long as he can see the road and safely steer the vehicle, the driver could freely choose his or her seating on the platform. Figure 11.4 shows pictures from the General Motors 'HyWire' concept that has built on this concept. This gets as near to a revolution of private vehicle handling as today's imagination lets us. Taking it even a step further, vehicle users could swap the hull of the vehicle for any kind of 'skin' that currently suits their transport needs best. Businesses could for instance be built on the concept of 'renting out' hulls in the form of pick-ups, family transport, goods transport, recreational needs *etc.* thus increasing the value of the basic vehicle (chassis) due to its flexibility and easy adaptability to the driver's needs.

One aspect, though, needs further consideration. A major advantage of fuel cells in public awareness is the issue of environmental protection and freedom

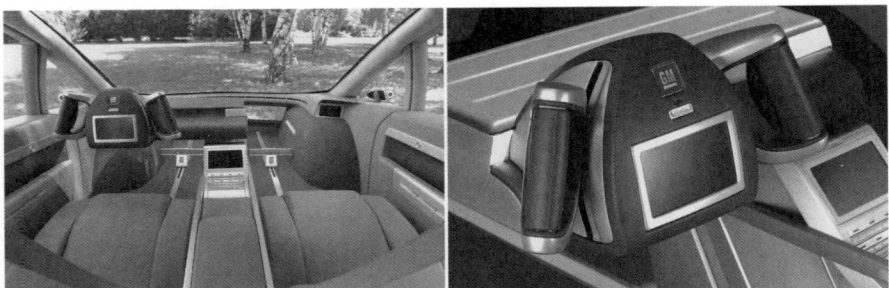

Figure 11.4 General Motors 'HyWire' full electric driving and steering interface. View of the driving console from the front (left) and from the driver's perspective (right). Reproduction courtesy of GM.

from emissions (of any type). This is reflected in the importance fuel cell technology has in advertising for power companies and in the technology portfolio for instance of car manufacturers worldwide. But, public awareness of environmental issues is very critical as the examples of green electricity and other 'green' products show. The public views products and conveniences marketed under 'environmental' labels a great deal more critically than conventional products (which everybody expects to be environmentally inferior, anyway).

Some regulation for certification will therefore be necessary in order to clearly indicate the environmental performance of fuel cell products along the whole line form primary energy to product usage. This applies for instance to hydrogen fuel with the clear need to declare the carbon footprint it has accumulated over the manufacturing chain, as well as to residential fuel cell combined heat and power (CHP) units which need to prove that their operation actually reduces the carbon footprint of the energy supply system. Regulation could, for instance, consist of a minimum allowable conversion efficiency in electricity generation. This would systematically call in high-efficiency generation to offset the technically inferior equipment.

11.8   Outlook

Fuel cells are a technology on the verge of market entry and aiming at replacing long-established conventional energy conversion processes like engines and turbines. Upon entry into the first markets the necessarily higher costs need to be offset by the 'added value' consumers allocate to fuel cell products. The highest additional marginal costs will be achievable from lifestyle oriented issues including recreational applications and remote power. Some market potential can be expected from environmentally oriented clientele. In this case the environmental aspects are of supreme importance and environmental superiority needs to be clearly proven, including all aspects from fuel supply to operation and recycling. Only if this performance is convincing will fuel cells be able to claim a substantial part of their potential market. Due to the relatively 'abstract' nature of the environmental benefits, government regulation, for instance on the efficiency of overall electricity power generation and/or emission levels, will be helpful in market introduction, especially of fuel cells for stationary applications.

Acknowledgements

Thanks go to Herbert Wancura of NTDA Energia for the discussions leading towards the development of Figure 11.2 in the context of the European Fuel Cells and Hydrogen Joint Undertaking (FCH JU) 2010 annual programme preparation.

References

1. W. Schiebel and S. Pöchträger, in *Proceedings of the Second Austrian Technology Assessment Conference*, Vienna, May 2002, ITA newsletter, June 2002, p. 9.
2. R. Steinberger-Wilckens, *Int. J. Hydrogen Energy*, 2003, **28**, 763–770.
3. G. Barudio, *Die Weltgeschichte des Erdöls*, Stuttgart, Klett-Cotta, 2001, ISBN 3-608916-80-6.
4. Dietmar Abt, Die Erklärung der Technikgenese des Elektroautomobils [*Explaining the Technical Genesis of the Electric Automobile*]. Frankfurt am Main/New York, Peter Lang, 2008, ISBN 3-631-33085-5.
5. ARAL GmbH press release, 29 April 2004.
6. P. C. Flynn, *Energy Policy*, 2002, **30**, 613–619.
7. D. Asendorpf, in *Die Zeit*, 2 May 2002.
8. *cf.* for instance web pages with tariff comparisons like www.billiger-telefonieren.de *etc.*
9. Regulierungsbehörde f. Telekommunikation und Post, Jahresbericht 2001, Bonn, 2002.
10. Th. Feck, Ökobilanzierung unterschiedlicher Kraftstofflebenszyklen für Wasserstoffahrzeuge [*Life Cycle Analysis of Various Fuels for Hydrogen Vehicles*], Master's thesis, Carl-von-Ossietzky Universität Oldenburg, Germany, July, 2001.
11. M. Machat and K. Werner, Entwicklung der spezifischen Kohlendioxidemissionen des deutschen Strommix [*Development of the specific carbon dioxide emissions of the German electricity mix*], UBA, Dessau, 2007, ISSN 1862-4359.
12. N. Rosenberg, *Perspectives on Technology*, Cambridge University Press, Cambridge, 1976.
13. J. Howell, *J. Management Studies*, 2002, **39**, 887–907.

14. NASA Research and Technology Program and Project Management Requirements, NASA Procedural Requirements 7120.8, Appendix J, Technology Readiness Levels (TRLs), Washington DC, 5 February 2008.
15. J. P. Strakey, in *3rd Annual SECA Workshop*, Washington DC, March 2002, NETL, Pittsburgh, PA, 2002.
16. European Commission, *Energy, Environment and Sustainable Development. Work Programme Update: August 2001. Part B: Energy – Priorities and roadmaps 2001–2002*, Brussels, 2001.
17. Smart Fuel Cells: Nine Months Report 2009, Brunnthal, October 2009.

Subject Index